Switching Theory and Logic Design

Switching Theory and Logic Design

Editor:

Gerard Ian Prudhomme

ArclerPress

www.arclerpress.com

Switching Theory and Logic Design

Editor: Gerard Ian Prudhomme

708 3rd Avenue, 6th Floor
New York
NY 10017
United States of America
www.arclerpress.com

ISBN: 978-1-68094-446-4

Library of Congress Control Number: 2016951153

About the Editor

Gerard Ian Prudhomme

Gerard I. Prudhomme has a graduate degree (M.S.) for Computer Science from University College London (UCL). He has also worked as a software programmer and tech writer for different Fortune 500 companies, and studied at UCL, Harvard, and Oxford.

Contents

List of Contributors

Chengjian Wen
Department of Computer Science and Engineering, Beihang University, Beijing, China

Yifen Mu
Key Laboratory of Systems and Control, Institute of Systems Sciences, Academy of Mathematics and Systems Science, Chinese Academy of Sciences, Beijing, China

Michael Alexander Nugent
M. Alexander Nugent Consulting, Santa Fe, New Mexico, United States of America
KnowmTech LLC, Albuquerque, New Mexico, United States of America
Xeiam LLC, Santa Fe, New Mexico, United States of America

Timothy Wesley Molter
M. Alexander Nugent Consulting, Santa Fe, New Mexico, United States of America
KnowmTech LLC, Albuquerque, New Mexico, United States of America
Xeiam LLC, Santa Fe, New Mexico, United States of America

Ning An
State Key Laboratory of Software Development Environment, Beihang University, Beijing, China

Rui Wang
State Key Laboratory of Software Development Environment, Beihang University, Beijing, China

Yuan Gao
School of Computer Science and Engineering, Beihang University, Beijing, China

Hailong Yang
School of Computer Science and Engineering, Beihang University, Beijing, China

Depei Qian
School of Computer Science and Engineering, Beihang University, Beijing, China

Bin Shao
The State Key Laboratory for Artificial Microstructures and Mesoscopic Physics, School of Physics, Peking University, Beijing, China
The Center for Quantitative Biology and Peking-Tsinghua Center for Life Sciences, Peking University, Beijing, China

Xiang Liu
The Center for Quantitative Biology and Peking-Tsinghua Center for Life Sciences, Peking University, Beijing, China

Dongliang Zhang
The State Key Laboratory for Artificial Microstructures and Mesoscopic Physics, School of Physics, Peking University, Beijing, China

Jiayi Wu
The State Key Laboratory for Artificial Microstructures and Mesoscopic Physics, School of Physics, Peking University, Beijing, China

Qi Ouyang
The State Key Laboratory for Artificial Microstructures and Mesoscopic Physics, School of Physics, Peking University, Beijing, China
The Center for Quantitative Biology and Peking-Tsinghua Center for Life Sciences, Peking University, Beijing, China

Antoni Morro
Electronic Engineering Group, Physics Department, Universitat de les Illes Balears, Palma de Mallorca, Balears, Spain

Vincent Canals
Electronic Engineering Group, Physics Department, Universitat de les Illes Balears, Palma de Mallorca, Balears, Spain

Antoni Oliver
Electronic Engineering Group, Physics Department, Universitat de les Illes Balears, Palma de Mallorca, Balears, Spain

Miquel L. Alomar
Electronic Engineering Group, Physics Department, Universitat de les Illes Balears, Palma de Mallorca, Balears, Spain

Josep L. Rossello
Electronic Engineering Group, Physics Department, Universitat de les Illes Balears, Palma de Mallorca, Balears, Spain

M. Fernando Gonzalez-Zalba
Hitachi Cambridge Laboratory, Cambridge CB3 0HE, United Kingdom

Chiara Ciccarelli
Cavendsih Laboratory, University of Cambridge, Cambridge, United Kingdom

Liviu P. Zarbo
Institute of Physics, The Academy of Science of the Czech Republic, v.v.i, Cukrovarnick'a 10, 162 53 Praha 6, Czech Republic

Andrew C. Irvine
Cavendsih Laboratory, University of Cambridge, Cambridge, United Kingdom

Richard C. Campion
School of Physics and Astronomy, University of Nottingham, Nottingham NG7 2RD, United Kingdom

Bryan L. Gallagher
School of Physics and Astronomy, University of Nottingham, Nottingham NG7 2RD, United Kingdom

Tomas Jungwirth
Institute of Physics, The Academy of Science of the Czech Republic, v.v.i, Cukrovarnick'a 10, 162 53 Praha 6, Czech Republic
School of Physics and Astronomy, University of Nottingham, Nottingham NG7 2RD, United Kingdom

Andrew J. Ferguson
Cavendsih Laboratory, University of Cambridge, Cambridge, United Kingdom

Joerg Wunderlich
Hitachi Cambridge Laboratory, Cambridge CB3 0HE, United Kingdom
Institute of Physics, The Academy of Science of the Czech Republic, v.v.i, Cukrovarnick'a 10, 162 53 Praha 6, Czech Republic

José Portêlo
INESC-ID, Lisbon, Portugal
Instituto Superior Técnico, Universidade de Lisboa, Lisbon, Portugal

Bhiksha Raj
Language Technologies Institute, Carnegie Mellon University, Pittsburgh, Pennsylvania, United States of America

Isabel Trancoso
INESC-ID, Lisbon, Portugal
Instituto Superior Técnico, Universidade de Lisboa, Lisbon, Portugal

Javier Macia
ICREA-Complex Systems Lab, Universitat Pompeu Fabra, Barcelona, Spain
Institut de Biologia Evolutiva, UPF-CSIC, Barcelona, Spain

Ricard Sole
ICREA-Complex Systems Lab, Universitat Pompeu Fabra, Barcelona, Spain
Institut de Biologia Evolutiva, UPF-CSIC, Barcelona, Spain
Santa Fe Institute, Santa Fe, New Mexico, United States of America

Edward H. Hellen
Department of Physics and Astronomy, University of North Carolina Greensboro, Greensboro, North Carolina, United States of America

Syamal K. Dana
Council of Scientific and Industrial Research-Indian Institute of Chemical Biology, Kolkata, India

Ju¨ rgen Kurths
Potsdam Institute for Climate Impact Research, Potsdam, Germany

Elizabeth Kehler
Department of Physics and Astronomy, University of North Carolina Greensboro, Greensboro, North Carolina, United States of America

Sudeshna Sinha
Indian Institute of Science Education and Research Mohali, SAS Nagar, Mohali, Punjab, India

Angel Gon˜ i-Moreno
Systems Biology Program, Centro Nacional de Biotecnologı´a CSIC, Cantoblanco-Madrid, Spain

Martyn Amos
School of Computing, Mathematics and Digital Technology, Manchester Metropolitan University, Manchester, United Kingdom

Fernando de la Cruz
Departamento de Biologı´a Molecular e Instituto de Biomedicina y Biotecnologı´a de Cantabria (IBBTEC), Universidad de Cantabria-CSIC-SODERCAN, Santander, Spain

Jon Borresen
School of Computing, Mathematics and Digital Technology, Manchester Metropolitan University, Manchester, United Kingdom

Stephen Lynch
School of Computing, Mathematics and Digital Technology, Manchester Metropolitan University, Manchester, United Kingdom

Laszlo Bela Kish
Department of Electrical and Computer Engineering, Texas A&M University, College Station, Texas, United States of America

Claes-Go¨ ran Granqvist
Department of Engineering Sciences, The A˚ngstro¨m Laboratory, Uppsala University, Uppsala, Sweden

Mahesh Yaragatti
Biotechnology Program, University of Pennsylvania, Philadelphia, Pennsylvania, United States of America

Qi Wen
Department of Physics, University of Pennsylvania, Philadelphia, Pennsylvania, United States of America

Yajie Wang
The Higher Educational Key Laboratory for Measuring & Control Technology and Instrumentations of Heilongjiang Province, School of Measurement–Control Technology and Communications Engineering, Harbin University of Science and Technology, Harbin, China

Yunbo Shi
The Higher Educational Key Laboratory for Measuring & Control Technology and Instrumentations of Heilongjiang Province, School of Measurement–Control Technology and Communications Engineering, Harbin University of Science and Technology, Harbin, China

Xiaoyu Yu
The Higher Educational Key Laboratory for Measuring & Control Technology and Instrumentations of Heilongjiang Province, School of Measurement–Control Technology and Communications Engineering, Harbin University of Science and Technology, Harbin, China

Yongjie Liu
The Higher Educational Key Laboratory for Measuring & Control Technology and Instrumentations of Heilongjiang Province, School of Measurement–Control Technology and Communications Engineering, Harbin University of Science and Technology, Harbin, China

Manuel Caldas-Morgan
Department of Mechanical Engineering, University of São Paulo, São Paulo, SP, Brazil

Alexander Alvarez-Rosario
Department of Mechanical Engineering, University of São Paulo, São Paulo, SP, Brazil

Linilson Rodrigues Padovese
Department of Mechanical Engineering, University of São Paulo, São Paulo, SP, Brazil

Preface

Switching circuit theory is the analytical investigation of the attributes of systems. Sequential organizational behavior is utilized to layout, and consequently develop, finite state machines. Boolean logic is broken down into combinational logic, together with sequential logic. In electronic components, a logic gate is an idealized, or, alternatively, an actual apparatus, employing Boolean functionality. It carries out a rational procedure upon inputs.

Sequential logic is a type of Boolean logic, in which the performance is a product of both the present inputs together with past information. Combinational logic circuits, on the other hand, rely only on current input. These functions are unable to rely on overall parameters, or other supplementary factors, such as memory. Combinational logic is utilized to carry out Boolean procedures on binary access messages together with binary information. Both styles of logic are often used to implement various electronic components presently utilized.

Combinational Logic Circuits are produced up from straightforward logic NOT, NAND, or NOR gates, which have been joined together with each other to come up with more advanced switching circuits. Due to the fact that combinational logic circuits are produced exclusively from independent logic gates, they are able to, at the same time, be regarded as a form of circuit which produces decisions. Therefore, combinational logic ultimately concerns itself with bringing together multiple logic gates alongside one another, to handle multiple messages, and enables them to manufacture one or more end result signals in accordance with the rational functionality of the logic gates.

Editor
Gerard Ian Prudhomme
Hawaii, US

CHAPTER 1 POWER AND PERFORMANCE MANAGEMENT IN NONLINEAR VIRTUALIZED COMPUTING SYSTEMS VIA PREDICTIVE CONTROL

Chengjian Wen[1], Yifen Mu[2]

[1]Department of Computer Science and Engineering, Beihang University, Beijing, China

[2]Key Laboratory of Systems and Control, Institute of Systems Sciences, Academy of Mathematics and Systems Science, Chinese Academy of Sciences, Beijing, China

ABSTRACT

The problem of power and performance management captures growing research interest in both academic and industrial field. Virtulization, as an advanced technology to conserve energy, has become basic architecture for most data centers. Accordingly, more sophisticated and finer control are desired in virtualized computing systems, where multiple types of control actions exist as well as time delay effect, which make it complicated to formulate and solve the problem. Furthermore, because of improvement on chips and reduction of idle power, power consumption in modern machines shows significant nonlinearity, making linear power models(which is commonly adopted in previous work) no longer suitable. To deal with this, we build a discrete system state model, in which all control actions and time delay effect are included by state transition and performance and power can be defined on each state. Then, we design the predictive controller, via which the quadratic cost function integrating performance and power can be dynamically optimized. Experiment results show the effectiveness of the controller. By choosing a moderate weight, a good balance can be achieved between performance and power: 99.76% requirements can be dealt with and power consumption can be saved by 33% comparing to the case with open loop controller.

INTRODUCTION

In large scale computing systems, like data centers, huge power has been consumed and the consumption is increasing greatly each year([1]). This gives rise to heavy pressure on environment and resource. Thus performance and power management has become more and more important and captured growing research attention in both industrial and academic fields.

Citation: Wen C, Mu Y (2015) Power and Performance Management in Nonlinear Virtualized Computing Systems via Predictive Control. PLoS ONE 10(7): e0134017. doi:10.1371/journal.pone.0134017.

As an advanced technology, virtualization can provide a promising approach to save power ([2]) and has become a basic architecture in most data centers today. In the virtualized systems, the services are enclosed in the virtual machines (VM). Many VMs can be hosted on one physical machine (PM) and the VMs can be created or destroyed or migrated between different PMs at little cost. Then, by consolidating many VMs to few PMs, the PMs with low workload, utilization or frequency can be shut down and power consumption can be reduced. Hence, in the virtualized computing systems, more control actions is feasible, including turning on/off the PMs, turning on/off the VMs, migrating the VMs between PMs, regulating the frequency of processor etc.. Additionally, control actions often have time delay. For example, turning on/off the PM needs several minutes and thus will influence performance and power consumption greatly. These factors make it complicated to design sophisticated and fine control to manage performance and power consumption.

Because of the significance and difficulty, researchers have been trying to solve the problem from different aspects. Usually, the first step is to build a model of energy consumption(called power model below for simplicity). In the literature, linear power model is often adopted. However, modern machines show significantly nonlinear features, which can be caught from data. Other modelling include fuzzy model etc., see e.g. [3, 4]. A review for power metering can be seen in [5]. With power model, the problem can be formulated in several ways. For example, we can maximize performance under power budget(e.g. [6]) or minimize power consumption when tracking performance or load balance between VMs (e.g. [7]), or optimize a newly defined objective, which integrates performance, power and the balance between different machines(e.g. [3][8]).

Based on the formulations, different kinds of solutions have been proposed, see [9] for a review of energy-efficient algorithms. Next we describe several specific ways. In [3] [10], the authors use reinforcement learning, and in [3] the reinforcement learning method is used together with fuzzy rule bases to achieve a defined objective, which shows robust performance improvement. In [11], the authors take advantage of the min-max and shares features inherent in virtualization technologies to allocate resources to a VM based on available resources, power costs, and application utilities. In [12], a temperature-aware workload placement is proposed in data centers. In [13], game theory is applied to formulate the problem and optimize power and performance at each level of the hierarchy while maintaining scalability.

Since it can provide a unified framework as well as rigorous controller design and can deal with dynamic and uncertain environment, control theory has been applied more and more to solve the problem. For example, in [7], the self-tuning regulator(which is an adaptive controller) is utilized to track performance and then optimize the energy assumption based on linear power model; in [8] the authors designs optimal controller by integrating the SLA function, and introduces a two-level control with one level being faster and the other being slower; in [14] Kalman filter is introduced to track the CPU utilizations and update the allocations of CPU resources to VMs accordingly; in [15] PID controller is proposed to manage power consumption and CPU utilization; in [16] MPC controller is designed; in [6] both PID controller and MPC are adopted at the same time; in [17] the authors compare effects of different controllers and find that predictive controller performs better and has some self-learning behavior. Other recent related papers are referred to [18–22].

This paper aims to provide a predictive controller for the virtualized computing systems with significantly nonlinear features on power consumption. First, significant nonlinearity on energy consumption of newer servers are discovered from data. Especially, we find that the max utilization/workload does not correspond to the max performance-power ratio. This announces invalidation of linear power model, which was commonly adopted in the literature. To deal with the nonlinear features, we build a discrete system state model as well as the state transition graph. Then, performance and power, especially power, is defined on each state, which is more accurate than linear power model. Furthermore, turning on/off a PM becomes feasible control actions, and the time-delay effect can also be taken into account. Based on the discrete model, we design a predictive controller naturally and minimize the quadratic cost function (which integrates both performance and power) dynamically by searching on the feasible state space. During the control process, the time-varying workload is predicted using an AR model. By choosing a moderate prediction horizon, the search approach becomes feasible. Experiment results validate effectiveness of the controller: by choosing a moderate weight, performance requirement can be satisfied for more than 99% times; meanwhile, the power consumption can be saved by 33% compared to the system without predictive controller. Part of this paper is presented in [23].

The paper is organized as follows. After a brief review on power models, Section 2 shows novel features on energy consumption in modern machines, which make linear power model not suitable any more. Section 3 formulates the problem by building a discrete state model. Section 4 describes the design of predictive controller. Section 5 validates the effectiveness of the controller by experiments. And finally, Section 6 concludes the paper with some remarks.

NOVEL FEATURES ON POWER CONSUMPTION

Linear Power Models in the Literature

In order to save power, a suitable power model, which catches the relationship between energy consumption per time unit (power, kilo · hour)) and physical parameters when running the machine, needs to be built. In the literature, linear power model is commonly adopted for simplicity and convenience, in which power consumption is modeled as a linear function with respect to certain system parameters, like CPU utilization, CPU frequency or number of VMs. Below we give some typical examples.

In [6], the VM's power P_{VM} is assumed to be $P_{VM} = c_{freq} * u_{cpu}$, where P_{VM} denotes VM's power, u_{cpu} is CPU utilization of the VM, and c_{freq} is a model parameter which is dependent on processor frequency.

In [8], the power (kilo · hour) is assumed to be Power = $P_{idle} + \alpha \cdot U_{cpu}$, where P_{idle} denotes the power consumption at idle state, U_{cpu} denotes CPU utilization and α is a parameter which is dependent on the specific machine and application.

In [24], the speed(or frequency) of the server s(GHz) is assumed to be $s = s_b + \alpha(P - b)$, where s_b(Hertz) denotes the speed of a fully utilized server running at b Watts, P (Watts) denotes the power allocated to the server, while the coefficient α (units of GHz per Watt) is the slope of the power-to-frequency curve, which can be obtained by experiment.

In [25], the processor power is modeled as an approximately linear function with respect to the DVFS level within the limited DVFS adaptation range available in real multi-core processors, i.e., $cp(k) = a(f(k + 1) - f(k)) + cp(k - 1)$, in which cp(k) denotes power consumption of the entire chip in the k-th control period, f(k) denotes total aggregated frequency of all cores on the chip in the kth control period and a denotes a generalized parameter that may vary for different chips and applications.

In most practical cases, linear power model is suitable and can provide convenience for utilization, see [8]. This causes nonlinear power models rare to be studied in the literature. However, linear power model has a defect, which can be seen from Fig 1. For the server Sun Netra x4250 and IBM x3450, power can be approximated by a linear function with respect to CPU load, where the positive power value when load equals zero denotes the idle power. However, when the machine is turned off, the system load is also zero, but the power value now is zero rather than idle power. Thus, linear power model cannot distinguish the state when the PM is shut down from the state when it holds zero workload, which makes turning on/off a PM deleted from the optional control action set in most previous work. As we have stated in Section 1, this does not fit the virtualized computing systems.

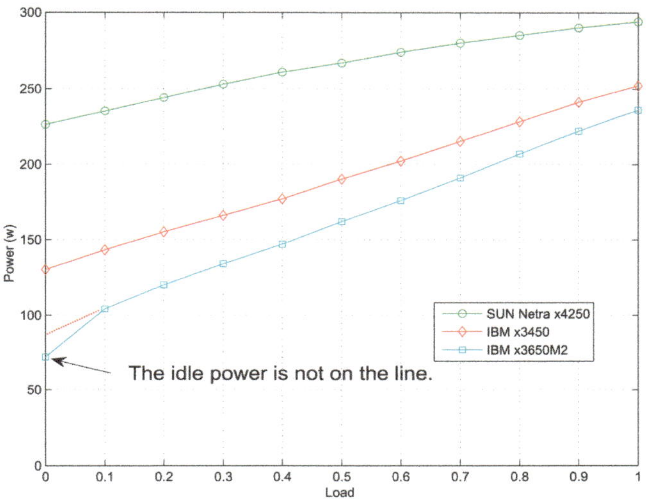

Figure 1: Linear model cannot reflect the reduced idle power of new machine.

http://dx.doi.org/10.1371/journal.pone.0134017.g001

Nonlinear Features of New Machines

In recent years, server technology has developed swiftly, making processors more and more advanced. Along with the progress, linear power model encounters new problems, making it unable to reflect novel features of new machines, as stated below.

First, a significant progress is the considerable reduction of idle power. Meanwhile, other parameters, such as power at big load, do not change with it. This makes linear power model no longer accurate, as shown in Fig 1.

In Fig 1, power curves for the servers Sun Netra x4250 and IBM x3450 are basically in accordance with a line. However, for the server IBM x3650M2, the idle power is reduced significantly (from 86 to 75 in fact), which is far below the line while other points still fit with it. Apparently, linear power model is no longer suitable now and thus management based on linear model is no longer optimal.

On the other hand, power saving is far from being the sole objective or even a prior objective. A good performance-power ratio(performance obtained per unit of power) makes more sense, since in practice it is more important to complete a job in the given time or satisfying other requirements.

From this point of view, in a linear power model, the performance-power ratio peaks at 100% load: assume

$$Power = P_{idle} + \alpha \cdot Load,$$

then the performance-power ratio

$$\frac{\text{Perf}}{\text{Power}} =: \frac{\text{Load}}{\text{Power}} = \frac{1}{\alpha + \frac{P_{\text{idle}}}{\text{Load}}}$$

is a strictly increasing function with respect to Load and peaks when Load = 100%.

However, the real case is different for new servers, as shown in Fig 2 below. Fig 2 draws the curve of performance-power ratio(defined as the requirements number that can be dealt with using one unit of power) with respect to load for the server IBM x3300M4 (data come from [26]).

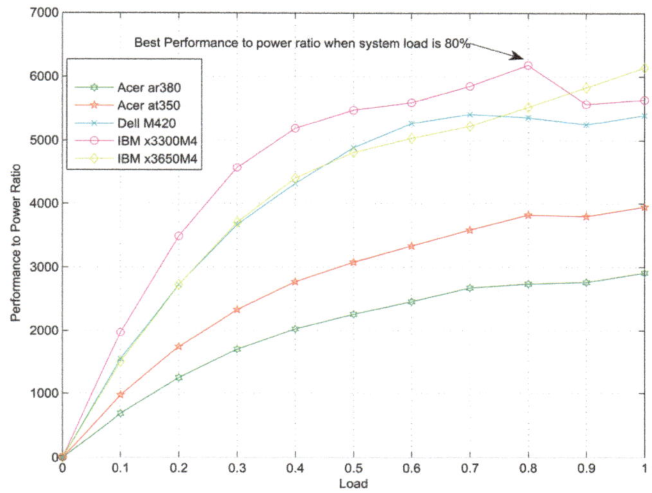

Figure 2: Performance-power ratio not peaks at 100% load.

http://dx.doi.org/10.1371/journal.pone.0134017.g002

From Fig 2, we can see that the biggest performance-power ratio occurs when the load is nearly 80% for some servers. Apparently, linear power model is not suitable any more and if we still use linear power model to simulate a modern server, we will miss the best choice. To deal with this, we will build a discrete-state-based power model in next section, which can include these nonlinear features.

DISCRETE STATE MODEL

In a virtualized computing system, there are some physical machines(PMs), denoted by PM_i, i= 1, 2, ..., n, which are homogeneous or heterogeneous. Virtual machines(VMs) are hosted on each PM and can be immigrated between different PMs. In this paper, we assume that at most one VM can be hosted on one processor of a PM.

Generally, the state of a PM describes whether the PM is turned on or off, the running frequency of its processors, and the VM number hosted on it. Thus, we define the state of PM_i as $S_i = [on, vms, freq]$, in which the variable on is taken as on or off, corresponding to PM_i being turned on or turned off, vms stands for the number of VMs running on PM_i, and freq represents the current frequencies of the running processors(so freq here is not a scalar).

In most situations, jobs allocated to one PM has the same type. Then, all influential factors are included in the state vector of a PM to complete the job. When different jobs are allocated to a PM, more factors, such as the architecture among the cores, must be considered, which is beyond the focus of this paper.

The time varying strength of a job is called workload and denoted by $\lambda(k)$ at time k. Suppose at time k, the state is $S_i(k)$, then the performance of PM_i, denoted by $perf_i(k)$, is defined to be themaximal workload which can be dealt with by PM_i. Obviously, if the allocated computing resource is not sufficient, it can happen that $perf_i(k) < \lambda(k)$. Here we assume that a requirement will be abandoned if it cannot be dealt with immediately, i.e., no aggregation effect exists here.

On the other hand, at time k, the power consumption of PM_i, denoted by $power_i(k)$, also is mainly determined by $S_i(k)$. In the experiment, $power_i(k)$ can be measured at each time.

Apparently, there is a tradeoff between high performance and low power consumption. Thus in practice, we usually aim to get a good balance between them. To this end, we will build a discrete state model of the system and then design predictive controller.

To simplify but without loss of generality, we describe the discrete model as we do in experiment. Let For PM_i, S_i=[on, vms, freq] to be discrete and taken out of the following 6 typical values:

$s^0 = [off, 0, null], s^1 = [on, 0, null],$

$s^2 = [on, 1, 1VM\ on\ low\ frequency], s^3 = [on, 1, 1VM\ on\ high\ frequency],$

$s^4 = [on, 2, 2VMs\ both\ on\ low\ frequency], s^5 = [on, 2, 2VMs\ both\ on\ high\ frequency]$

which are abbreviated as Turn off, Turn on, 1VM low frequency, 1VM high frequency, 2VM low frequency, 2VM high frequency respectively. Then, by regulating resource, one state can be transformed to another, as shown in Fig 3, in which each directed line represents a transition between the states with positive time delay. Transition without time delay, such as the transition between s^2 and s^3, between s^4 and s^5, are omitted in the figure.

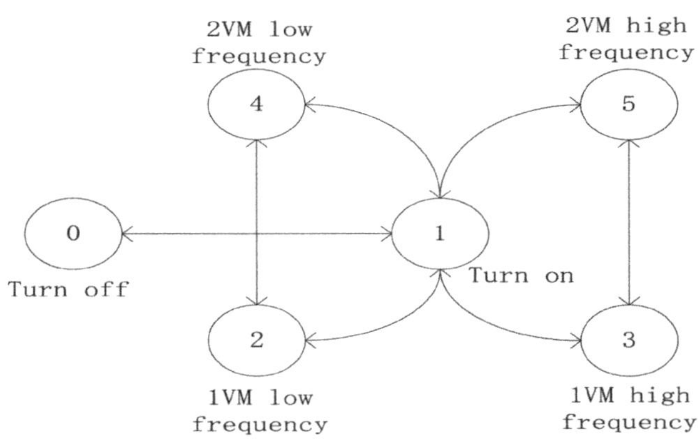

Figure 3: State transition graph for a single PM.

http://dx.doi.org/10.1371/journal.pone.0134017.g003

Now, assume that a job J is hosted on some PMs. Then, the performance and power at time kis defined on the state as

$$perfiJ_i^J(k) = function(S_i(k)) \in \{f^J(s^q), q = 0, 1, ..., 5\}, \quad (1)$$

$$poweriJ_i^J(k) = function(S_i(k)) \in \{g^J(s^q), q = 0, 1, ..., 5\}. \quad (2)$$

In practice, the value of $perf_i^J(k), power_i^J(k)$ can be learnt by prior experiments.

For a system with n PMs, define the performance and power at time k as

$$perf(k) = \sum_{i=1}^{n} perf_i(k), \quad (3)$$

$$power(k) = \sum_{i=1}^{n} power_i(k), \quad (4)$$

which are reasonable in view of the meanings of $perf_i(k)$ and $power_i(k)$.

There are several advantages of the discrete state model

- By defining discrete states and transition between them, turning on/off a PM can be treated as a control action. To be specific, s^0 to s^1 stands for turning on the machine and the reverse stands for turning it off.
- The idle state is also included in the model by state s^1, and thus it can be taken into account. Accordingly, the performance-power ratio is hidden in the model naturally and can be reflected in the cost function. Thus the discrete state model overcomes the drawbacks of linear power model and is suitable for new servers.
- Time delay effect, which exists during turning on/off a PM, now is taken into consideration. Time delay effect were often neglected in the literature, partly because it is hard to add into the model, partly because the PMs are always turned on in the past. Here, by modelling the system state discretely, time delay can be studied in a straight way.

In fact, the state transition graph builds a Markovian chain between states. This property, together with time delay effect, leads us to design predictive controller to regulate resources.

Finally, we note that this model is a coarse approximation to the real nonlinear relationship between performance, power and the system resource allocation. Putting it in a broader background, the method is commonly used to deal with a significantly nonlinear function. In the future, it is necessary and possible to make fine solution in which the frequency of each processor, the share of each virtual CPU, the memory and computing load will be added into the state.

MODEL BASED PREDICTIVE CONTROLLER

Controller Design

As we reviewed in Section 1, performance and power management problem can be handled in different ways: we can minimize power consumption under performance requirement, or we can maximize the performance within energy budget. Here we will define a quadratic cost function which combines performance and power together. By optimizing the cost function, we try to get a good balance between performance and power. To take time delay effect into consideration, the cost function must cover a time interval, which acquires the controller to predict the state for several time steps. The model predictive controller can just fit our scenario, see [24][25] for more introduction on the theory and applications of MPC.

Define the cost function at time k as below:

$$cost(k) = cost(S(k), u(k+1), \ldots, u(k+P))$$
$$\triangleq \sum_{j=1}^{P} \| perf(k+j|k) - \hat{\lambda}(k+j) \|_{Q_j}^2 + \sum_{j=1}^{M} \| power(k+j|k) \|_{R_j}^2$$
$$\triangleq \sum_{j=1}^{P} \| perf(\hat{S} + (k+j)|S(k)) - \hat{\lambda}(k+j) \|_{Q_j}^2 + \sum_{j=1}^{M} \| power(\hat{S} + (k+j)|S(k)) \|_{R_j}^2 \quad (5)$$

where j = 1, …P, P > 0 is prediction horizon, 0 < M ≤ P is control horizon, u(k + j) is a feasible control action at time k + j; $\hat{\lambda}(k+j)$ is predicted workload at time k + j; $\hat{S}(k+j)$ is predicted state at time k + j according to Fig 3 if the control actions are take as u(k + 1), …, u(k + j) sequentially from the state S(k); perf(k + j|k) and power(k + j|k) are predicted performance and power at time k + j; Q_j and R_j are the weights(matrices if perf(k + j|k) and power(k + j|k) are vectors) between performance and power. Note that here we assume that there is only one PM and omit the subscript i in $S_i(k)$, $perf_i(k)$, $power_i(k)$.

In Eq (5), the first part represents the quality of performance. $\hat{\lambda}(k+j)$, the prediction of workload, is the required performance at time k + j. Because of time delay and uncertainty, it is possible that current control cannot create the desired effect immediately. Therefore, the difference $perf(k+j \mid k) - \hat{\lambda}(k+j)$ can be treated as performance tracking error. And the quadratic term $\|perf(k+j \mid k) - \hat{\lambda}(k+j)\|_{Q_j}^2$ is the error with weight Q_j. $\|power(k+j \mid k)\|_{R_j}^2$ is the power value with its weight R_j. Here power(k + j|k) contains the switching cost caused by time delay. For example, the idle power is actually included in Eq (5). The larger Q_j and smaller R_j mean that the controller cares performance more, and vice versa. Obviously, appropriate choice of Q_j and R_j is significant for control, otherwise it cannot lead to a good balance between performance and power. This point will be discussed in Section 5 with more details.

The predictive controller aims to minimize the cost function at each time while state transition graph defines constraints for control action u(k + j):

$$\min_{\{u(k+1),\ldots,u(k+P)\}} cost(k),$$
$$\text{subject to that } u(k+j), j = 1, \ldots, P, \text{ coincides } Fig.3 \quad (6)$$

and it works as below. At each time k, the system state S(k) now is known. Then by choosing control action sequence u(k + 1), …, u(k + P), a predictive controller can predict the virtual state path $\hat{S}(k+1), \ldots, \hat{S}(k+P)$ in the future P steps according to the state model, i.e. Fig 3. Different control sequence u(k + 1), …, u(k + P) will lead to different state path $\hat{S}(k+1), \ldots, \hat{S}(k+P)$, which will lead to different performance and power prediction too. Given the cost function defined above, the optimal state path can be found and thus the optimal control sequence can be determined. Then, the optimal control action u(k + 1), …, u(k+ M), M ≤ P, in the future M steps can be determined, which will be adopted by PM at time k.

When there are more jobs and more machines, performance and power in Eq (5) can be defined as a vector. Then Q_j and R_j are weight matrices and the element implies the weight between each job and each machine, which makes the calculation of cost(k) more complicated. However, the predictive controller can be designed in a similar way.

Prediction for Workload

From description above, prediction of future workload $\hat{\lambda}(k+j)$ is needed for the controller to optimize the cost function. Here we assume that workload cannot wait and accumulate, thus its prediction is independent of controller. There are certain statistical models in the literature, in which the typical ones are ANOVA and AR model, or models combined with them. We refer to [17] for more details.

Briefly speaking, ANOVA model is suitable to analyze data for long time which have repeatable pattern, while AR model is suitable for prediction of data for shorter time. Since we only need the instant prediction and do not focus on data's pattern, we will apply the AR(Auto-Regression) model to predict workload in this paper.

For a sequence {x(k)}, an AR model with order r, simplified as AR(r) model, is to utilize linear combination of history data to fit the sequence and predict future values, i.e., to assume that $\hat{x}(k) = \sum_{i=1}^{r} a_i \cdot x(k-i) + e_k$ where e_k is noise and parameters a_i can be estimated off line or online.

By Eq (5), at each time k, we need to get the prediction $\hat{\lambda}(k+1), \ldots, \hat{\lambda}(k+P)$. Take r = 2 in the model as an example. Then we have $\hat{\lambda}(k) = a_1 \cdot \lambda(k-1) + a_2 \cdot \lambda(k-2)$. The parameters a_1, a_2 can be estimated by data fitting offline from history data. Then, at timek + 1, suppose we have λ(1), …λ(k), predictions for the future P steps are obtained by

$$\begin{aligned} \hat{\lambda}(k+1) &= a_1 \cdot \lambda(k) + a_2 \cdot \lambda(k-1); \\ \hat{\lambda}(k+2) &= a_1 \cdot \hat{\lambda}(k+1) + a_2 \cdot \lambda(k); \\ &\ldots\ldots \\ \hat{\lambda}(k+P) &= a_1 \cdot \hat{\lambda}(k+P-1) + a_2 \cdot \hat{\lambda}(k+P-2) \end{aligned} \quad (7)$$

in the order.

Optimization

At time k, suppose that we have predicted the future workload, we are at the place to choose control action sequence u(k + 1), …, u(k + P) to solve the Eq (6), i.e., to optimize cost(k). Still the workload is assumed to be abandoned if it is not dealt with at current time. Note that the optimization can not be finished directly by a Matlab function, as [18] did.

Due to the discrete state model, choice of u(k + 1), …, u(k + P) seems a tough task. It is true for the worst case when the state transition graph Fig 3 is a complete directed graph, i.e., all the states can be transited to all the states in one step. At this case, the state of a single PM can be taken from 6 different values, so the state space for P times has a volume of 6^P, which will increase exponentially with state number and thus will be very huge when there are many states of a PM or CPU. However, in the scenario of this paper, the state transition must satisfy physical constraints, which causes that Fig 3 is not a complete digraph and u(k + 1) can only be transited to certain states. This will shrink the state space greatly. On the other hand, considering the performance requirement in practice, the control actions which are far from being able to satisfy performance requirement, will also be abandoned first.

In this paper, we will search on the state space in the future P times. Then performance and power and thus the cost function cost(k) for each state path in the prediction horizon can be calculated according to the model. By choosing the path with minimal cost function, we can determine the control actions for the future M times.

Remark 4.1: For optimization Eq (6), another feasible sophisticated approach is to use dynamic programming. To do this, we need to draw the state transition graph for future P steps originating from current state, which is actually a tree. When the state space is huge and P is big, dynamic programming might be computationally efficient.

Remark 4.2: For large scale computing systems covering a large number of processors, e.g., more than 1000, the state space will amplify sharply which will result in great challenge on search algorithms. To deal with this, the usual method is to compute off line based on history data. On the other hand, there are often several batches of homogeneous machines in real systems, which can also simplify the problem highly by coarse grained modelling. Surely this brings about another tradeoff between coarse grained and finer grained modelling, which actually is a tradeoff between high-cost-optimal-outcome and low-cost-suboptimal-outcome.

EXPERIMENT VALIDATION

In this section, experiments will be implemented to check the effectiveness of the predictive controller based on the discrete state model.

Test Bed

We will take n = 3, m = 2, i.e., the system is composed of 3 physical machines PM_1, PM_2,PM_3, and at most 2 virtual machines can be hosted on each physical machine. Configuration of the PMs are listed in Table 1.

Table 1: Configuration of physical machines

PM name	Core/thread	DVFS frequency	Idle power(w)	Max power(w)
PM_3 (s-i7)	4/8	1.60, 2.93GHz	177	229
PM_2 (s-cf)	4/4	1.99, 2.49GHz	138	178
PM_1 (s-quad)	4/4	2.00, 2.83GHz	115	176

doi:10.1371/journal.pone.0134017.t001

http://dx.doi.org/10.1371/journal.pone.0134017.t001

The processors of PM_1, PM_2, PM_3 are Core i7, Xeon 5320, Core 2 Quad respectively, the OS is Fedora 16, and the virtualization software is KVM. Note that these PMs or processors are chosen to be different, which can happen usually in practice. The heterogeneity on PMs or processors can be easily handled in the experiment by choosing appropriate parameters in performance and power models; we will see that they will not influence the effectiveness of our controller, which implies extendibility of the method.

The parameters pertinent to performance and power during control process are measured from chosen servers and listed in Table 2 below, in which it is set 1T = 5 min, and perf, freq are abbreviations for performance, frequency.

Table 2: Parameters of control actions about performance and power.

http://dx.doi.org/10.1371/journal.pone.0134017.t002

Possible control action	Time delay	Power	Perf delay
turning off a PM	1T	max power	0
turning on a PM	2T	max power	0
changing freq by DVFS	0	determined by current freq	determined by new freq
turning on a VM	1T	power after VM turning on	0
turning off a VM	0	power before VM turning down	0
VM immigration	0	determined by new PM	determined by new PM

doi:10.1371/journal.pone.0134017.t002

Then the flow chart with predictive controller in the virtualized computer system is drawn in Fig 4.

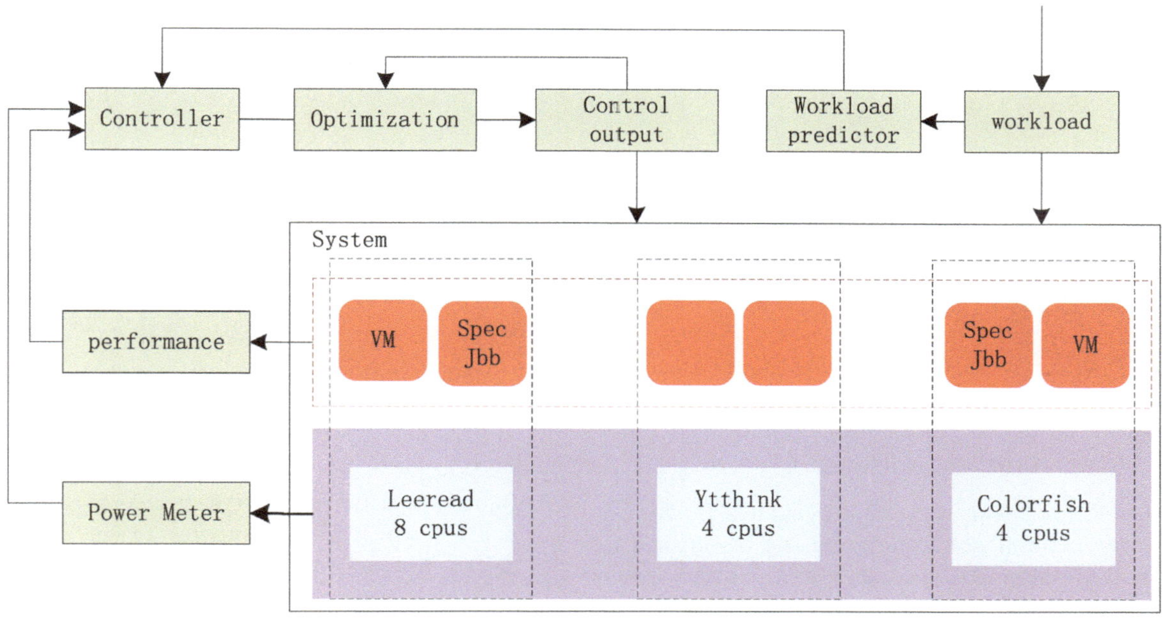

Figure 4: The architecture and control flow in the virtualized computing system.

http://dx.doi.org/10.1371/journal.pone.0134017.g004

Choice of Control Parameters

By the definition of cost(k) in Eq (5), four parameters will influence its value: the integers P, Mand the weights Q, R. Hence, they will influence the solution of optimization Eq (6) and thus the effect of controller. Additionally, P, M influence complexity of the problem more heavily while Q, R will influence the choice of control actions more significantly.

Now, we decide the prediction and control horizon P and M first. As we have stated above, long prediction horizon will lead to exponentially increasing search space for predicted state, which is not bearable for large scale servers. So P cannot be very large. On the other hand, Pmust be chosen according to the characteristics of workload so that prediction accuracy can be assured. With deeper prediction steps, prediction of workload will become less accurate. Based on these points, we take P = 3 in the experiment. M is taken as 1 as in the literature to avoid the conflict between control actions at time k and k + 1.

Then, we decide the weights Q_j and R_j. First, we set $Q_j = 1, j = 1, 2, \ldots, P$, by which we mean that the cost resulting from performance tracking at each time are the same.

As to the power saving weight R (denoted by R since M = 1), it is important to choose a moderate value for it: different R leads to different cost function and thus different optimal control action, and then different performance quality and power consumption. Moreover, it can be imagined that a moderate value of R is dependent of jobs' types. In the experiment, we will choose different R for an identical workload sequence, and compare the results under each R.

Experiment Results

To test the effectiveness of predictive controller, we will compare performance and power of the system under control with the case without controller(or equally, with open loop controller). For the system with predictive controller, we will compare experiment results under different choice of R. When the system is managed without controller, all physical machines will be always turned on so as to guarantee the performance requirement, thus turning on/off a PM is not an option of control actions in this case. This is just the practice for many data centers today.

We define several criteria as below:

$$PoU = \frac{|\{k : perf(k) < \lambda(k)\}|}{|k|} \cdot 100\%,$$

$$NNR = mean((\lambda(k) - perf(k)) \cdot 1_{\{perf(k)<\lambda(k)\}}),$$

$$PoS = \left(1 - \frac{\sum_k power_{olc}(k)}{\sum_k power_{pc}(k)}\right) \cdot 100\%, \quad (8)$$

where mean(x(k)) is average of sequence x(k), $power_{olc}(k)$ is the measured power without controller, and $power_{pc}(k)$ is the measured power with predictive controller. It is easy to see that PoU stands for the percent of times under performance requirement(i.e., the performance is not satisfied), NNR stands for the number of non-handled-request, and PoS stands for thepercent of power saving compared with the open-loop-controller case.

Basic Results.

Table 3 below roughly shows the experiments results with different R:

Table 3: Performance and power criteria under different R.

http://dx.doi.org/10.1371/journal.pone.0134017.t003

R	PoU	NNR	PoS
5	0.23%	1000	21%
50	0.24%	3478	33%
100	38.00%	7911	35%
200	50.20%	11854	38%
500	74.48%	35582	51%

doi:10.1371/journal.pone.0134017.t003

From Table 3, we can see that the predictive controller can save power considerably compared to the system with open-loop controller while the performance tracking result is very different under different R. To be specific,

- When R = 5, PoU is very small, being 0.23%, meaning that the performance requirement can be satisfied for almost all the times. Averagely, 1000 requests cannot be handled at each time. However, in this case, the power can be saved only by 21%, which is not bad but can be improved still.
- When R = 50, PoU is also very small, being 0.24%, which is almost the same with R = 5 case. And now there are 3478 requests which will be abandoned at each time. On the other hand, power consumption can be saved by 33%, which is great progress to R = 5 case.
- When R = 100, PoU now increases pretty highly, being 38%, implying that for more than one-third time, the performance requirement cannot be satisfied, which is very bad. And now the amount of requests that cannot be handled at each time is high, being 7911. At the same time, power is saved by only 35%, which is very near that of R = 50 case.
- When R = 200, PoU now becomes 50.2%, implying that performance cannot be satisfied for more than half of the time, while power is saved by 38%, which is still near that of R = 50 case.
- When R = 500, PoU now becomes 74.48%, implying a terrible performance. Power consumption is saved by 51%, which is a great progress to smaller R. Obviously, we cannot choose such a decision since performance is prior to power.

With respect to increasing R, results in Table 3 coincide with the physical meaning of R: with Rincreasing, power saving holds a bigger weight in the cost function, and thus power saving can become better, while performance becomes worse. We can find that R = 50 is a suitable weight, which can achieve a good balance between performance tracking (99.76% requirement can be deal with) and power saving (energy consumption can be saved by 33%).

Control Process in Details

Next, we will present the detailed process of the predictive controller. To this end, curves of the state, performance and power for three PMs are shown in each of Fig 5, Fig 6, Fig 7 as below, which contains five pictures in them. The data is measured for 24 hours every 5 minutes and the weight R is taken as 5, 50, 500 respectively in three figures.

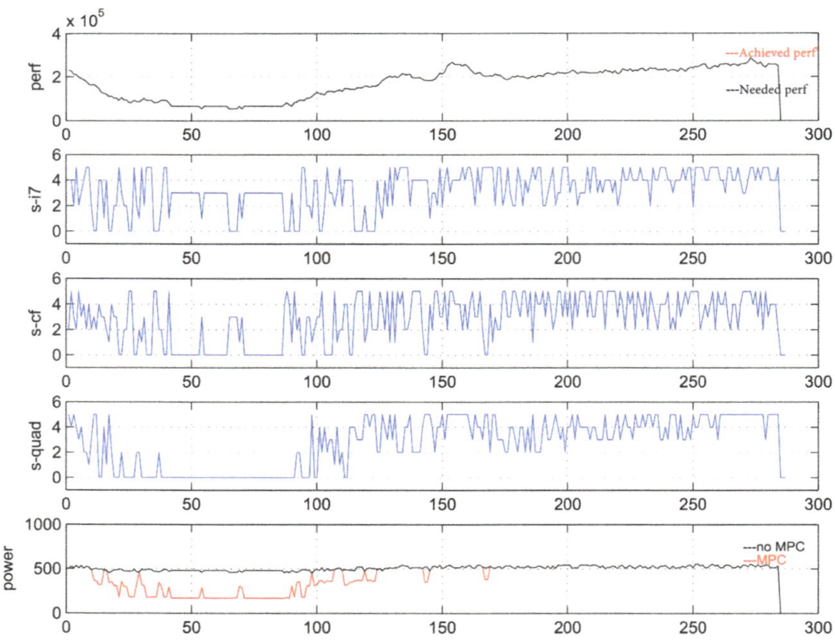

Figure 5: Performance, power and state transition in one day at R = 5.

http://dx.doi.org/10.1371/journal.pone.0134017.g005

In Fig 5, Fig 6, Fig 7, the horizontal axis denotes time; the vertical axis of the first picture denotes number of requirements that are coming and handled; the vertical axes of the second, third and fourth pictures denote the frequencies of three PMs; the vertical axis of the fifth picture denotes total power of the system. In the first picture, black curve denotes the coming requirement while red curve denotes the handle requirement at each time. In the fifth curve, black curve denotes the power without controller while red curve denotes the power with predictive controller.

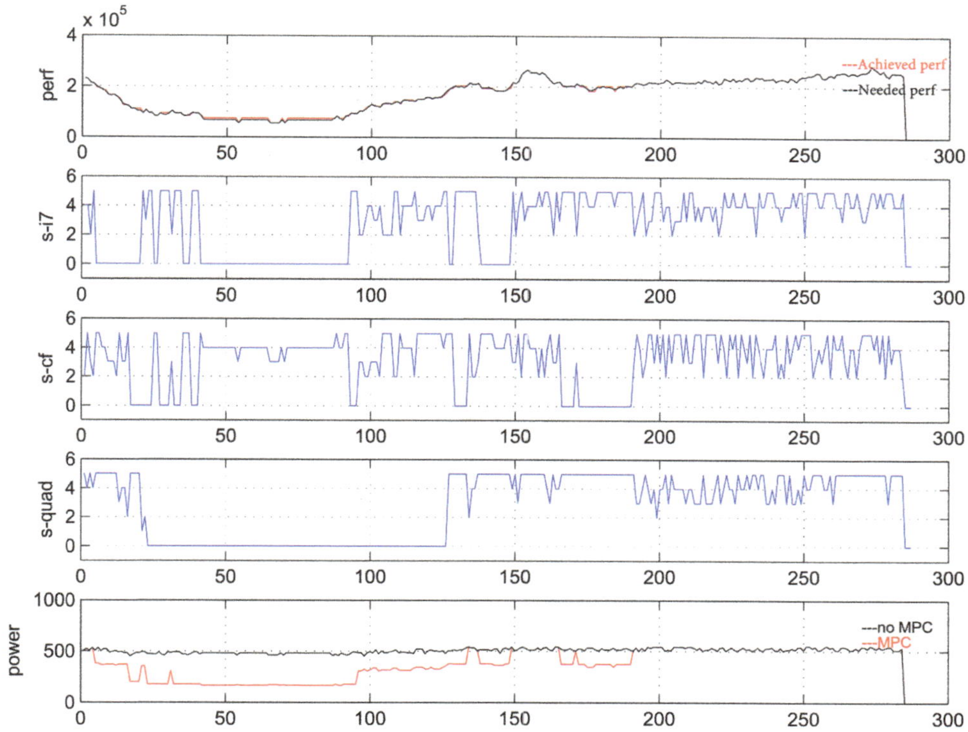

Figure 6: Performance, power and state transition in one day at R = 50.

http://dx.doi.org/10.1371/journal.pone.0134017.g006

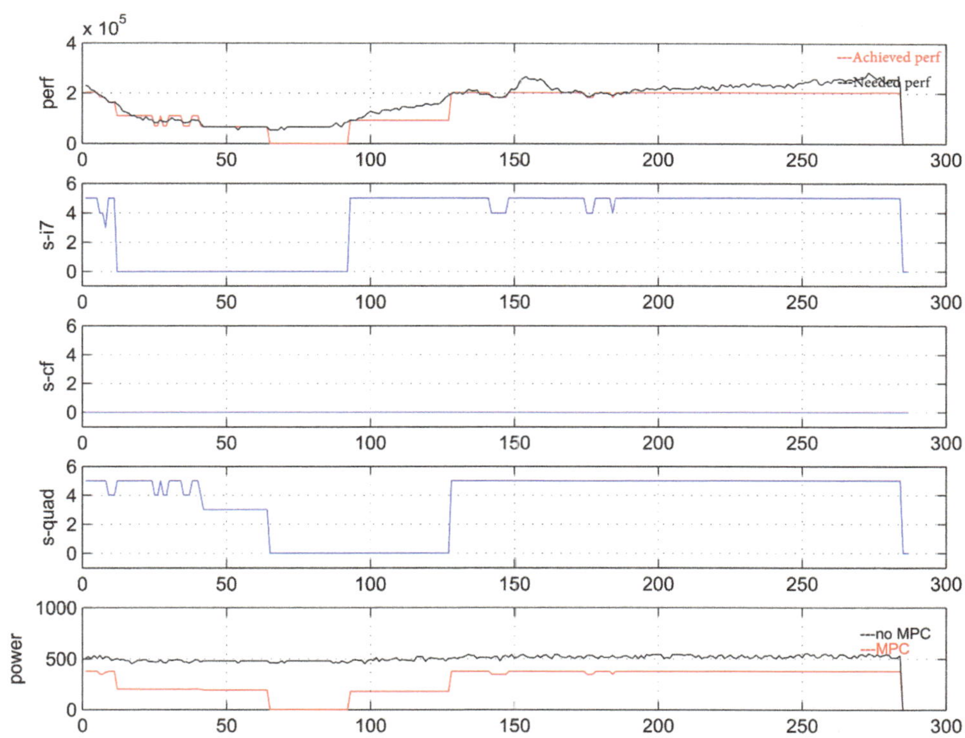

Figure 7: Performance, power and state transition in one day at R = 500.

http://dx.doi.org/10.1371/journal.pone.0134017.g007

From the curves, control actions can be distinguished from the change of frequencies and power clearly. When the frequency is zero and the power is smaller than the sum of idle powers, we can see that PM under observation is indeed shut down. And we can also see that under open loop control, the power is always maximal.

From Fig 5, Fig 6, Fig 7, when R = 5, the action of turning on/off the PM is used rarely, and the performance can be satisfied well while power saving is not very good. When R = 500, one PM is always being turned off, the power saving is good, while the performance tracking is bad. R= 50 is a moderate choice, which can lead to a good balance between performance and power. Meanwhile, at R = 50, VMs can be migrated between PMs, and power cycling is an effective control action.

To sum up, the choice of R is very crucial for control. If R can be taken moderately, such as R= 50, then a good balance between performance and power can be achieved: on one hand, the performance can be satisfied well; on the other hand, the power consumption can be saved considerably. In practice, R can be obtained by learning from experience, which depends on the specific system and jobs.

CONCLUSIONS AND FUTURE WORK

In this paper, the power saving problem in the nonlinear virtualized computing system is studied. First, we present some novel nonlinear characteristics of newer servers which are found from data. Such nonlinear features make linear power model not suitable again for fine control. Then, we build a discrete system state model, in which all control actions such as turning on/off a PM (or a VM) can be included and time delay effect can be reflected too. Then, by defining a quadratic cost function which involves both performance and power, the predictive controller based on the discrete state model can be designed in a natural way. Thus the cost function can be optimized by regulating computing resources dynamically. Experimental results show that with an appropriate weight of performance and power in the cost function, the predictive controller can achieve efficient performance and power management: almost 99.76% requirements can be dealt with and 33% power consumption can be saved compared with the case without controller.

In practice, the architecture of processors is also important and needs to be taken into account. Additionally, as we have noted before, the proposed discrete state model in this paper is still very rough. In order to get better effect, more sophisticated models need to be built so that more details can be included such as the share of VCPU, the frequency values of processors and so on. These will be left in the future work.

Finally, new technologies have been booming in control theory, such as the theory on multi-agent systems, in which massive agents with individual dynamics are contained and the consensus or special formations are often control objectives([26][27]). It can be surprisingly correlated with distributed computation, see [28][29]. Will such theory provide some inspirations to problems under study in the current paper? We do not know now while it might be interesting if new relations can be built.

AUTHOR CONTRIBUTIONS

Conceived and designed the experiments: CW. Performed the experiments: CW. Analyzed the data: YM. Contributed reagents/materials/analysis tools: YM. Wrote the paper: YM.

REFERENCES

1. Koomey J. Growth in data center electricity use 2005 to 2010. 2011, Oakland, CA: Analytics Press. August 1. Aavailable:http://www. analyticspress. com/datacenters. html.
2. Wen CJ, He J, Zhang J, Long X. PCFS: Power Credit Based Fair Scheduler Under DVFS for Muliticore Virtualization Platform. Proceedingoint of view, in a linear power model, the ps of the 2010 IEEE/ACM Int'l Conference on Green Computing and Communications & Int'l Conference on Cyber, Physical and Social Computing. 2010: 163–170.
3. Vengerov D. A reinforcement learning approach to dynamic resource allocation. Journal Engineering Applications of Artificial Intelligence. 2007, 20(3): 383–390. doi: 10. 1016/j. engappai. 2006. 06. 019.
4. Xu J, Zhao M, Fortes J, Carpenter R, Yousif M. On the use of fuzzy modeling in virtualized data center management. Proceedings of the IEEE Int'l Confonrefence on Autonomic Computing 2007. 2007: 25ï¿½C35. 10. 1109/ICAC. 2007. 28.
5. Krishman B, Amur H, Gavrilovska A, Schwan K. VM power metering: feasibility and challenges. ACM SIGMETRICS Performance Evaluation Review archive. 2010, 38(3): 56–60. doi: 10. 1145/1925019. 1925031.
6. Lim H, Kansal A, Liu J. Power Budgeting for Virtualized Data Centers. Proceedings of the 2011 USENIX conference on USENIX annual technical conference. 2011: 5–15.
7. Wen CJ, Long X, Mu YF. Adaptive Controller for Dynamic Power and Performance Management in the Virtualized Computing Systems. PLoS ONE. 8(2): e57551. 10. 1371/journal. pone. 0057551. doi: 10. 1371/journal. pone. 0057551. pmid:23451241
8. Kusic D, Kephart JO, Hanson JE, Kandasamy N, Jiang GF. Power and Performance Management of Virtualized Computing Environments Via Lookahead Control. Journal Cluster Computing. 2009, 12(1): 1–15. doi: 10. 1007/s10586-008-0070-y.
9. Albers S. Energy-efficient algorithms. Communications of the ACM. 2010: 53(5): 86–96. doi: 10. 1145/1735223. 1735245.
10. Tesauro G, Jong NK, Das R, Bennani MN. A hybrid reinforcement learning approach to autonomic resource allocation. Proceedings of the IEEE Int'l Conference on Autonomic Computing 2006. 2006: 65–73, 10. 1109/ICAC. 2006. 1662383.
11. Cardosa M, Korupolu MR, Singh A. Shares and utilities based power consolidation in virtualized server environments. Proceedings of the 11th IFIP/IEEE International Symposium on Integrated Network Management. 2009: 327–334. doi: 10. 1109/INM. 2009. 5188832

12. Moore JD, Chase J, Ranganathan P, Sharma R. Making scheduling 'cool': temperature-aware workload placement in data centers. Proceedings of the Annual Conference on USENIX Annual Technical Conference 2005. 2005: 5–5.

13. Khargharia B, Hariri S,Yousif MS. Autonomic power and performance management for computing systems. Cluster Computing. 2008, 11(2): 167–181. doi: 10. 1007/s10586-007-0043-6.

14. Kalyvianaki E, Charalambous T, Hand S. Self-adaptive and self-configured cpu resource provisioning for virtualized servers using kalman filters. Proceedings of the 6th International Conference on Autonomic Computing, ACM. 2009: 117–126.

15. Kjaer MA, Kihl M, Robertsson A. Resource allocation and disturbance rejection in web servers using SLAs and virtualized servers. IEEE Trans. Network and Service Management. 2009, 6(4): 226C239. doi: 10. 1109/ TNSM. 2009. 04. 090403.

16. Xu CZ, Liu BJ, Wei JB. Model predictive feedback control for QoS assurance in webservers. Computer. 2008, 41(3): 66C72. doi: 10. 1109/MC. 2008. 93.

17. Xu W, Zhu Xy, Singhal S, Wang Zk. Predictive Control for Dynamic Resource Allocation in Enterprise Data Centers. The 10th IEEE/IFIP Network Operations and Management Symposium, Vancouver, Canada. 2006: 115–126.

18. Wang YF, Wang XR, Chen M, Zhu XY. Power-Efficient Response Time Guarantees for Virtualized Enterprise Servers. Proceedings of the 2008 Real-Time Systems Symposium. 2008: 303–312. doi: 10. 1109/RTSS. 2008. 20.

19. Ardagna D, Trubian M, Zhang L. SLA based resource allocation policies in autonomic environments. Journal of Parallel and Distributed Computing. 2007, 67(3): 259–270. doi: 10. 1016/j. jpdc. 2006. 10. 006.

20. Raghavendra R, Ranganathan P, Talwar V, Wang ZK, Zhu XY. No power struggles: Coordinated multi-level power management for the data center. ACM SIGARCH Computer Architecture News. 2008, 36(1): 48–59. doi: 10. 1145/1353534. 1346289.

21. Govindan S, Choi J, Urgaonkar B, Sivasubramaniam A, Baldni A. Statistical profiling-based techniques for effective power provisioning in data centers. Proceedings of the 4th ACM European Conference on Computer Systems. 2009: 317–330.

22. Kusic D,Kandasamy N, Jiang GF. Combined power and performance management of virtualized computing environments serving session-based workloads. IEEE Transactions on Network and Service Management. 2011, 8(3): 245–258. doi: 10. 1109/TNSM. 2011. 0726. 100045.

23. Wen CJ, Long X, Mu YF. Power and performance management via model predictive control for virtualized cluster computing. Proceedings of the 32nd Chinese Control Conference. 2013: 2591–2596.

24. Gandhi A, Harchol-Balter M, Das R, Lefurgy C. Optimal power allocation in server farms. Proceedings of the 11th International Joint Conference on Measurement and Modeling of Computer Systems, ACM. 2009: 157–168.

25. Ma K, Li X, Chen M, Wang XR. Scalable power control for many-core architectures running multi-threaded applications. ACM SIGARCH Computer Architecture News. 2011, 39(3): 449–460. doi: 10. 1145/2024723. 2000117.

26. Standard Performance Evaluation Corporation. The SPEC Powerssj2008 benchmark. available: http://www. spec. org/power_ssj2008/results/res2010q4/power_ ssj2008-20100921-00294. html

27. Richalet J. Industrial applications of model based predictive control. Automatica. 1993, 29(5): 1251–1274. doi: 10. 1016/0005-1098(93)90049-Y.

28. Qin SJ, Badgwell TA. A survey of industrial model predictive control technology. Control Engineering Practice. 2003, 11: 733C764. doi: 10. 1016/S0967-0661(02)00186-7.

29. Wang L, Chen ZQ, Liu ZX, Yuan ZZ. Finite time agreement protocol design of multi-agent systems with communication delays. Asian Journal of Control. 2009, 11(3):281–286. doi: 10. 1002/asjc. 104.

30. Wang L, Sun SW, Xia CY. Finite-time stability of multi-agent system in disturbed environment. Nonlinear Dynamics. 2012, 67(3): 2009–2016. doi: 10. 1007/ s11071-011-0125-0.

31. Tsitsiklis JN, Bertsekas DP, Athans M. Distributed asynchronous deterministic and stocahstic gradient optimization algorithms. IEEE Transactions on Automatic Control. 1986, 31(9): 803C812. doi: 10. 1109/TAC. 1986. 1104412.

32. Jadbabaie A, Lin J,Morse AS. Coordination of groups of mobile autonomous agents using nearest neighbor rules. IEEE Transactions on Automatic Control. 2003, 48(6):988C1001. doi: 10. 1109/TAC. 2003. 812781.

CHAPTER 2 AHAH COMPUTING–FROM METASTABLE SWITCHES TO ATTRACTORS TO MACHINE LEARNING

Michael Alexander Nugent[1,2,3] and Timothy Wesley Molter[1,2,3]

[1]M. Alexander Nugent Consulting, Santa Fe, New Mexico, United States of America
[2]KnowmTech LLC, Albuquerque, New Mexico, United States of America
[3]Xeiam LLC, Santa Fe, New Mexico, United States of America

ABSTRACT

Modern computing architecture based on the separation of memory and processing leads to a well known problem called the von Neumann bottleneck, a restrictive limit on the data bandwidth between CPU and RAM. This paper introduces a new approach to computing we call AHaH computing where memory and processing are combined. The idea is based on the attractor dynamics of volatile dissipative electronics inspired by biological systems, presenting an attractive alternative architecture that is able to adapt, self-repair, and learn from interactions with the environment. We envision that both von Neumann and AHaH computing architectures will operate together on the same machine, but that the AHaH computing processor may reduce the power consumption and processing time for certain adaptive learning tasks by orders of magnitude. The paper begins by drawing a connection between the properties of volatility, thermodynamics, and Anti-Hebbian and Hebbian (AHaH) plasticity. We show how AHaH synaptic plasticity leads to attractor states that extract the independent components of applied data streams and how they form a computationally complete set of logic functions. After introducing a general memristive device model based on collections of metastable switches, we show how adaptive synaptic weights can be formed from differential pairs of incremental memristors. We also disclose how arrays of synaptic weights can be used to build a neural node circuit operating AHaH plasticity. By configuring the attractor states of the AHaH node in different ways, high level machine learning functions are demonstrated. This includes unsupervised clustering, supervised and unsupervised classification, complex signal prediction, unsupervised robotic actuation and combinatorial optimization of procedures–all key capabilities of biological nervous systems and modern machine learning algorithms with real world application.

INTRODUCTION

How does nature compute? Attempting to answer this question naturally leads one to consider biological nervous systems, although examples of computation abound in other manifestations of life. Some examples include plants [1]–[5], bacteria [6], protozoan [7], and swarms [8], to name a few. Most attempts to understand biological nervous systems fall along a spectrum. One end of the spectrum attempts to mimic the observed physical properties of nervous systems. These models necessarily contain parameters that must be tuned to match the biophysical and architectural properties of the natural model. Examples of this approach include Boahen's neuromorphic

Citation: Nugent MA, Molter TW (2014) AHaH Computing–From Metastable Switches to Attractors to Machine Learning. PLoS ONE 9(2): e85175. doi:10.1371/journal.pone.0085175.

circuit at Stanford University and their Neurogrid processor [9], the mathematical spiking neuron model of Izhikevich [10] as well as the large scale modeling of Eliasmith [11]. The other end of the spectrum abandons biological mimicry in an attempt to algorithmically solve the problems associated with brains such as perception, planning and control. This is generally referred to as machine learning. Algorithmic examples include support vector maximization [12], *k*-means clustering [13] and random forests [14]. Many approaches fall somewhere along the spectrum between mimicry and machine learning, such as the CAVIAR [15] and Cognimem [16] neuromorphic processors as well as IBM's *neurosynaptic core* [17]. In this paper we consider an alternative approach outside of the typical spectrum by asking ourselves a simple but important question: How can a brain compute given that it is built of volatile components?

A brain, like all living systems, is a far-from-equilibrium energy dissipating structure that constantly builds and repairs itself. We can shift the standard question from "how do brains compute?" or "what is the algorithm of the brain?" to a more fundamental question of "how do brains build and repair themselves as dissipative attractor-based structures?" Just as a ball will roll into a depression, an attractor-based system will fall into its attractor states. Perturbations (damage) will be fixed as the system reconverges to its attractor state. As an example, if we cut ourselves *we heal*. To bestow this property on our computing technology we must find a way to represent our computing structures as attractors. In this paper we detail how the attractor points of a plasticity rule we call Anti-Hebbian and Hebbian (AHaH) plasticity are computationally complete logic functions as well as building blocks for machine learning functions. We further show that AHaH plasticity can be attained from simple memristive circuitry attempting to maximize circuit power dissipation in accordance with ideas in nonequilibrium thermodynamics.

Our goal is to lay a foundation for a new type of practical computing based on the configuration and repair of volatile switching elements. We traverse the large gap from volatile memristive devices to demonstrations of computational universality and machine learning. The reader should keep in mind that the subject matter in this paper is necessarily diverse, but is essentially an elaboration of these three points:

- AHaH plasticity emerges from the interaction of volatile competing energy dissipating pathways.
- AHaH plasticity leads to attractor states that can be used for universal computation and advanced machine learning
- Neural nodes operating AHaH plasticity can be constructed from simple memristive circuits.

The Adaptive Power Problem

Through constant dissipation of free energy, living systems continuously repair their seemingly fragile state. A byproduct of this condition is that living systems are intrinsically adaptive at all scales, from cells to ecosystems. This presents a difficult challenge when we attempt to simulate such large scale adaptive networks with modern von Neumann computing architectures. Each adaptation event must necessarily reduce to memory–processor communication as the state variables are modified. The energy consumed in shuttling information back and forth grows in line with the number of state variables that must be continuously modified. For large scale adaptive systems like the brain, the inefficiencies become so large as to make simulations impractical.

As an example, consider that IBM's recent cat-scale cortical simulation of 1 billion neurons and 10 trillion synapses [18] required 147,456 CPUs, 144 TB of memory, running at 1/83 real-time. At a power consumption of 20 W per CPU, this is 2.9 MW. Under perfect scaling, a real-time simulation of a human-scale cortex would dissipate over 7 GW of power. The number of adaptive variables under constant modification in the IBM simulation is orders of magnitude less than the biological counterpart and yet its power dissipation is orders of magnitude larger. Another example from Google to train neural networks on YouTube data roughly doubled the accuracy from previous attempts [19]. The effort took an array of 16,000 CPU cores working at full capacity for 3 days. The model contained 1 billion connections, which although impressive pales in comparison to biology. The average human neocortex contains 150,000 billion connections [20] and the number of synapses in the neocortex is a fraction of the total number of connections in the brain. At 20 W per core, Google's simulation consumed about 320 kW. Under perfect scaling, a human-scale simulation would dissipate 48 GW of power.

At the core of the adaptive power problem is the energy wasted during memory–processor communication. The ultimate solution to the problem entails finding ways to let memory configure itself, and AHaH computing is one such method.

The Adaptive Power Solution

Consider two switches, one non-volatile and the other volatile. Furthermore, consider what it takes to change the state of each of these switches, which is the most fundamental act of adaptation or reconfiguration. Abstractly, a switch can be represented as a potential energy well with two or more minima.

In the non-volatile case, sufficient energy must be applied to overcome the barrier potential. Energy must be dissipated in proportion to the barrier height once a switching event takes place. Rather than just the switch, it is also the electrode leading to the switch that must be raised to the switch barrier energy. As the number of adaptive variables increases, the power required to sustain the switching events scales as the total distance needed to communicate the switching events and the square of the voltage.

A volatile switch on the other hand cannot be read without damaging its state. Each read operation lowers the switch barriers and increases the probability of random state transitions. Accumulated damage to the state must be actively repaired. In the absence of repair, the act of reading the state is alone sufficient to induce state transitions. The distance that must be traversed between memory and processing of an adaptation event goes to zero as the system becomes intrinsically adaptive. The act of accessing the memory *becomes* the act of configuring the memory.

In the non-volatile case some process external to the switch (i.e. an algorithm on a CPU) must provide the energy needed to effect the state transition. In the volatile case an external process must *stop* providing the energy needed for state repair. These two antisymmetric conditions can be summarized as: "Stability for free, adaptation for a price" and "adaptation for free, stability for a price", respectively.

Not only does it make physical sense to build large scale adaptive systems from volatile components but furthermore there is no supporting evidence to suggest it is possible to do the contrary. A brain is a volatile dissipative out-of-equilibrium structure. It is therefore reasonable that a volatile solution to machine learning at low power and high densities exists. The goal of AHaH computing is to find and exploit this solution.

Historical Background

In 1936, Turing, best known for his pioneering work in computation and his seminal paper 'On computable numbers' [21], provided a formal proof that a machine could be constructed to be capable of performing any conceivable mathematical computation if it were representable as an algorithm. This work rapidly evolved to become the computing industry of today. Few people are aware that, in addition to the work leading to the digital computer, Turing anticipated connectionism and neuron-like computing. In his paper 'Intelligent machinery' [22], which he wrote in 1948 but was not published until well after his death in 1968, Turing described a machine that consists of artificial neurons connected in any pattern with modifier devices. Modifier devices could be configured to pass or destroy a signal, and the neurons were composed of NAND gates that Turing chose because any other logic function can be created from them.

In 1944, physicist Schrödinger published the book *What is Life?* based on a series of public lectures delivered at Trinity College in Dublin. Schrödinger asked the question: "How can the events in space and time which take place within the spatial boundary of a living organism be accounted for by physics and chemistry?" He described an aperiodic crystal that predicted the nature of DNA, yet to be discovered, as well as the concept of *negentropy* being the entropy of a living system that it exports to keep its own entropy low [23].

In 1949, only one year after Turing wrote 'Intelligent machinery', synaptic plasticity was proposed as a mechanism for learning and memory by Hebb [24]. Ten years later in 1958 Rosenblatt defined the theoretical basis of connectionism and simulated the *perceptron*, leading to some initial excitement in the field [25].

In 1953, Barlow discovered neurons in the frog brain fired in response to specific visual stimuli[26]. This was a precursor to the experiments of Hubel and Wiesel who showed in 1959 the existence of neurons in the primary visual cortex of the cat that selectively responds to edges at specific orientations [27]. This led to the theory of receptive fields where cells at one level of organization are formed from inputs from cells in a lower level of organization.

In 1960, Widrow and Hoff developed ADALINE, a physical device that used electrochemical plating of carbon rods to emulate the synaptic elements that they called *memistors* [28]. Unlike memristors, memistors are three terminal devices, and their conductance between two of the terminals is controlled by the time integral of the current in the third. This work represents the first integration of memristive-like elements with electronic feedback to emulate a learning system. In 1969, the initial excitement with perceptrons was tampered by the work of Minsky and Papert, who analyzed some of the properties of perceptrons and illustrated how they could not compute the XOR function using only local neurons [29]. The reaction to Minsky and Papert diverted attention away from connection networks until the emergence of a number of new ideas, including Hopfield networks (1982) [30], back propagation of error (1986) [31], adaptive resonance theory (1987) [32], and many other permutations. The wave of excitement in neural networks began to fade as the key problem of generalization versus memorization became better appreciated and the computing revolution took off.

In 1971, Chua postulated on the basis of symmetry arguments the existence of a missing fourth two terminal circuit element called a memristor (*memory resistor*), where the resistance of the memristor depends on the integral of the input applied to the terminals [33], [34].

VLSI pioneer Mead published with Conway the landmark text *Introduction to VLSI Systems* in 1980 [35]. Mead teamed with John Hopfield and Feynman to study how animal brains compute. This work helped to catalyze the fields of Neural Networks (Hopfield), Neuromorphic Engineering (Mead) and Physics of Computation (Feynman). Mead created the world's first neural-inspired chips including an artificial retina and cochlea, which was documented in his book *Analog VLSI Implementation of Neural Systems* published in 1989 [36].

Beinenstock, Cooper and Munro published a theory of synaptic modification in 1982 [37]. Now known as the BCM plasticity rule, this theory attempts to account for experiments measuring the selectivity of neurons in primary sensory cortex and its dependency on neuronal input. When presented with data from natural images, the BCM rule converges to selective oriented receptive fields. This provides compelling evidence that the same mechanisms are at work in cortex, as validated by the experiments of Hubel and Wiesel. In 1989 Barlow reasoned that such selective response should emerge from an unsupervised learning algorithm that attempts to find a factorial code of independent features [38]. Bell and Sejnowski extended this work in 1997 to show that the independent components of natural scenes are edge filters [39]. This provided a concrete mathematical statement on neural plasticity: Neurons modify their synaptic weight to extract independent components. Building a mathematical foundation of neural plasticity, Oja and collaborators derived a number of plasticity rules by specifying statistical properties of the neuron's output distribution as objective functions. This lead to the principle of *independent component analysis* (ICA) [40], [41].

At roughly the same time, the theory of support vector maximization emerged from earlier work on statistical learning theory from Vapnik and Chervonenkis and has become a generally accepted solution to the generalization versus memorization problem in classifiers [12], [42].

In 2004, Nugent et al. showed how the AHAH plasticity rule is derived via the minimization of a kurtosis objective function and used as the basis of self-organized fault tolerance in support vector machine network classifiers. Thus, the connection that margin maximization coincides with independent component analysis and neural plasticity was demonstrated [43], [44]. In 2006, Nugent first detailed how to implement the AHaH plasticity rule in memristive circuitry and demonstrated that the AHaH attractor states can be used to configure a universal reconfigurable logic gate [45]–[47].

In 2008, HP Laboratories announced the production of Chua's postulated electronic device, the memristor [48] and explored their use as synapses in neuromorphic circuits [49]. Several memristive devices were previously reported by this time, predating HP Laboratories [50]–[54], but they were not described as memristors. In the same year, Hylton and Nugent launched the Systems of Neuromorphic Adaptive Plastic Scalable Electronics (SyNAPSE) program with the goal of demonstrating large scale adaptive learning in integrated memristive electronics at biological scale and power. Since 2008 there has been an explosion of worldwide interest in memristive devices [55]–[59] device models [60]–[65], their connection to biological synapses[66]–[72], and use in alternative computing architectures [73]–[84].

THEORY

On the Origins of Algorithms and the 4th Law of Thermodynamics

Turing spent the last two years of his life working on mathematical biology and published a paper titled 'The chemical basis of morphogenesis' in 1952 [85]. Turing was likely struggling with the concept that algorithms represent structure, brains and life in general are clearly capable of creating such structure, and brains are ultimately a biological chemical process that emerge from chemical homogeneity. How does complex spatial-temporal structure such as an algorithm emerge from the interaction of a homogeneous collection of units?

Answering this question in a physical sense leads one straight into the controversial 4th law of thermodynamics. The 4th law is is attempting to answer a simple question with profound consequences if a solution is found: If the 2nd law says everything tends towards disorder, why does essentially everything we see in the Universe contradict this? At almost every scale of the Universe we see self-organized structures, from black holes to stars, planets and suns to our own earth, the life that abounds on it and in particular the brain. Non-biological systems such as Benard convection cells [86], tornadoes, lightning and rivers, to name just a few, show us that matter does not tend toward disorder in practice but rather does quite the opposite. In another example, metallic spheres in a non-conducting liquid medium exposed to an electric field will self-organize into fractal dendritic trees [87].

One line of argument is that ordered structures create entropy faster than disordered structures do and self-organizing dissipative systems are the result of *out of equilibrium thermodynamics*. In other words, there may not actually be a distinct 4th law, and all observed order may actually result from dynamics yet to be unraveled mathematically from the 2nd law. Unfortunately this argument does not leave us with an understanding sufficient to allow us to exploit the phenomena in our technology. In this light, our work with AHaH attractor states may provide a clue as to the nature of the 4th law in so much as it lets us construct useful self-organizing and adaptive computing systems.

One particularly clear and falsifiable formulation of the 4th law comes from Swenson in 1989:

"A system will select the path or assembly of paths out of available paths that minimizes the potential or maximizes the entropy at the fastest rate given the constraints [88]."

Others have converged on similar thoughts. For example, Bejan postulated in 1996 that:

"For a finite-size system to persist in time (to live), it must evolve in such a way that it provides easier access to the imposed currents that flow through it [89]."

Bejan's formulation seems intuitively correct when one looks at nature, although it has faced criticism that it is too vague since it does not say what particle is flowing. We observe that in many cases the particle is either directly a carrier of free energy dissipation or else it gates access, like a key to a lock, to free energy dissipation of the units in the collective. These particles are not hard to spot. Examples include water in plants, ATP in cells, blood in bodies, neurotrophins in brains, and money in economies.

More recently, Jorgensen and Svirezhev have put forward the *maximum power principle* [90]and Schneider and Sagan have elaborated on the simple idea that "nature abhors a gradient"[91]. Others have put forward similar notions much earlier. Morowitz claimed in 1968 that the flow of energy from a source to a sink will cause at least one cycle in the system [91] and Lotka postulated the *principle of maximum energy flux* in 1922 [92].

The Container Adapts

Hatsopoulos and Keenan's *law of stable equilibrium* [93] states that:

"When an isolated system performs a process, after the removal of a series of internal constraints, it will always reach a unique state of equilibrium; this state of equilibrium is independent of the order in which the constraints are removed."

The idea is that a system erases any knowledge about how it arrived in equilibrium. Schneider and Sagan state this observation in their book *Into the Cool: Energy Flow, Thermodynamics, and Life* [91] by claiming: "These principles of erasure of the path, or past, as work is produced on the way to equilibrium hold for a broad class of thermodynamic systems." This principle has been illustrated by connected rooms, where doors between the rooms are opened according to a particular sequence, and only one room is pressurized at the start. The end state is the same regardless of the path taken to get there. The problem with this analysis is that it relies on an external agent: the door opener.

We may reformulate this idea in the light of an adaptive container, as shown in Figure 1. A first*replenished* pressurized container P_0 is allowed to diffuse into two non-pressurized empty containers P_1 and P_2 though a region of matter M. Let us presume that the initial fluid conductance G_a between P_0 and P_1 is less than G_b. Competition for limited resources within the matter (conservation of matter) enforces the condition that the sum of conductances is constant:

$$G_a + G_b = k. \tag{1}$$

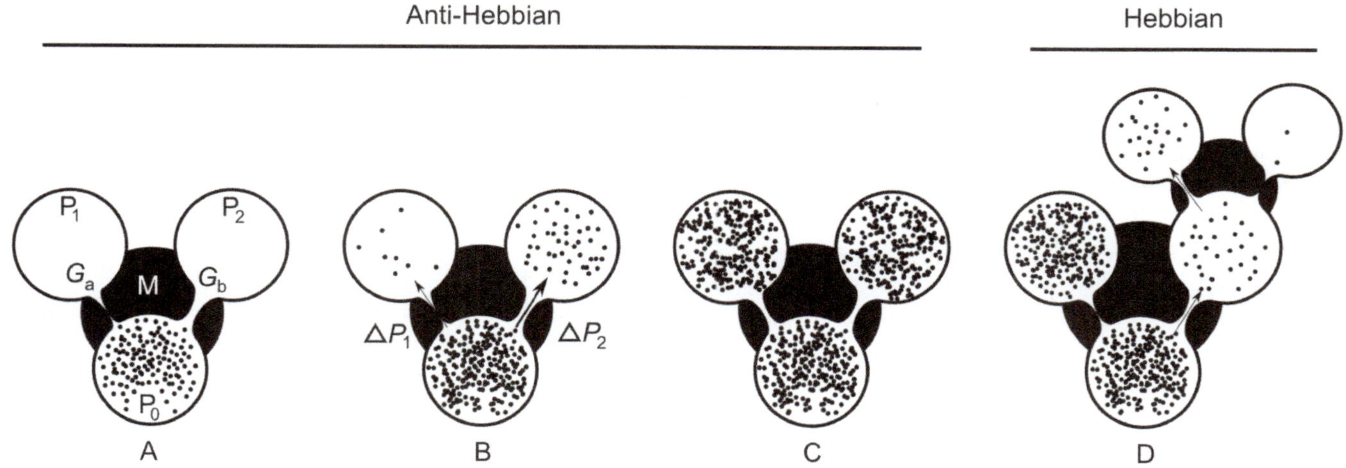

Figure 1: AHaH process.

A) A first replenished pressurized container P_0 is allowed to diffuse into two non-pressurized empty containers P_1 and P_2 though a region of matter M. B) The gradient ΔP_2 reduces faster than the gradient ΔP_1 due to the conductance differential. C) This causes G_a to grow more than G_b, reducing the conductance differential and leading to anti-Hebbian learning. D) The first detectable signal (work) is available at P_2 owing to the differential that favors it. As a response to this signal, events may transpire in the environment that open up new pathways to particle dissipation. The initial conductance differential is reinforced leading to Hebbian learning.

http://dx.doi.org/10.1371/journal.pone.0085175.g001

Now we ask how the container adapts as the system attempts to come to equilibrium. If it is the *gradient* that is driving the change in the conductance, then it becomes immediately clear that the container will adapt in such a way as to erase any initial differential conductance:

$$\Delta G = \lambda \Delta P \Delta t. \quad (2)$$

The gradient ΔP_2 will reduce faster than the gradient ΔP_1 and G_a will grow more than G_b. When the system comes to equilibrium we will find that the conductance differential, $G_a - G_b$has been reduced.

The sudden pressurization of P_2 may have an effect on the environment. In the moments right after the flow sets up, the first detectable signal (work) will be available at P_2 owing to the differential that favors it. As a response to this signal, any number of events could transpire in the environment that open up new pathways to particle dissipation. The initial conductance differential will be reinforced as the system rushes to equalize the gradient in this newly discovered space. Due to conservation of adaptive resources (Equation 1), an increase in $G_b$will require a drop in G_a, and vice versa. The result is that as $\Delta P_1 \to 0$, $G_a \to 0$, $G_b \to k$ and the system selects one pathway over another. The process illustrated in Figure 1 creates structure so long as new sinks are constantly found and a constant particle source is available. We now map this thermodynamic process to anti-Hebbian and Hebbian (AHaH) plasticity and show that the resulting attractor states support universal algorithms and broad machine learning functions. We furthermore show how AHaH plasticity can be implemented via physically adaptive memristive circuitry.

Anti-Hebbian and Hebbian (AHaH) Plasticity

The thermodynamic process outlined above can be understood more broadly as: (1) particles spread out along all available pathways through the environment and in doing so erode any differentials that favor one branch over the other, and (2) pathways that lead to dissipation (the flow of the particles) are stabilized. Let us first identify a synaptic weight, w, as the differential conductance formed from two energy dissipating pathways:

$$w = G_a - G_b. \quad (3)$$

We can now see that the synaptic weight possess state information. If $G_a > G_b$ the synapse is positive and if $G_a < G_b$ then it is negative. With this in mind we can explicitly define AHaH learning:

- Anti-Hebbian (erase the path): Any modification to the synaptic weight that reduces the probability that the synaptic state will remain the same upon subsequent measurement.
- Hebbian (select the path): Any modification to the synaptic weight that increases the probability that the synaptic state will remain the same upon subsequent measurement.

Our use of Hebbian learning follows a standard mathematical generalization of Hebb's famous postulate:

"When an axon of cell A is near enough to excite B and repeatedly or persistently takes part in firing it, some growth process or metabolic change takes place in one or both cells such that A's efficiency, as one of the cells firing B, is increased [24]."

Hebbian learning can be represented mathematically as $\Delta w \propto xy$, where x and y are the activities of the pre- and post-synaptic neurons and Δw is the change to the synaptic weight between them. Anti-Hebbian learning is the negative of Hebbian: $\Delta w \propto -xy$. Notice that intrinsic to this mathematical definition is the notion of state. The pre- and post-synaptic activities as well as the weight may be positive or negative. We achieve the notion of state in our physical circuits via differential conductances (Equation 3).

Linear Neuron Model

To begin our mapping of AHaH plasticity to computing and machine learning systems we use a standard linear neuron model. The choice of a linear neuron is motivated by the fact that they are ubiquitous in machine learning and also because it is easy to achieve the linear sum function in a physical circuit, since currents naturally sum.

The inputs x_i in a linear model are the outputs from other neurons or spike encoders (to be discussed). The weights w_i are the strength of the inputs. The larger w_i, the more x_i affects the neuron's output. Each input x_i is multiplied by a corresponding weight w_i and these values, combined with the bias b, are summed together to form the output y:

$$y = b + \sum_{i=0}^{N} x_i w_i. \tag{4}$$

The weights and bias change according to AHaH plasticity, which we further detail in the sections that follow. The AHaH rule acts to *maximize the margin* between positive and negative classes. In what follows, *AHaH nodes* refer to linear neurons implementing the AHaH plasticity rule.

AHaH Attractors Extract Independent Components

What we desire is a mechanism to extract the underlying building blocks or *independent components* of a data stream, irrespective of the number of discrete channels those components are communicated over. One method to accomplish this task is independent component analysis. The two broadest mathematical definitions of independence as used in ICA are (1) minimization of mutual information between competing nodes and (2) maximization of non-Gaussianity of the output of a single node. The non-Gaussian family of ICA algorithms uses negentropy and kurtosis as mathematical objective functions from which to derive a plasticity rule. To find a plasticity rule capable of ICA we can minimize a kurtosis objective function over the node output activation. The result is ideally the opposite of a peak: a bimodal distribution. That is, we seek a hyperplane that separates the input data into two classes resulting in two distinct *positive* and *negative* distributions. Using a kurtosis objective function, it can be shown that a plasticity rule of the following form emerges [43]:

$$\Delta w_i = x_i\left(\alpha y - \beta y^3\right), \tag{5}$$

where α and β are constants that control the relative contribution of Hebbian and anti-Hebbian plasticity, respectively. Equation 5 is one form of many that we call the *AHaH rule*. The important functional characteristics that Equation 5 shares with all the other forms is that as the magnitude of the post-synaptic activation grows, the weight update transitions from Hebbian to anti-Hebbian learning.

AHaH Attractors Make Optimal Decisions

An AHaH node is a hyperplane attempting to bisect its input space so as to make a binary decision. There are many hyperplanes to choose from and the question naturally arises as to which one is best. The generally agreed answer to this question is "the one that maximizes the separation (margin) of the two classes." The idea of *maximizing the margin* is central to support vector machines, arguably one of the more successful machine learning algorithms. As demonstrated in [43], [44], as well as the results of this paper, the attractor states of the AHaH rule coincide with the maximum-margin solution.

AHaH Attractors Support Universal Algorithms

Given a discrete set of inputs and a discrete set of outputs it is possible to account for all possible transfer functions via a logic function. Logic is usually taught as small two-input gates such as NAND and OR. However, when one looks at a more complicated algorithm such as a machine learning classifier, it is not so clear that it is performing a logic function. As demonstrated in following sections, AHaH attractor states are computationally complete logic functions. For example, when robotic arm actuation or prediction is demonstrated, self-configuring logic functions is also being demonstrated.

In what follows we will be adopting a *spike encoding*. A spike encoding consists of either a spike (1) or no spike (z). In digital logic, the state '0' is opposite or complimentary to the state '1' and it can be communicated. One cannot communicate a pulse of *nothing* (z). For this reason, we refer to a spike as '1' and no spike as a 'z' or *floating* to avoid this confusion. Furthermore, the output of an AHaH node can be positive or negative and hence possess a*state*. We can identify these positive and negative output states as logical outputs, for example the standard logical '1' is *positive* and '0' is *negative*.

Let us analyze the simplest possible AHaH node; one with only two inputs. The three possible input patterns are:

$$(x_0, x_1) = (z,1),(1,z),(1,1). \tag{6}$$

Stable synaptic states will occur when the sum over all weight updates is zero. We can plot the AHaH node's stable decision boundary on the same plot with the data that produced it. This can be seen in Figure 2, where decision boundaries A, B and C are labeled. Although the D state is theoretically achievable, it has been difficult to achieve in circuit simulations, and for this reason we exclude it as an available state. Note that every state has a corresponding anti-state. The AHaH plasticity is a local update rule that is attempting to maximize the margin between opposing positive and negative data distributions. As the positive distribution pushes the decision boundary away (making the weights more positive), the magnitude of the positive updates decreases while the magnitude of the opposing negative updates increases. The net result is that strong attractor states exist when the decision boundary can cleanly separate a data distribution.

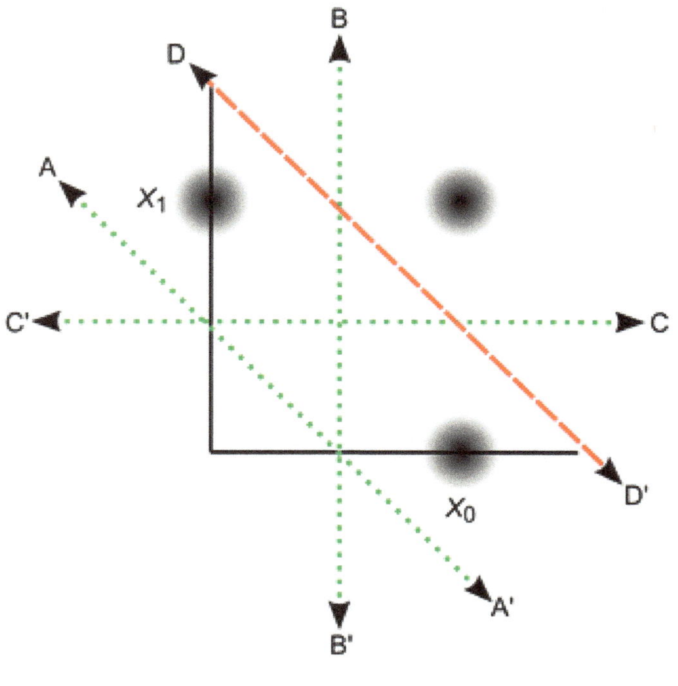

Figure 2: Attractor states of a two-input AHaH node.

The AHaH rule naturally forms decision boundaries that maximize the margin between data distributions (black blobs). This is easily visualized in two dimensions, but it is equally valid for any number of inputs. Attractor states are represented by decision boundaries A, B, C (green dotted lines) and D (red dashed line). Each state has a corresponding anti-state: $\psi = -\psi'$. State A is the null state and its occupation is inhibited by the bias. State D has not yet been reliably achieved in circuit simulations.

http://dx.doi.org/10.1371/journal.pone.0085175.g002

We refer to the A state as the null state. The null state occurs when an AHaH node assigns the same weight value to each synapse and outputs the same state for every pattern. The null state is mostly useless computationally, and its occupation is inhibited by bias weights. Through strong anti-Hebbian learning, the bias weights force each neuron to split the output space equally. As the neuron *locks on* to a stable bifurcation, the effect of the bias weights is minimized and the decision margin is maximized via AHaH learning on the input weights.

Recall Turing's idea of a network of NAND gates connected by *modifier devices* as mentioned in the Historical Background section. The AHaH nodes extract independent component states, the *alphabet* of the data stream. As illustrated in Figure 3, by providing the sign of the output of AHaH nodes to static NAND gates, a universal reconfigurable logic gate is possible. Configuring the AHaH attractor states, ψ_i, configures the logic function. We can do even better than this however.

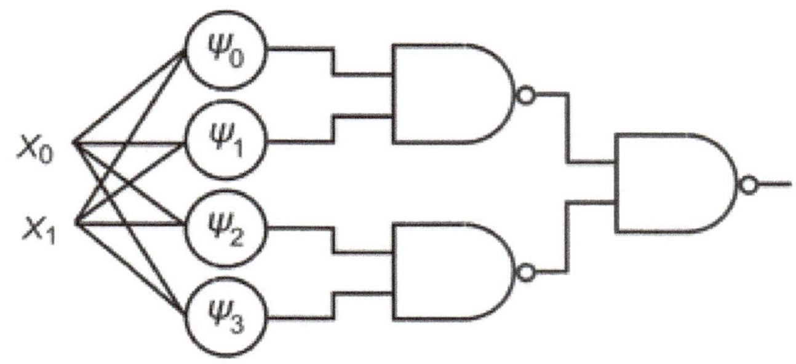

Figure 3: Universal reconfigurable logic.

By connecting the output of AHaH nodes (circles) to the input of static NAND gates, one may create a universal reconfigurable logic gate by configuring the AHaH node attractor states (ψ_i). The structure of the data stream on binary encoded channels X_0 and X_1 support AHaH attractor states $\psi_i = \{A,B,C,D\}$ (Figure 2). Through configuration of node attractor states the logic function of the circuit can be configured and all logic functions are possible. If inputs are represented as a spike encoding over four channels then AHaH node attractor states can attain all logic functions without the use of NAND gates.

http://dx.doi.org/10.1371/journal.pone.0085175.g003

We can achieve all logic functions directly (without NAND gates) if we define a *spike logic*code, where $0 = (1, z)$ and $1 = (z, 1)$, as shown in Table 1. As any algorithm or procedure can be attained from combinations of logic functions, AHaH nodes are building blocks from which any algorithm can be built. This analysis of logic is necessary to prove that AHaH attractor states can support any algorithm, not that AHaH computing is intended to replace modern methods of high speed digital logic.

Table 1: Spike logic patterns

http://dx.doi.org/10.1371/journal.pone.0085175.t001

Logic Pattern	Spike Logic Pattern
(0, 0)	(1, z, 1, z)
(0, 1)	(1, z, z, 1)
(1, 0)	(z, 1, 1, z)
(1, 1)	(z, 1, z, 1)

Digital logic states '0' and '1' across two input lines are converted to a spike encoding across four input lines. A spike encoding consists of either spikes (1) or no spikes (z). This encoding insures that the number of spikes at any given time is constant.

doi:10.1371/journal.pone.0085175.t001

AHaH Attractors are Bits

Every AHaH attractor consists of a state/anti-state pair that can be configured and therefore appears to represent a *bit*. In the limit of only one synapse and one input line activation, the state of the AHaH node is the state of the synapse just like a typical bit. As the number of simultaneous inputs grows past one, the *AHaH bit* becomes a collective over all interacting synapses. For every AHaH attractor state that outputs a '1', for example, there exists an equal and opposite AHaH attractor state that will output a '−1'. The state/anti-state property of the AHaH attractors follows mathematically from ICA, since ICA is in general not able to uniquely determine the sign of the source signals. The AHaH bits open up the possibility of configuring populations to achieve computational objectives. We take advantage of AHaH bits in the AHaH clustering and AHaH motor controller examples presented later in this paper. It is important to understand that AHaH attractor states are a reflection of the underlying statistics of the data stream and cannot be fully understood as just the collection of synapses that compose it. Rather, it is both the collection of synapses and also the structure of the information that is being processed that result in an AHaH attractor state. If we equate the data being processed as a sequence of measurements of the AHaH bit's state, we arrive at an interesting observation: the act of measurement not only effects the state of the AHaH bit, it actually*defines* it. Without the data structure imposed by the sequence of measurements, the state would simply not exist. This bears some similarity to ideas that emerge from quantum mechanics.

AHaH Memristor Circuit

Although we discuss a *functional* or *mathematical* representation of the AHaH node, AHaH computing necessarily has its foundation in a physical embodiment or circuit. The AHaH rule is achievable if one provides for competing adaptive dissipating pathways. The modern memristor provides us with just such an adaptive pathway. Two memristors provide us with two competing pathways. While some neuromorphic computing research has focused on exploiting the synapse-like behavior of a single memristor [68], [83] or using two serially connected memristive devices with different polarities [67], we implement synaptic weights via a differential pair of memristors with the same polarities (Figure 4) [45]–[47] acting as competing dissipation pathways.

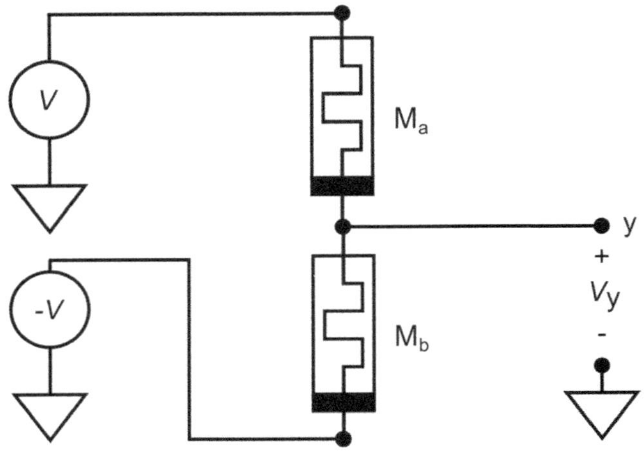

Figure 4: A differential pair of memristors forms a synapse.

A differential pair of memristors is used to form a synaptic weight, allowing for both a sign and magnitude. The bar on the memristor is used to indicate polarity and corresponds to the lower potential end when driving the memristor into a higher conductance state. M_a and M_b form a voltage divider causing the voltage at node y to be some value between V and $-V$. When driven correctly in the absence of Hebbian feedback a synapse will evolve to a symmetric state where $V_y=0$ V, alleviating issues arising from device inhomogeneities.

http://dx.doi.org/10.1371/journal.pone.0085175.g004

The circuits capable of achieving AHaH plasticity can be broadly categorized by the electrode configuration that forms the differential synapses as well as how the input activation (current) is converted to a feedback voltage that drives unsupervised anti-Hebbian learning [46], [47]. Synaptic currents can be converted to a feedback voltage statically (resistors or memristors), dynamically (capacitors), or actively (operational amplifiers). Each configuration requires unique circuitry to drive the electrodes so as to achieve AHaH plasticity, and multiple driving methods exist. The result is that a very large number of AHaH circuits exist, and it is well beyond the scope of this paper to discuss all configurations. Herein, a '2-1' two-phase circuit configuration is introduced because of its compactness and because it is amenable to mathematical analysis.

The functional objective of the AHaH circuit shown in Figure 5 is to produce an analog output on electrode y, given an arbitrary spike input of length N with k active inputs and $N-k$inactive (floating) inputs. The circuit consists of one or more memristor pairs (synapses) sharing a common electrode labeled y. Driving voltage sources are indicated with circles and labeled with an S, B or F, referring to spike, bias, or feedback respectively. The individual driving voltage sources for spike inputs of the AHaH circuit are labeled $S^0, S^1 \ldots, S^n$. The driving voltage sources for bias inputs are labeled $B^0, B^1 \ldots, B^m$. The driving voltage source for supervised and unsupervised learning is labeled F. The subscript values a and b indicate the positive and negative dissipative pathways, respectively.

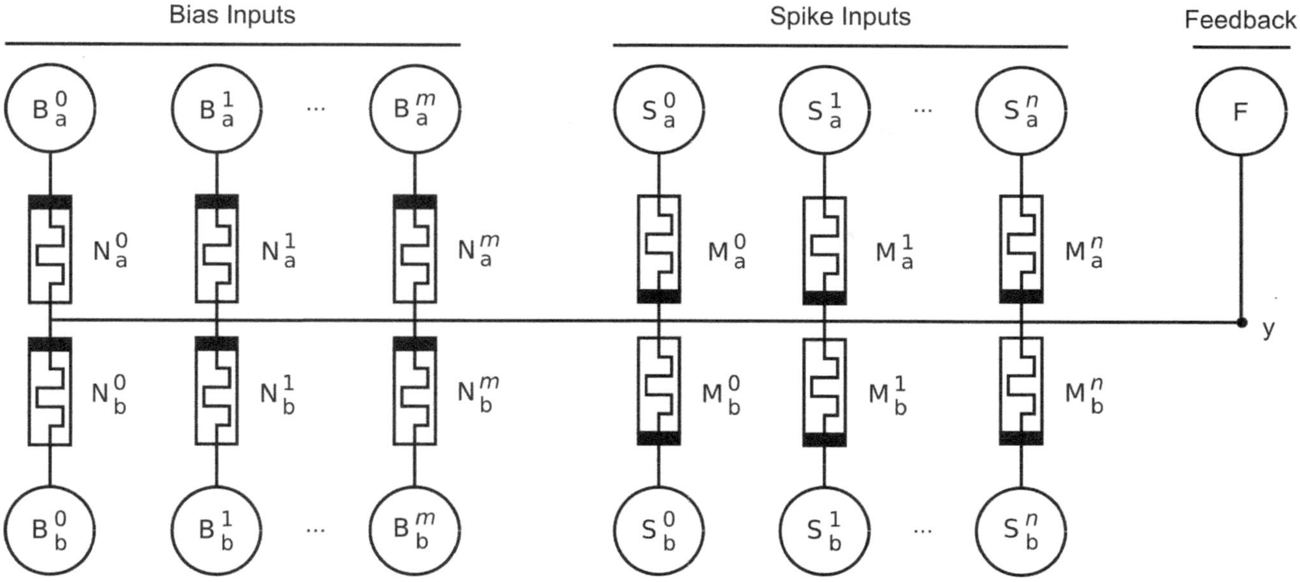

Figure 5: AHaH 2-1 two-phase circuit diagram.

The circuit produces an analog voltage signal on the output at node y given a spike pattern on its inputs labeled S^0, S^1. . ., S^n. The bias inputs B^0, B^1. . ., B^m are equivalent to the spike pattern inputs except that they are always active when the spike pattern inputs are active. F is a voltage source used to implement supervised and unsupervised learning via the AHaH rule. The polarity of the memristors for the bias synapse(s) is inverted relative to the input memristors. The output voltage, V_y, contains both state (positive/negative) and confidence (magnitude) information.

http://dx.doi.org/10.1371/journal.pone.0085175.g005

During the read phase, driving voltage sources S_a and S_b are set to $+V$ and $-V$respectively for all k active inputs. Inactive S inputs are left floating. The number of bias inputs to drive, m, is fixed or a function of k and driving voltage sources B_a and B_b are set to $+V$ and $-V$ respectively for all bias pairs. The combined conductance of the active inputs and biases produce an output voltage on electrode y. This analog signal contains useful confidence information and can be digitized via the $\mathrm{sgn}()$ function to either a logical 1 or a 0, if desired.

During the write phase, driving voltage source F is set to either $V_y^{\mathrm{write}} = V\overline{\mathrm{sgn}}(V_y^{\mathrm{read}})$(unsupervised) or $V_y^{\mathrm{write}} = V\overline{\mathrm{sgn}}(s)$ (supervised), where s is an externally applied teaching signal. The polarity of the driving voltage sources S and B are inverted to $-V$ and $+V$. The polarity switch causes all active memristors to be driven to a less conductive state, counteracting the read phase. If this dynamic counteraction did not take place, the memristors would quickly saturate into their maximally conductive states, rendering the synapses useless.

A more intuitive explanation of the above feedback cycle is that "the winning pathway is rewarded by not getting decayed." Each synapse can be thought of as two competing energy dissipating pathways (positive or negative evaluations) that are building structure (differential conductance). We may apply reinforcing Hebbian feedback by (1) allowing the winning pathway to dissipate more energy or (2) forcing the decay of the losing pathway. If we chose method (1) then we must at some future time ensure that we decay the conductance before device saturation is reached. If we chose method (2) then we achieve both decay and reinforcement at the same time.

AHaH Rule from Circuit Derivation

Without significant demonstrations of utility there is little motivation to pursue a new form of computing. Our functional model abstraction is necessary to reduce the computational overhead associated with simulating circuits and enable large scale simulations that tackle benchmark problems with real world utility. In this section, we derive the AHaH plasticity rule again, but instead of basing it on statistical independent components as in the derivation ofEquation 5, we derive it from simple circuit physics.

During the read phase, simple circuit analysis shows that the voltage on the electrode labeled y in the circuit shown in Figure 5 is:

$$V_y^{\mathrm{read}} = V\frac{\sum_i (G_a^i - G_b^i)}{\sum_i (G_a^i + G_b^i)}, \tag{7}$$

where G_a^i and G_b^i are the conductances of the ith memristors for the positive and negative dissipative pathways, respectively. The driving voltage sources S_a and S_b as well as B_a and B_b are set to $+V$ and $-V$ for all i active inputs and bias pairs.

During the write phase the driving voltage source F is set according to either a supervisory signal or in the unsupervised case, the anti-signum of the previous read voltage:

$$V_y^{\mathrm{write}} = V\overline{\mathrm{sgn}}(V_y^{\mathrm{read}}) = \begin{cases} +V & : V_y^{\mathrm{read}} < 0 \\ 0 & : V_y^{\mathrm{read}} = 0. \\ -V & : V_y^{\mathrm{read}} > 0 \end{cases} \tag{8}$$

We may adapt Equation 2 by replacing pressure with voltage:

$$\Delta G = \lambda \Delta V \Delta t. \quad (9)$$

Using Equation 9, the change to memristor conductances over the read and write phases is given in Table 2 and corresponds to the circuits of Figure 6. There are a total of four possibilities because of the two phases and the fact that the polarities of the bias memristors are inverted relative to the spike input memristors. Driving voltage source F is set to $V = V\overline{\text{sgn}}(V_y^{\text{read}})$ during the write phase for both spike and bias inputs. The terms in Table 2can be combined to show the total update to the input memristors over the read and write cycle:

$$\begin{aligned} \Delta G_a &= \beta\lambda V - \beta\lambda V_y^{\text{read}} - \alpha\lambda V - \alpha\lambda V\overline{\text{sgn}}(V_y^{\text{read}}) \\ \Delta G_b &= \beta\lambda V + \beta\lambda V_y^{\text{read}} + \alpha\lambda V\overline{\text{sgn}}(V_y^{\text{read}}) - \alpha\lambda V \quad , \\ \Delta G &= \Delta G_a - \Delta G_b = -2\beta\lambda V_y^{\text{read}} + 2\alpha\lambda V\text{sgn}(V_y^{\text{read}}) \end{aligned} \quad (10)$$

and likewise for the bias memristors:

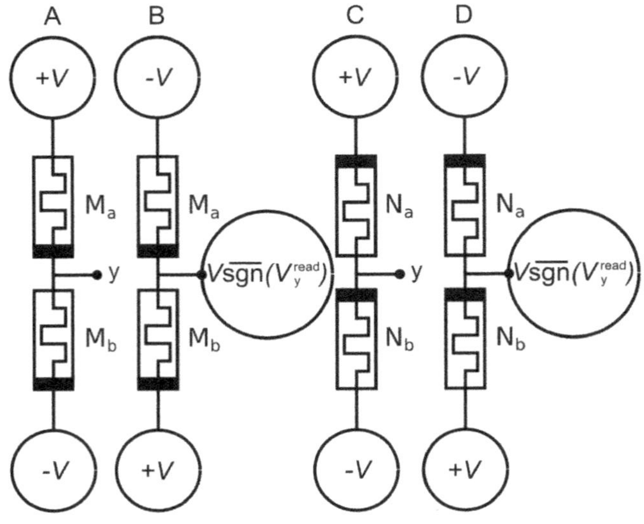

Figure 6: Circuit voltages across memristors during the read and write phases.

A) Voltages during read phase across spike input memristors. B) Voltages during write phase across spike input memristors. C) Voltages during read phase across bias memristors. D) Voltages during write phase across bias memristors.

http://dx.doi.org/10.1371/journal.pone.0085175.g006

Table 2: Memristor conductance updates during the read and write cycle

http://dx.doi.org/10.1371/journal.pone.0085175.t002

	Input Memristors		Bias Memristors	
	Read	**Write**	**Read**	**Write**
	$\Delta t = \beta$	$\Delta t = \alpha$	$\Delta t = \beta$	$\Delta t = \alpha$
	Accumulate	Decay	Decay	Accumulate
ΔG_a	$\beta\lambda\left(V - V_y^{\text{read}}\right)$	$-\alpha\lambda\left(V + V\overline{\text{sgn}}(V_y^{\text{read}})\right)$	$\beta\lambda\left(V_y^{\text{read}} - V\right)$	$\alpha\lambda\left(V\overline{\text{sgn}}(V_y^{\text{read}}) + V\right)$
ΔG_b	$\beta\lambda\left(V + V_y^{\text{read}}\right)$	$\alpha\lambda\left(V\overline{\text{sgn}}(V_y^{\text{read}}) - V\right)$	$-\beta\lambda\left(V + V_y^{\text{read}}\right)$	$\alpha\lambda\left(V - V\overline{\text{sgn}}(V_y^{\text{read}})\right)$

Both input and bias memristors are updated during one read/write cycle. During the read phase the active input memristors increase in conductance (accumulate) while the bias memristors decrease in conductance (decay). During the write phase the active input memristors decrease in conductance while the bias memristors increase in conductance. The changes in memristor conductances, ΔG_a and ΔG_b, for the memristor pairs are listed for all four cases.
doi:10.1371/journal.pone.0085175.t002

$$\begin{aligned} \Delta G_a &= -\beta\lambda V + \beta\lambda V_y^{\text{read}} + \alpha\lambda V + \alpha\lambda V\overline{\text{sgn}}(V_y^{\text{read}}) \\ \Delta G_b &= -\beta\lambda V - \beta\lambda V_y^{\text{read}} - \alpha\lambda V\overline{\text{sgn}}(V_y^{\text{read}}) + \alpha\lambda V\,. \\ \Delta G &= \Delta G_a - \Delta G_b = 2\beta\lambda V_y^{\text{read}} - 2\alpha\lambda V\text{sgn}(V_y^{\text{read}}) \end{aligned} \quad (11)$$

The quantity W^+, which we call the weight conjugate, remains constant due to competition for limited feedback:

$$W^+ = \sum_i \left(G_a^i + G_b^i\right) = k. \quad (12)$$

The output voltage during the read phase reduces to:

$$V_y^{\text{read}} = \frac{1}{k} V W^-, \quad (13)$$

where we have used the substitution:

$$W^- = \sum_i \left(G_a^i - G_b^i\right). \quad (14)$$

We identify the quantity VW^- as the standard linear sum over the active weights of the node (Equation 4). Furthermore, we identify the change of the ith weight as:

$$\Delta w^i = \Delta w_a^i - \Delta w_b^i = -2\beta\lambda V_y^{\text{read}} + 2\alpha\lambda V\text{sgn}(V_y^{\text{read}}). \quad (15)$$

By absorbing k, λ and the two constant 2s into the α and β constants we arrive at the functional form *Model A* of the AHaH rule:

$$\begin{aligned} y &= \sum_i w_i + \sum_{j=0}^{M} b_j \\ \Delta w_i &= -\beta y + \alpha\text{sgn}(y) + \eta - (1-\delta)w_i \; , \\ \Delta b_j &= \beta y - \alpha\text{sgn}(y) + \eta - (1-\delta)b_j \end{aligned} \quad (16)$$

where w_i is the ith spike input weight, b_j is the jth bias weight, and M is the total number of biases. To shorten the notation we make the substitution $V_y^{\text{read}} \rightarrow y$. Also note that the quantity

$\sum w_i$ is intended to denote the sum over the active (spiking) inputs. The noise variable η (normal Gaussian) and the decay variable δ account for the underlying stochastic nature of the memristive devices.

Model A is an approximation that is derived by making simplifying assumptions that include linearization of the update and non-saturation of the memristors. However, when a weight reaches saturation, $|w_a - w_b| \rightarrow \max$, it becomes resistant to Hebbian modification since the weight differential can no longer be increased, only decreased. This has the desirable effect of inhibiting null state occupation. However, it also means that functional Model A is not sufficient to account for these anti-Hebbian forces that grow increasingly stronger as weights near saturation. The result is that Model A leads to strange attractor dynamics and weights that can (but may not) grow without bound, a condition that is clearly unacceptable for a functional model and is not congruent with the circuit.

To account for the growing effect of anti-Hebbian forces we can make a modification to the bias weight update, and we call the resulting form functional *Model B*:

$$
\begin{aligned}
y &= \sum_i w_i + \sum_{j=0}^{M} b_j \\
\Delta w_i &= -\beta y + \alpha \operatorname{sgn}(y) + \eta - (1-\delta) w_i \\
\Delta b_j &= -\beta y + \eta - (1-\delta) b_j
\end{aligned}
\tag{17}
$$

The purpose of a functional model is to capture equivalent function with minimal computational overhead so that we may pursue large scale application development on existing technology without incurring the computational cost of circuit simulations. We justify the use of Model B because simulations prove it is a close functional match to the circuit, and it is computationally less expensive than Model A. However, it can be expected that better functional forms exist. Henceforth, any reference to the *functional* model refers to Model B.

Finally, in cases where supervision is desired, the sign of the Hebbian feedback may be modulated by an external supervisory signal, s, rather than the evaluation state y:

$$\Delta w_i = -\beta y + \alpha \operatorname{sgn}(s) + \eta - (1-\delta) w_i. \tag{18}$$

Compare Equation 17 to Equation 5. Both our functional models as well as the form ofEquation 5 converge to functionally similar attractor states. The common characteristic between both forms is a transition from Hebbian to anti-Hebbian learning, as the magnitude of node activation, y, grows large. This transition insures stable AHaH attractor states.

Generalized Memristive Device Model

Note that AHaH computing is not constrained to just one particular memristive device; any memristive device can be used as long as it meets the following criteria: (1) it is incremental and (2) its state change is voltage dependent. In order to simulate the proposed AHaH node circuit shown in Figure 5, a memristive device model is therefore needed. An effective memristive device model for our use should satisfy several requirements. It should accurately model the device behavior, it should be computationally efficient, and it should model as many different devices as possible. Many memristive device models exist, but we felt compelled to create another one which modeled a wider range of devices and, in particular, shows a transition from stochastic binary to incremental analog properties. Any device that can be manufactured to have electronic behavioral characteristics fitting to our model should be considered a viable component for building AHaH computing devices.

In our proposed semi-empirical model, the total current through the device comes from both a memory-dependent current component, I_m, and a Schottky diode current, I_s in parallel:

$$I = \phi I_m(V,t) + (1-\phi) I_s(V), \tag{19}$$

where $\phi \in [0,1]$. A value of $\phi = 1$ represents a device that contains no Schottky diode effects.

The Schottky component, $I_s(V)$, follows from the fact that many memristive devices contain a Schottky barrier formed at a metal–semiconductor junction [48], [63], [68], [94]. The Schottky component is modeled by forward bias and reverse bias components as follows:

$$I_s = \alpha_f e^{\beta_f V} - \alpha_r e^{-\beta_r V}, \tag{20}$$

where α_f, β_f, α_r, and β_r are positive valued parameters setting the exponential behavior of the forward and reverse biases exponential current flow across the Schottky barrier.

The memory component of our model, I_m, arises from the notion that memristors can be represented as a collection of conducting channels that switch between states of differing resistance. The channels could be formed from molecular switches, atoms, ions, nanoparticles or more complex composite structures. Modification of device resistance is attained through the application of an external voltage gradient that causes the channels to transition between conducting and non-conducting states. As the number of channels increases, the memristor will become more incremental as it acquires the ability to access more states. By modifying the number of channels we may cover a range of devices from binary to incremental. We treat each channel as a *metastable switch* (MSS) and the conductance of a collection of metastable switches capture the memory effect of the memristor.

An MSS possesses two states, A and B, separated by a potential energy barrier as shown inFigure 7. Let the barrier potential be the reference potential $V = 0$. The probability that the MSS will transition from the B state to the A state is given by P_A, while the probability that the MSS will transition from the A state to the B state is given by P_B. The transition probabilities are modeled as:

$$P_A = \alpha \frac{1}{1 + e^{-\beta(V - V_A)}} = \alpha \Gamma(V, V_A) \tag{21}$$

And

$$P_B = \alpha (1 - \Gamma(V, -V_B)), \tag{22}$$

Where

$$\beta = \frac{q}{kT} = (V_T)^{-1}$$

Here, V_T is the thermal voltage and is equal to approximately $26\,\text{mV}^{-1}$ at $T=300$ K, $\alpha=\frac{\Delta t}{t_c}$ is the ratio of the time step period Δt to the characteristic time scale of the device, t_c, and V is the voltage across the switch. The probability P_A is defined as the positive-going direction, so that a positive applied voltage increases the chances of occupying the A state. An MSS possesses utility in an electrical circuit as an adaptive element so long as these conductances differ. Each state has an intrinsic electrical conductance given by G_A and G_B. The convention is that $G_B > G_A$. Note that the logistic function $\frac{1}{1+e^{-x}}$ is similar to the hyperbolic-sign function used in other memristive device models including the nonlinear ion-drift, the Simmons tunnel barrier, the threshold adaptive models, and physics-based models [64], [95]–[98]. Our use of the logistic function follows simply from the requirement that probabilities must be bounded between 0 and 1.

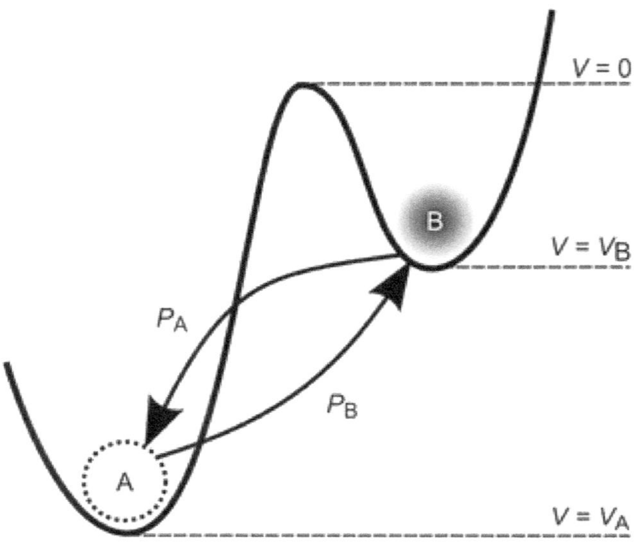

Figure 7: Generalized Metastable Switch (MSS).

An MSS is an idealized two-state element that switches probabilistically between its two states as a function of applied voltage bias and temperature. The probability that the MSS will transition from the B state to the A state is given by P_A, while the probability that the MSS will transition from the A state to the B state is given by P_B. We model a memristor as a collection of N MSSs evolving over discrete time steps.

http://dx.doi.org/10.1371/journal.pone.0085175.g007

We model a memristor as a collection of N MSSs evolving in discrete time steps, Δt. The total memristor conductance is given by the sum over each MSS:

$$G_m = N_A G_A + N_B G_B = N_B(G_B - G_A) + N G_A, \tag{23}$$

where N_A is the number of MSSs in the A state, N_B is the number of MSSs in the B state and $N = N_A + N_B$.

At each time step some subpopulation of the MSSs in the A state will transition to the B state, while some subpopulation in the B state will transition to the A state. The probability that kMSSs will transition out of a population of n MSSs is given by the binomial distribution:

$$P(n,k) = \frac{n!}{k!(n-1)!} p^k (1-p)^{n-k}, \tag{24}$$

where p is the probability a MSS will transition states. As n becomes large we may approximate the binomial distribution with a normal distribution:

$$\mathcal{N}(\mu,\sigma^2) = \frac{e^{\frac{-(x-\mu)^2}{2\sigma^2}}}{\sqrt{2\pi\sigma^2}}, \tag{25}$$

where $\mu = np$ and $\sigma^2 = np(1-p)$.

We model the change in conductance of a memristor as a probabilistic process where the number of switches that transition between A and B states is picked from a normal distribution with a center at np and variance $np(1-p)$, and where the state transition probabilities are given by Equations 21 and 22.

The update to the memristor conductance is given by the contribution from two random variables picked from two normal distributions:

$$\Delta N_B = \mathcal{N}(N_A P_A, N_A P_A(1-P_A)) - \mathcal{N}(N_B P_B, N_B P_B(1-P_B)). \tag{26}$$

The final update to the conductance of the memristor is then given by:

$$\Delta G_m = \Delta N_B (G_B - G_A). \tag{27}$$

Reducing the number of MSSs in the model will reduce the averaging effects and cause the memristor to behave in a more stochastic way. As the number of MSSs becomes small, the normal approximation to the binomial distribution breaks down. However, our desired operating regime of many metastable switches, and hence incremental behavior, is within the acceptable bounds of the approximation.

METHODS

All experiments are software based, and they involve the simulation of AHaH nodes in various configurations to perform various adaptive learning tasks. The source code for the experiments is written in the Java programming language and can be obtained from a Git repository linked to from Xeiam LLC's main web page at http://xeiam.com under the AHaH! project. The code used for the experiments in this paper is tagged as *PLOS_AHAH* on the *master* branch giving a pointer to the exact code used for this paper. The specific programs for each experiment are clearly identified at the end of each experiment description in the methods section. Further details about the programs and the relevant program parameters can be found in the source code itself in the form of comments.

There are two distinct models used for the simulation experiments: functional and circuit. The simulations based on the functional model use functional Model B as described above. The simulations based on the circuit model use ideal electrical circuit components and the generalized model for memristive devices. Nonideal behaviors such as parasitic impedances are not included in the circuit simulation experiments. We want to emphasize that at this stage we are attempting to cross the considerable divide between memristive electronics and general machine learning by defining a theoretical methodology for computing with dissipative attractor states. By focusing on nonideal circuit behavior at this stage we risk obfuscating what is otherwise a theory with minimal complexity.

Generalized Memristive Device Model

By adjusting the free variables in the generalized memristive device model and comparing the subsequent current-voltage hysteresis loops to four real world memristive device I–V data, matching model parameters were determined as shown in Table 3. The devices include the Ag-chalcogenide [55], AIST [99], GST [70], and WO_x [63] devices, and they represent a wide spectrum of incremental memristive devices found in recent publications exhibiting diverse characteristics. All simulations in this paper involving AHaH node circuitry use the memristor model parameters of the Ag-chalcogenide device, unless otherwise noted. The remaining three are presented in support of our general model.

Table 3: General memristive device model parameters fit to various devices

http://dx.doi.org/10.1371/journal.pone.0085175.t003

Device	t_c [ms]	G_A [mS]	G_B [mS]	V_A [V]	V_B [V]	ϕ	α_f	β_f	α_r	β_r
Ag-chalc	0.32	8.7	0.91	0.17	0.22	1	–	–	–	–
AIST	0.15	40	10	.23	.25	1	–	–	–	–
GST	0.42	.12	1.2	.9	0.6	0.7	5×10^{-3}	3.0	5×10^{-3}	3.0
WO_x	0.80	.025	0.004	0.8	1.0	.55	1×10^{-9}	8.5	22×10^{-9}	6.2

The devices used to test our general memristive device model include the Ag-chalcogenide, AIST, GST, and WO_x devices. The parameters in this table were determined by comparing the model response to a simulated sinusoidal or triangle-wave voltage to real I–V data of physical devices.
doi:10.1371/journal.pone.0085175.t003

Figure 8A shows the hysteresis curve of the model and raw Ag-chalcogenide device data driven at 100 Hz with a sinusoidal voltage of 0.25 V amplitude. Additional 1000 Hz and 10 kHz simulations are also shown. The predicted behavior of the model shows a good fit to the physical Ag-chalcogenide device. In fact the model is arguably better than other models (linear ion drift and nonlinear ion drift) tested for a similar device in [61]. Figure 8B shows the predicted response of two series-connected arbitrary memristive devices with differing parameters driven by the sinusoidal voltage as in 8A. The simulation of two devices in series (Figure 4) as shown in Figure 8B also displayed expected characteristics and agrees with results in [100] where the linear ion drift model was used. Experiments have not yet been carried out on physical devices to verify this. Figure 8C shows the incremental pulsed resistance change of a single Ag-chalcogenide modeled device for three different pulse train configurations. The three different pulse trains were chosen to show that by changing both the pulse width or the pulse voltage, the modeled behavior is predicted as expected. Figure 8Dshows the time response of the Ag-chalcogenide modeled device at frequencies of 100 Hz, 150 Hz, and 200 Hz. Figure 8E shows the simulated response of the Ag-chalcogenide modeled device to a triangle wave of both +0.1 V and −0.1 V amplitude at 100 Hz designed to show the expected incremental prediction of the model. Figure 8F shows additional model fits to the AIST, GST, and WO_x devices. As demonstrated, our model can be applied to a wide range of memristive devices from Chalcogenides to metal-oxides and more. The source code for these simulations is in *AgChalcogenideHysteresisPlotA.java*, *AgChalcogenideHysteresisPlotB.java*,*AgChalcogenidePulseTrainPlotC*, *AgChalcogenideTimePlotD*, *AgChalcogenideTrianglePlotE*,*AgInSbTeHysteresisPlot.java*, *GSTHysteresisPlot.java*, and *PdWO3WHysteresisPlot.java*.

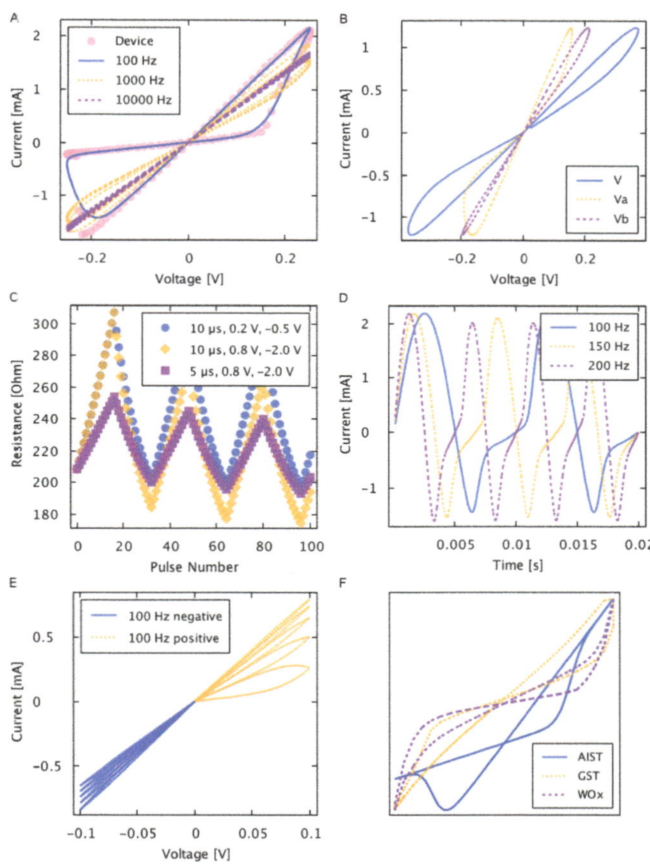

Figure 8: Generalized memristive device model simulations.

A) Solid line represents the model simulated at 100 Hz and dots represent the measurements from a physical Ag-chalcogenide device from Boise State University. Physical and predicted device current resulted from driving a sinusoidal voltage of 0.25 V amplitude at 100 Hz across the device. B) Simulation of two series-connected arbitrary devices with differing model parameter values. C) Simulated response to pulse trains of {10 μs, 0.2 V, −0.5 V}, {10 μs, 0.8 V, −2.0 V}, and {5 μs, 0.8 V, −2.0 V} showing the incremental change in resistance in response to small voltage pulses. D) Simulated time response of model from driving a sinusoidal voltage of 0.25 V amplitude at 100 Hz, 150 Hz, and 200 Hz. E) Simulated response to a triangle wave of 0.1 V amplitude at 100 Hz showing the expected incremental behavior of the model. F) Simulated and scaled hysteresis curves for the AIST, GST, and WO_x devices (not to scale).

http://dx.doi.org/10.1371/journal.pone.0085175.g008

When it comes time to manufacture AHaH node circuitry, an ideal memristor will be chosen taking into consideration many properties. It is likely that some types of memristors will be better candidates, some will not be suitable at all, and that the best device has yet to be fabricated. Based on our current understanding, the ideal device would have low thresholds of adaptation (<0.2 V), on-state resistance of ~100 kΩ or greater, high dynamic range, durability, the capability of incremental operation with very short pulse widths and long retention times of a week or more. However, even devices that deviate considerably from these parameters will be useful in more specific applications. As an example, short retention times on the order of seconds are perfectly compatible with combinatorial optimizers.

AHaH Circuit Simulation

Circuit simulations were carried out by solving for the voltage at node y in each AHaH node (Figure 5) using Kirchhoff's Current law (KCL) during the read phase followed by updating all memristor conductance values according to the generalized MSS model given the voltage drop across each memristor and the read period length. During the write phase, the memristor conductance values were individually updated according to the generalized MSS model given the voltage drop across each memristor and the write period length. The source code for the circuit is available in AHaH21Circuit.java. Parameters for operation of the circuit were set as follows: V_{dd}=0.5 V, V_{ss}=−0.5 V, read period (α)=1 μs, and write period (β)=1 μs. The number of input and bias memristors differed depending on the simulation task, as noted in each section below or in the source code.

Spike Encoding

All machine learning applications built from AHaH nodes have one thing in common: the inputs to the AHaH nodes take as input a spike pattern. A spike pattern is a set of integers that specify which synapses in the AHaH node are coactive. In terms of a circuit, this is a description of what physical input lines are being driven by the driving voltage (V). All other inputs remain floating (z). Any data source can be converted into a spike encoding with a spike encoder. As an example, the eye converts electromagnetic radiation into spikes, the ear converts sound waves into spikes, and the skin converts pressure into spikes. Each of these may be considered a spike encoder and each is optimized for a specific data source.

A simple example makes spike encoding for an AHaH node clear. Suppose a dataset is available where the colors of a person's clothes are associated with the sex of the person. The entire dataset consists of several colors → sex associations. For each person, the colors are mapped to an integer and added to a vector of variable length:

$$\begin{aligned} \{\text{red,blue,black}\} &\rightarrow \{1,2,5\} \\ \{\text{red,yellow,white}\} &\rightarrow \{1,3,4\}, \\ \{\text{white,black}\} &\rightarrow \{4,5\} \end{aligned} \tag{28}$$

where red maps to 1, blue maps to 2, yellow maps to 3, etc. The spike patterns for this dataset are then $\{1,2,5\}$, $\{1,3,4\}$, and $\{4,5\}$. In order to accommodate the range of spikes, the AHaH nodes would require at least five inputs or a spike space of five.

In the case of real-value numbers, a simple recursive method for producing a spike encoding can also conveniently be realized through strictly anti-Hebbian learning via a binary decision tree with AHaH nodes at each tree node. Starting from the root node and proceeding to the leaf node, the input x is summed with a bias b, $y=x+b$. Depending on the sign of the result y, it is routed in one direction or another toward the leaf nodes. The bias is updated according to anti-Hebbian learning, the practical result being a subtraction of an adaptive average:

$$\Delta b = -\beta y. \tag{29}$$

If we then assign a unique integer to each node in the decision tree, the path that was taken from the root to the leaf becomes the spike encoding. This process is an adaptive analog to digital conversion. The source code used to generate this spike encoding is in *AHaHA2D.java*. This adaptive binning procedure can be extended to sparse-spike encoded patterns if

$$y = \sum_i w_i + b, \tag{30}$$

where w_i is sampled randomly from the set $\{-1,1\}$ with equal frequency.

Circuit and Functional Model Correspondence

We demonstrate that both the functional and circuit implementation of the AHaH node are equivalent and functioning correctly in order to establish a link between our benchmark results and the physical circuit. The source code for these experiments can be found in*AHaHRuleFunctionalApp.java* and *AHaHRuleCircuitApp.java* for both the functional and circuit form respectively. In both applications a four-input AHaH node receives the spike patterns from the set $\{(1,z),(z,1)\}$, and the change in the synaptic weights is measured as a function of the output activation, y. Recall that we must encode the nonlinearly separable two-input channels into four-input linearly separable *spike logic* channels so that we can achieve all logic functions (XOR) directly with AHaH attractor states. For both the functional and circuit form of the AHaH node, a bias synapse is included in addition to the normal inputs.

In the derivation of the functional model, the assumption was made that the quantity W^+ was constant (Equation 12). This enabled the treatment of the output voltage as a sum over the input and bias weights. This condition of conservation of adaptive resources is also required in the thermodynamic model (Equation 1). To demonstrate we have attained this conservation, the quantities W^+ and W^- (Equations 12 and 14) are plotted for five different four-input AHaH nodes receiving the spike patterns from the set $\{(1,z),(z,1)\}$ for 1100 time steps.

The source code for this experiment is in *DifferentialWeightApp.java*.

AHaH Logic

A two input AHaH node will receive three possible spike patterns $\{(1,z),(z,1),(1,1)\}$ and converge to multiple attractor states. Each decision boundary plotted in Figure 2 represents a state and its anti-state (i.e. an AHaH bit), since two solutions exist for each stable decision boundary. The 6 possible states are labeled A, A′, B, B′, C, and C′. Fifty two-input AHaH nodes with Ag-chalcogenide memristors were simulated. All AHaH nodes were initialized with random weights picked from a Gaussian distribution with low weight saturation. That is, the memristors were initialized close to their minimally conductive states. Each node was given a stream of 500 inputs randomly picked with equal probability from the set $\{(1,z),(z,1),(1,1)\}$. The source code for this experiment is in a file called *TwoInputAttractorsApp.java*, and there exists a functional form and a circuit form version to show correspondence between the two.

As stated earlier, the attractor states A, B, and C can be viewed as logic functions. It was earlier demonstrated how NAND gates can be used to make these attractor states computationally complete. It was also described how a spike encoding consisting of two input lines per channel can be used to achieve completeness directly with AHaH attractor states. To investigate this, 5000 AHaH nodes were initialized with random weights with zero mean. Each AHaH node was driven with 1000 spikes randomly selected from the set $\{(1,z),(z,1),(1,1)\}$. Finally, each AHaH node's logic function was tested, and the distribution of logic functions was measured. The source code for this experiment is in*SpikeLogicStateOccupationFrequencyApp.java*, and there exists functional form and circuit form versions to show correspondence between the two.

To demonstrate that the attractor states and hence logic functions are stable over time, the above experiment can be repeated. However, the number of time steps can be significantly increased and the logic state of each AHaH node can be recorded at each time step. For this experiment, 100 AHaH nodes were randomly initialized, and their logic functions were tested over 50,000 time steps. The source code for this experiment is in*SpikeLogicFuntionVsTimeApp.java*, and there exists functional form and circuit form versions to show correspondence between the two.

AHaH Clustering

Clustering is a method of knowledge discovery which automatically tries to find hidden structure in data in an unsupervised manner [101]. Centroid-based clustering methods like *k*-means [102] require that the user define the number of cluster centers ahead of time. Density-based methods can be used without predefining cluster centers, but can fail if the clusters are of various densities [103]. Methods such as OPTICS [104] address the problem of variable cluster densities, but introduce the problem that they expect some kind of density drop, which leads to arbitrary cluster borders. On datasets consisting of a mixture of known cluster distributions, density-based clustering algorithms are outperformed by distribution-based methods such as expectation maximization (EM) clustering [105]. However, EM clustering assumes that the data is a mixture of a known distribution and as such is not able to model density-based clusters. It is furthermore prone to over-fitting.

An AHaH node converges to attractor states that cleanly partition its input space by maximizing the margin between opposing data distributions. The set of AHaH attractor states are furthermore computationally complete. These two properties enable a sufficiently large collective of AHaH nodes to assign unique labels to unique input data distributions while maintaining a high level of tolerance to noise. If a collective of AHaH nodes are allowed to randomly fall into attractor states, the binary output vector is a label for the input feature. For example, a four node collective with outputs (0,0,0,1) would encode the output '0001' and, if converted to base-10 integers, be assigned the cluster ID '1'. The collective node output (1,1,1,1) would encode the output '1111' and be assigned the cluster ID '15'. Such a collective is called an AHaH clusterer.

The total number of possible output labels from the AHaH collective is 2^N, where N is the number of AHaH nodes in the collective. The collective may output the same label for different spike patterns if N is small and/or the number of patterns, F, is high. However, as the number of AHaH nodes increases, the probability of this occurring drops exponentially. Under the assumption that all attractor states are equally likely, the probability that any two unique spike patterns, F, will be assigned the same binary label is:

$$P = \frac{1}{2^N} + \frac{2}{2^N} + \cdots + \frac{F}{2^N} = \frac{F^2+F}{2^{N+1}}. \tag{31}$$

For example, given 64 spike patterns and 16 AHaH nodes, the probability of the collective assigning the same label is 3%. By increasing N to 32, the probability falls to less than one in a million.

We developed a quantitative metric to characterize the performance of our AHaH clusterer. Given a unique spike pattern P we would ideally like a unique label L ($P \to L$). This is complicated by the presence of noise, occlusion, and non-stationary data or drift. Failure can occur in two ways. First, if the same underlying pattern is given more than one label, we may say that the AHaH clusterer is *diverging*. We measure the divergence, D, as the inverse of the average labels per pattern. Second, if two different patterns are given the same label, we may say that it is *converging*. We measure convergence, C, as the inverse of the average patterns per label.

Divergence and convergence may be combined to form a composite measure we call*vergence*, V:

$$V = \frac{D+C}{2}. \tag{32}$$

Perfect clustering extraction will occur with a vergence value of 1. The code used to encapsulate the vergence measurement can be found in *VergenceEvaluator.java*.

To investigate the AHaH clusterer's performance as measured by our vergence metric, we swept the following parameters individually while holding the others constant: learning rate (α, β), number of AHaH nodes, number of noise bits, spike pattern length, and number of spike patterns. The applications used to perform the sweeps can be found in the files*SweepLearningRateApp.java*, *SweepNumAhahNodesApp.java*,*SweepNumNoiseBitsVsSpikePatternLengthApp.java*, *SweepSpikePatternLengthApp.java*, and*SweepNumSpikePatternsApp.java*, respectively.

The number of inputs to the AHaH nodes making up the AHaH clusterer was 256. Synthetic spike patterns were created with a random spike pattern generator. Given a spike pattern length, the number of inputs available on the AHaH nodes and the number of unique spike patterns, a set of spike patterns was generated. Noise is generated by taking random input lines and activating them, or, if the input line is already active, deactivating it. The number of patterns that can be distinguished by the AHaH clusterer before vergence falls is a function of the input pattern sparsity, number of total patterns and the pattern noise. Both functional-based and circuit-based AHaH clusterers were investigated and showed good correspondence.

While the vergence experiments provide a quantitative measure of the characteristics of the AHaH clusterer, we also designed a program to qualitatively visualize the clustering capabilities. The basic idea is to create several spatial clusters in two-dimensional space and let the clusterer automatically determine the boundaries between clusters in an unsupervised manner. We used a *k*-nearest neighbor algorithm to translate the spatial location of cluster points into a spike representation, although other spike encoding methods are of course possible. The AHaH clusterer converges to attractor states that map spike patterns to integer, which is in turn mapped to a unique color. The visualizations give the observer a sense of how tolerant the AHaH clusterer is to variations in cluster type, size and temporal stability. The code for the clustering visualization is in *ClusteringApp.java*. There are several different visualizations including clusters of various sizes, arrangements, and numbers, both stationary and non-stationary.

AHaH Classification

Linear classification is a useful tool used in the field of machine learning to characterize and apply labels to samples from datasets. State of the art approaches to classification include algorithms such as decision trees, random forests, support vector machines (SVM) and naïve Bayes [106]. These approaches are used in real world applications such as image recognition, data mining, spam filtering, voice recognition, and fraud detection. Our AHaH-based linear classifier is different from these techniques mainly in that it is not just another algorithm; it can be realized as a physically adaptive circuit. This presents several competitive advantages, the main one being that such a circuit would increase the speed and reduce power consumption dramatically while eliminating the problems associated with disk I/O bottlenecks experienced in large scale data mining applications [107].

The AHaH classifier consists of one or more AHaH nodes, each node assigned to a classification label and each operating the supervised form of the AHaH rule of Equation 17. In cases where a supervisory signal is not available, the unsupervised form of the rule (Equation 18) may be used. Larger magnitude AHaH node output activations are interpreted as a higher confidence. There are multiple ways to interpret the output of the classifier depending on the situation. First, one can order all AHaH node outputs and choose the most positive. This method is ideal when only one label per pattern is needed and an output must always be generated. Second, one can choose all AHaH node outputs that exceed a confidence threshold. This method can be used when multiple labels exist for each input pattern. Finally, only the most positive AHaH node output is chosen if it exceeds a threshold, otherwise nothing is returned. This method can be used when only one label per pattern is needed, but rejection of a pattern is allowed.

To compare the AHaH classifier to other state of the art classification algorithms, we chose four popular classifier benchmark data sets: the Breast Cancer Wisconsin (Original), Census Income, MNIST Handwritten Digits, and the Reuters-21578 data sets. The source code for these classification experiments is found in *BreastCancerFunctionalApp.java*,*CensusIncomeApp.java*, *MnistApp.java*, and *Reuters21578App.java*, respectively.

We scored the classifiers' performance using standard classification metrics: precision, recall, F1, and accuracy. Information on these metrics and how they are used is widely available. The standard training and test sets were used for learning and testing respectively. More information about these benchmark datasets is widely available, and a large amount of classification algorithms have been benchmarked against them including SVM, naïve Bayes, and decision trees.

To further validate an AHaH classifier implemented with circuit AHaH nodes against functional AHaH nodes, we use the Breast Cancer Wisconsin (Original) benchmark dataset. This dataset is relatively small allowing the circuit level simulations to complete quickly. Each sample is either labeled *benign* or *malignant*. There were a total of 683 samples. The first 500 were designated as the training set and the last 183 as the test set. Our spike encoder for this data set produced a total of 70 unique spikes requiring 70 inputs for this particular classifier. The source code for the circuit form of the Breast Cancer Wisconsin experiment is in*BreastCancerCircuitApp.java*.

Continuous valued inputs were converted using the adaptive decision tree method of Equation 29. Text was converted to a bag-of-words representation where each unique word was representative of a unique spike. MNIST image data was first thresholded to produce a spike list of active pixels. The spike list in each 8×8 image patch was converted to a single spike via the method of Equation 30. The image patch was convolved and pooled over an 8×8 pixel region. The result of this procedure is a list of spikes with moderate translational invariance, which was fed to the AHaH classifier. The source code for this procedure is available in*MnistSpikeEncoder.java*.

The AHaH classifier is capable of unsupervised learning by evoking Equation 17. If no supervised labels are given but the classifier is able to output labels with high confidence, the output can be assumed to be correct and used as the supervised signal. The result is a continued convergence into the attractor state, which represent a point of maximal margin. This has application in any domain where large volumes of unlabeled data exist such as image recognition. By allowing the classifier to process these unlabeled examples, it can continue to improve its performance (bootstrap) without supervised labels.

To demonstrate this unsupervised learning capability we used the Reuters-21578 dataset. The entire training and test sets were lumped together and the classifier was given the first 25% inputs in a supervised manner. For the remaining 75%

of the news articles, the classifier was run in an unsupervised manner. Only when the confidence was 1.0, which indicates high certainty of a correct answer, did the classifier use its own classification as a supervised training signal. The F1 score was recorded after each story for the following most frequent labels: *earn*, *acq*, *money-fx*, *grain*, *crude*, *trade*, *interest*, *ship*, *wheat*, and *corn*, a common label set used in many benchmarking exercises using this dataset. The source code for this experiment is in *Reuters21578SemiSupervisedApp.java*.

AHaH Signal Prediction

Complex signal prediction involves using the prior history of a signal or group of signals to predict the future state of the signal. Signal prediction, also known as signal forecasting, is used in adaptive filters, resource planning and action selection. Some real world examples include production estimating, retail inventory planning, inflation prediction, insurance risk assessment, and weather forecasting. Current prediction algorithms include principle component analysis and regression and Kalman filtering [108], artificial neural networks [109]and Bayesian model averaging [110].

By posing signal prediction as a multi-label classification problem, complex signals can be learned and predicted using the AHaH classifier. As a simplified proof of concept exercise to demonstrate this, a complex temporal signal prediction experiment was designed. For each moment of time, the real-valued signal $F(S(t))$ is converted into a sparse feature representation using the method of Equation 29. These features are buffered to form a feature set:

$$\{F(S(t-N)),F(S(t-N+1)),\cdots,F(S(t-1))\}. \quad (33)$$

This feature set is then used to make predictions of the current feature $F(S(t))$, and the spikes of the current feature are used as supervised labels. After learning, the output prediction may be used in lieu of the actual input and run forward recursively in time. In this way, extended predictions about the future are possible. The source code for the experiment is available in *ComplexSignalPredictionApp.java*. The signal was generated from the summation of five sinusoidal signals with randomly chosen amplitudes, periods, and phases. The experiment ran for a total of 10,000 time steps. During the last 300 time steps, recursive prediction occurred.

AHaH Motor Control

Motor control is the process by which sensory information about the world and the current state of the body is used to execute actions to generate movement. Stabilizing Hebbian feedback applied to an AHaH node can occur any time after the Anti-Hebbian read, which opens the interesting possibility of using AHaH nodes for reinforcement-based learning. Here we show that a small collective of AHaH nodes, an AHaH motor controller, can be used in autonomous robotic control. As a proof-of-concept experiment we use an AHaH motor controller to guide a multi-jointed robotic arm to a target based on a value signal or cost function.

A virtual environment in which an AHaH motor controller controls the angles of N connected fixed length rods in order to make contact with a target was created as shown in Figure 9. The arm rests on a plane with its base anchored at the center, and all the joints have 360 degrees of freedom to rotate. New targets are dropped randomly within the robotic arm's reach radius after it captures a target. The robotic arm virtual environment is part of an open source project called Proprioceptron, also available at http://xeiam.com. Proprioceptron builds upon a 3D gaming library and offers virtual worlds and challenges for testing motor control algorithms. The robotic arm challenge offers 5 levels of difficulty starting with stationary targets, increasing target lateral speed as the levels increase.

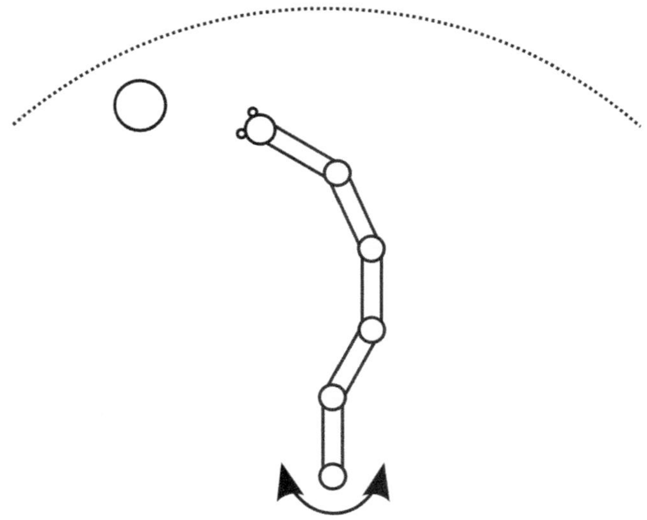

Figure 9: **Unsupervised robotic arm challenge.**

The robotic arm challenge involves a multi-jointed robotic arm that moves to capture a target. Each joint on the arm has 360 degrees of rotation, and the base joint is anchored to the floor. Using only a value signal relating the distance from the head to the target and an AHaH motor controller taking as input sensory stimuli in a closed-loop configuration, the robotic arm autonomously learns to capture stationary and moving targets. New targets are dropped within the arm's reach radius after each capture, and the number of discrete angular joint actuations required for each catch is recorded to asses capture efficiency.

http://dx.doi.org/10.1371/journal.pone.0085175.g009

Sensors measure the relative joint angles of each segment of the robot arm as well as the distance from the target ball to each of two "eyes" located on the side of the arm's "head". Sensor measurements are converted into a sparse spiking representation using the method ofEquation 29. A value signal is computed as the inverse distance of the head to the target:

$$V=\frac{1}{1+d}. \quad (34)$$

Opposing "muscles" actuate each joint. Each muscle is formed of many "fibers" and a single AHaH node controls each fiber. The number of discrete angular steps that move each joint, ΔJ, is given by:

$$\Delta J=\sum_{i=0}^{N}(\mathrm{H}(y_i^0)-\mathrm{H}(y_i^1)), \quad (35)$$

where N is the number of muscle fibers, y_i^0 is the post-synaptic activation of the ith AHaH node controlling the ith muscle fiber of the primary muscle, y_i^1 is the post-synaptic activation of the ith AHaH node controlling the ith muscle fiber of the opposing muscle, and H is the Heaviside step function. The number of discrete angular steps moved in each joint at each time step is then given by the difference in these two values.

Given a movement we can say if a fiber (AHaH node) acted for or against it. We can further determine if the movement was good or bad by observing the change in the value signal. If, at a later time, the value increased after a movement, then each fiber responsible for the movement receive rewarding Hebbian feedback. Likewise, if the fiber acted in support of a movement and later the value signal dropped, then the fiber is denied a Hebbian update. As the duration of time between movement and reward increases, so does the difficulty of the problem since many movements can be taken during the interval. A reinforcement scheme can be implemented in a number of ways over a number of timescales and may even be combined. For example, we may integrate over a number of time scales to determine if the value increased or decreased.

Experimental observation led to constant values of $\alpha=0.1$ and $\beta=0.5$ for the AHaH rule, although generally good performance was observed for a wide range of values. The choice of these parameters is influenced by the complexity of the problem and the need to learn complex compound sequences, as well as the duration between action (anti-Hebbian) and reward (Hebbian).

We measured the robotic arm's efficiency in catching targets by summing the total number of discrete angular joint actuations from the time the target was placed until capture. As a control, the same challenge was carried out using a simple random actuator. The challenge was carried out for both AHaH-controlled and random-controlled robotic arm actuation for different robotic arm lengths ranging from 3 to 21 joints in increments of three. The total joint actuation is the average amount of discrete joint actuation over the 100 captured targets. The source code for this experiment is available in *RoboticArmApp.java*.

AHaH Combinatorial Optimization

An AHaH node will descend into a probabilistic output state if the Hebbian feedback is withheld. As the magnitude of the synaptic weight falls closer to zero, the chance that state transitions will occur rises from 0% to 50%. This property can be exploited in probabilistic search and optimization tasks. Consider a combinatorial optimization task such as the traveling salesman problem where the city-to-city path is encoded as a binary vector $P=(b_0,b_1,\cdots,b_N)$. The space of all possible paths can be visualized as the leaves of a binary tree of depth N. The act of constructing a path can be seen as a routing procedure traversing the tree from trunk to leaf. By allowing prior attempted solutions to modify the routing probabilities, an initial uniform routing distribution can collapse into a subspace of more optimal solutions.

This can be accomplished by utilizing an AHaH node with a single input as a node within a virtual routing tree. As a route progresses from the trunk to a leaf, each AHaH node is evaluated for its state and receives the anti-Hebbian update. Should the route result in a solution that is better than the average solution, all nodes along the routing path receive a Hebbian update. By repeating the procedure over and over again, a positive feedback loop is created such that more optimal routes result in higher route probabilities that, in turn, result in more optimal routes. The net effect is a collapse of the route probabilities from the trunk to the leaves as a path is locked in. The process is intuitively similar to the formation of a lighting strike searching for a path to ground and as such we call it a *strike search*.

To evaluate the AHaH combinatorial optimizer, we used the functional model (Equation 17), setting $\alpha=\beta$ and making it a free parameter we call the *learning rate*, L:

$$L=\alpha=\beta. \tag{36}$$

The experiment consists of 200 strike searches, where L is set to a value chosen randomly from between 0.00015 and 0.0035 at the start of each trial. The noise variable, η, is picked from a random Gaussian distribution with zero mean and 0.025 variance. After every 10,000 solution attempts, branches with synaptic weight magnitudes less than 0.01 are pruned. A 64-city network is created where each city is directly connected to every other city, and the city coordinates are picked from a random Gaussian distribution with zero mean and a variance of one. The city path is encoded as a bit sequence such that the first city is encoded with 6 bits, and each successive city with only as many bits needed to resolve the remaining cities such that the second-to-last city requires one bit. The value of the solution is the path distance, which we are attempting to minimize. The strike process is terminated when the same solution is generated 5 successive times, indicating convergence. A random search is used as a control, where each new solution attempt is picked from a uniform random distribution. The code for this experiment is in *StrikeSearchApp.java*.

RESULTS AND DISCUSSION

AHaH Rule

The AHaH rule reconstructions for the functional and circuit forms of the AHaH node are shown in Figures 10A and 10B respectively. In both cases, the AHaH rule is clearly represented and there is congruence between both forms. However, Figure 10B hides complexity in the circuit that arises from the differential aspect of the weights and their limited dynamic range. Because of this, depending on the saturation state of a weight, the form of weight update may change over time. The AHaH rule reconstruction of Figure 10B is thus for a specific weight initialization for a specific time interval.

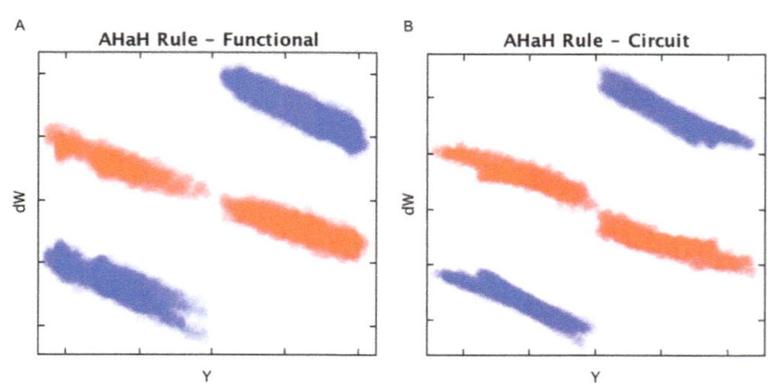

Figure 10: The AHaH rule reconstructed from simulations.

Each data point represents the change in a synaptic weight

as a function of AHaH node activation, y. Blue data points correspond to input synapses and red data points to bias inputs. There is good congruence between the A) functional and B) circuit implementations of the AHaH rule.

http://dx.doi.org/10.1371/journal.pone.0085175.g010

As part of our functional model derivation (Equation 12) and the connection to thermodynamics (Equation 1) we require that the quantity W^+ remains constant. As can be seen in Figure 11, the quantity W^+ does indeed asymptote and remains constant.

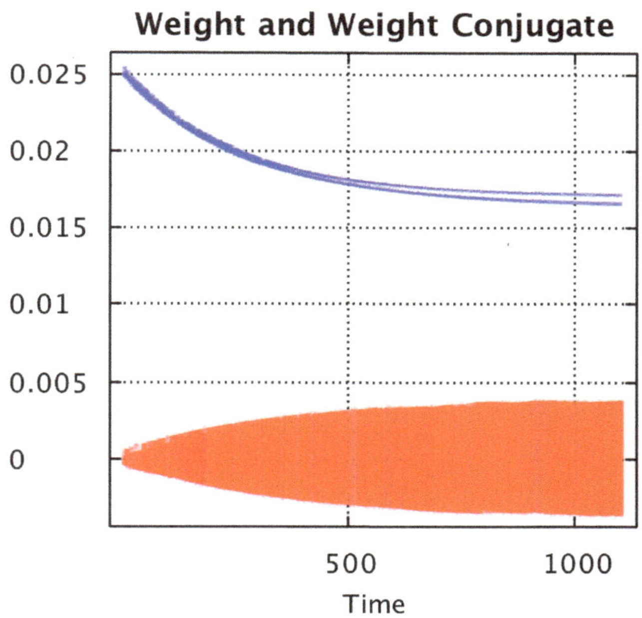

Figure 11: Justification of constant weight conjugate.

Multiple AHaH nodes receive spike patterns from the set $\{(1,z),(z,1)\}$ while the weight and weight conjugate is measured. Blue=weight conjugate (W^+), Red=weight (W^-). The quantity W^+ has a much lower variance than the quantity W^- over multiple trials, justifying the assumption that W^+ is a constant factor.

http://dx.doi.org/10.1371/journal.pone.0085175.g011

AHaH Logic

The 2-input AHaH node receiving 500 consecutive inputs randomly chosen from the set $\{(1,z),(z,1),(1,1)\}$ at 50 different initial synaptic weights evolves into one of the six attractor basins as shown in Figure 12. Labels A, A′, B, B′, C, and C′ indicate the attractor basins in these weight-space plots and correspond to the equivalent decision boundaries shown inFigure 2. The same experiment was performed with the functional form and the circuit form of the AHaH node (Figure 12A and 12B respectively) and close correspondence can be seen.

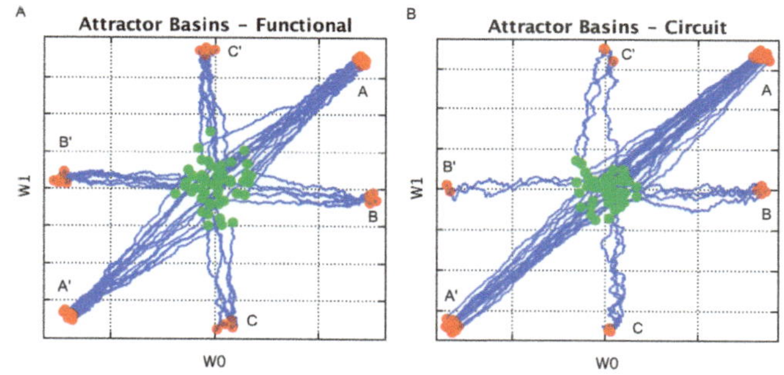

Figure 12: Attractor states of a two-input AHaH node under the three-pattern input.

The AHaH rule naturally forms decision boundaries that maximize the margin between data distributions. Weight space plots show the initial weight coordinate (green circle), the final weight coordinate (red circle) and the path between (blue line). Evolution of weights from a random normal initialization to attractor basins can be clearly seen for both the functional model (A) and circuit model (B).

http://dx.doi.org/10.1371/journal.pone.0085175.g012

After being initialized with random synaptic weights, the occupation of logic states of AHaH nodes receiving the spike logic patterns of Table 1 are shown in Figure 13 for both functional and circuit models. Each logic function was assigned a unique integer value as in Table 4. Experimental results show congruence between the functional form and circuit form simulations. All linear functions are represented by distinct AHaH attractor states. Absent are the expected nonlinear XOR functions 6 and 9. These functions are possible through combinations of other logic functions, meaning a multi-stage AHaH node network is capable of achieving any logic function. Since any algorithm or program can be reduced to successive utilizations of logic gates, the attractor states of AHaH nodes support universal computation. Logic functions remain stable over time, as indicated by Figure 13B.

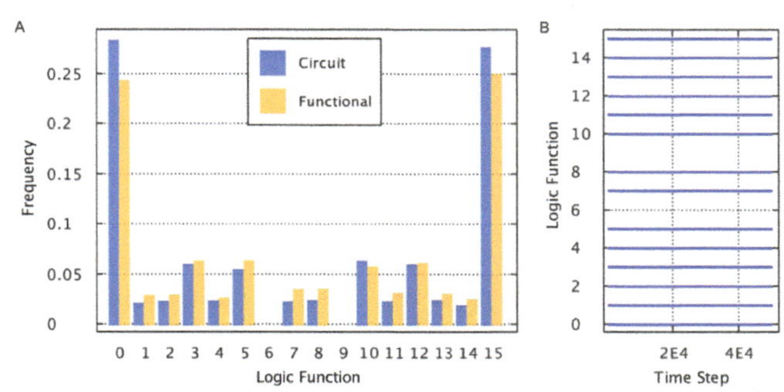

Figure 13: AHaH attractor states as logic functions.

A) Logic state occupation frequency after 5000 time steps for both functional model and circuit model. All logic functions can be attained directly from attractor states except for XOR functions, which can be attained via multi-stage circuits. B) The logic functions are stable over time for both functional model and circuit model, indicating stable attractor dynamics.

http://dx.doi.org/10.1371/journal.pone.0085175.g013

Table 4: Logic functions

http://dx.doi.org/10.1371/journal.pone.0085175.t004

SP⇓, LF⇒	15	14	13	12	11	10	9	8	7	6	5	4	3	2	1	0
(z, 1, z, 1)	1	1	1	1	1	1	1	1	0	0	0	0	0	0	0	0
(z, 1, 1, z)	1	1	1	1	0	0	0	0	1	1	1	1	0	0	0	0
(1, z, z, 1)	1	1	0	0	1	1	0	0	1	1	0	0	1	1	0	0
(1, z, 1, z)	1	0	1	0	1	0	1	0	1	0	1	0	1	0	1	0

The table defines all 16 possible logic functions (LF) for the four spike encoded input patterns (SP).
doi:10.1371/journal.pone.0085175.t004

Logic functions 0 and 15 represent the null state and their occupation is inhibited through the action of the bias. By increasing the number of bias inputs from 1 to 3 we can collapse the stable attractor logic states down to 3, 5, 10 and 12. These functions represent the pure independent component states and act to pass or invert each of the two input channels. Although these states are not computationally complete, they can be made so via the use of NAND gates as we discussed in the theory section. The advantage of using states 3, 5, 10 and 12 is that they are very stable. The disadvantage is that we must now rely on external circuitry (i.e. NAND gates) to achieve computational universality.

AHaH Clustering

The AHaH clusterer parameter sweep experiment results are summarized in Table 5. While setting the free parameters at their default values and sweeping the parameter under investigation, the range of that parameter that resulted in a vergence value greater than 0.90 was determined. The performance of the AHaH clusterer proved to be robust to input pattern noise. For example, the clusterer can achieve perfect performance with up to 18% noise under a 100% pattern load. A full pattern load occurs when the number of patterns (16) multiplied by the pattern size (16) is equal to the total number of input lines (256 in this case). The clusterer can achieve greater than 90% vergence with up to 44% noise, meaning 7 of the 16 spike input pattern's bits are reassigned random values.

Table 5: AHaH clusterer sweep results.

http://dx.doi.org/10.1371/journal.pone.0085175.t005

	Learning Rate	Number of AHaH nodes	Number of Noise Bits	Spike Pattern Length	Number of Spike Patterns
Default Value	0.0005	20	3	16	16
Range	.0002–.0012	>7	<=7	<=36	<=28

While sweeping each parameter of the AHaH clusterer and holding the others constant at their default values, the reported range is where the vergence remained greater than 90%.
doi:10.1371/journal.pone.0085175.t005

The results shown in Figure 14 illustrate that the performance as measured by vergence degrades as the number of spike patterns increase. This result is explained by the fact that AHaH plasticity is acting to maximize the margin between data distributions or patterns. As the number of patterns increases, the margin must decrease and hence becomes more susceptible to noise. For example, under a 200% pattern load (32 patterns), vergence falls below 90% after 12.5% noise (2 noise bits). Comparing Figure 14A and 14B further demonstrates that circuit and functional models produce similar results. Without noise, the clusterer has impressive capacity and can reliably assign labels to spike patterns with load factors that exceed 400%.

Functional (A) and circuit (B) simulation results of an AHaH clusterer formed of twenty AHaH nodes. Spike patterns were encoded over 16 active input lines from a total spike space of 256. The number of noise bits was swept from 1 (6.25%) to 10 (62.5%) while the vergence was measured. The performance is a function of the total number of spike patterns. Blue=16 (100% load), Orange=20 (125% load), Purple=24 (150% load), Green=32 (200% load), Red=64 (400% load).

http://dx.doi.org/10.1371/journal.pone.0085175.g014

Figure 15 shows screen shots of three different two-dimensional clustering visualizations. The AHaH clusterer performs well for clusters of various sizes and numbers as well as non-Gaussian clusters even though it does not need to know the number of clusters ahead of time or the expected cluster forms.

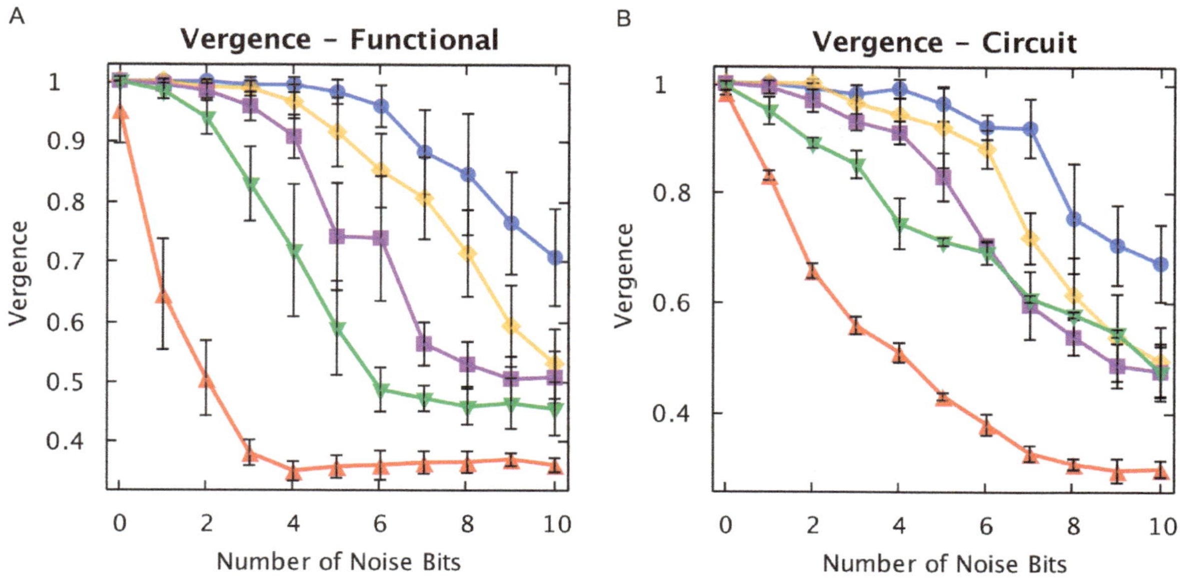

Figure 14: AHaH clusterer.

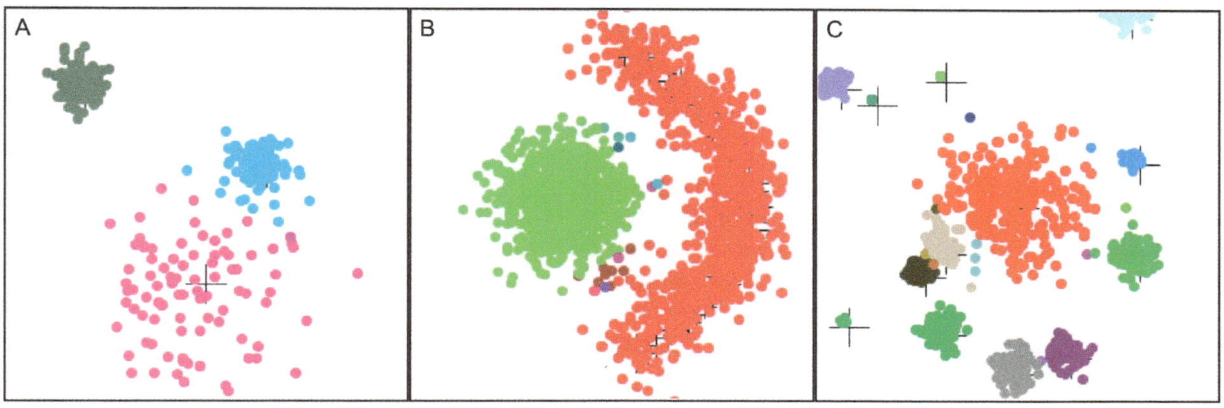

Figure 15: Two-dimensional spatial clustering demonstrations.

The AHaH clusterer performs well across a wide range of different 2D spatial cluster types, all without predefining the number of clusters or the expected cluster types. A) Gaussian B) non-Gaussian C) random Gaussian size and placement.

http://dx.doi.org/10.1371/journal.pone.0085175.g015

The results show that the AHaH clusterer is able to handle a spectrum of cluster types. We demonstrate the ability to detect Gaussian and non-Gaussian clusters, clusters of non-equal size, as well as non-stationary clusters. Whereas other methods have intrinsic failure modes for certain types of clusters, our method can apparently handle a wide range of cluster types. Although more work must be done to fully compare our methods to existing clustering methods, our results thus far indicate that our method offers a genuinely new clustering mechanism with a number of distinct advantages. The most significant advantage is that we can implement the AHaH clusterer in physically adaptive AHaH circuits. In other words, clustering can now become an adaptive hardware resource.

AHaH Classification

Our AHaH classifier benchmark scores for the Breast Cancer Wisconsin (Original), Census Income, MNIST Handwritten Digits, and the Reuters-21578 data sets are shown in Table 6along with results from other published studies using their respective classification methods. Our results compare well to published benchmarks and consistently match or exceed SVM performance. This is surprising given the simplicity of the approach, which amounts to simple sparse spike encoding followed by classification with independent AHaH nodes.

Table 6: Benchmark classification results

http://dx.doi.org/10.1371/journal.pone.0085175.t006

Breast Cancer Wisconsin (Original)		Census Income		MNIST Handwritten Digits		Reuters-21578	
AHaH	.997	AHaH	.853	AHaH	.98–.99	AHaH	.92
RS-SVM [115]	1.0	NBTree [116]	.86	deep convex net [117]	.992	SVM [118]	.864
SVM [119]	.972	naïve-Bayes [116]	.84	large conv. net [120]	.991	C4.5 [118]	.794
C4.5 [121]	.9474	C4.5 [116]	.858	polynomial SVM [42]	.986	naïve-Bayes [118]	.72

AHaH classifier classification scores for the Breast Cancer, Census Income, MNIST Handwritten Digits and Reuters-21578 classification benchmark datasets. The AHaH classifier results compare favorably with other methods. Higher scores on the MNIST dataset are possible by increasing the resolution of the spike encoding.
doi:10.1371/journal.pone.0085175.t006

In comparing our MNIST results with other methods, it is important to account for data preprocessing and artificial inflation of the training data set through transformations of training samples. We do not inflate the training set; our results are achievable with only one online training epoch. Both the training and test are completed on a standard desktop computer processor in a few minutes to less than an hour, depending on the resolution of the spike encoding. The current state of the art achieves a recognition rate of 99.65% and "took a few days" to train on a desktop computer with GPU acceleration [111]. Another study determined that human performance on this task is 97.27% [112].

The Reuters-21578, Census Income and Breast Cancer datasets cover a range of data types from strings to integers to continuous real-valued signals. The Census Income dataset furthermore contains mixed data types as well as exemplars with missing attributes. In all cases the AHaH classifier combined with the simple spike encoder of Equation 29 matched or exceeded state of the art classifiers. This is significant primarily for the reason that both spike encoding and classification functions can be attained via AHaH learning and support the idea that a generic adaptive learning hardware resource is possible.

Figure 16 provides a more detailed look at the individual classification experiments. Typical for all benchmark data sets, as the confidence threshold of the AHaH classifier is increased, the precision increases while recall drops (Figure 16A and 16B). In other words, the classifier makes fewer mistakes at the expense of not being able to answer some queries. The circuit-level simulation using the Ag-chalcogenide device model yielded a classification score as a function of confidence threshold similar to the functional simulations as shown in Figures 16C and 16D respectively. Similar circuit-level simulation results were obtained using the AIST and WO_x device models. The results of the MNIST experiment are shown in Figures 16E and 16F. While Figure 16E shows the average over all digits, Figure 16F shows the scores of the individual digits.

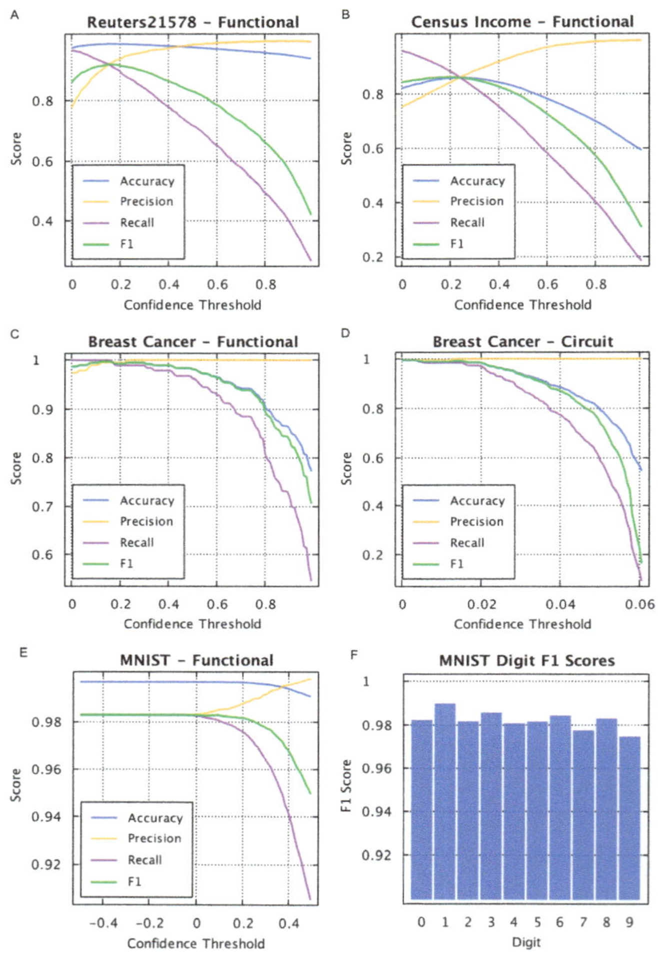

Figure 16: Classification benchmarks results.

A) Reuters-21578. Using the top ten most frequent labels associated with the news articles in the Reuters-21578 data set, the AHaH classifier's accuracy, precision, recall, and F1 score was determined as a function of its confidence threshold. As the confidence threshold increases, the precision increases while recall drops. An optimal confidence threshold can be chosen depending on the desired results and can be dynamically changed. The peak F1 score is 0.92. B) Census Income. The peak F1 score is 0.853 C) Breast Cancer. The peak F1 score is 0.997. D) Breast Cancer repeated but using the circuit model rather than the functional model. The peak F1 score and the shape of the curves are similar to functional model results. E) MNIST. The peak F1 score is 0.98–.99, depending on the resolution of the spike encoding. F) The individual F1 classification scores of the hand written digits.

http://dx.doi.org/10.1371/journal.pone.0085175.g016

Using the confidence threshold as a guide, the AHaH classifier can also be used in a semi-supervised mode. Starting in supervised mode and learning over a range of training data, the classifier can then switch to unsupervised mode. In unsupervised mode we may activate Hebbian learning if the confidence exceeds a value. Results are shown in Figure 17, which shows continued improved F1 score without supervision.

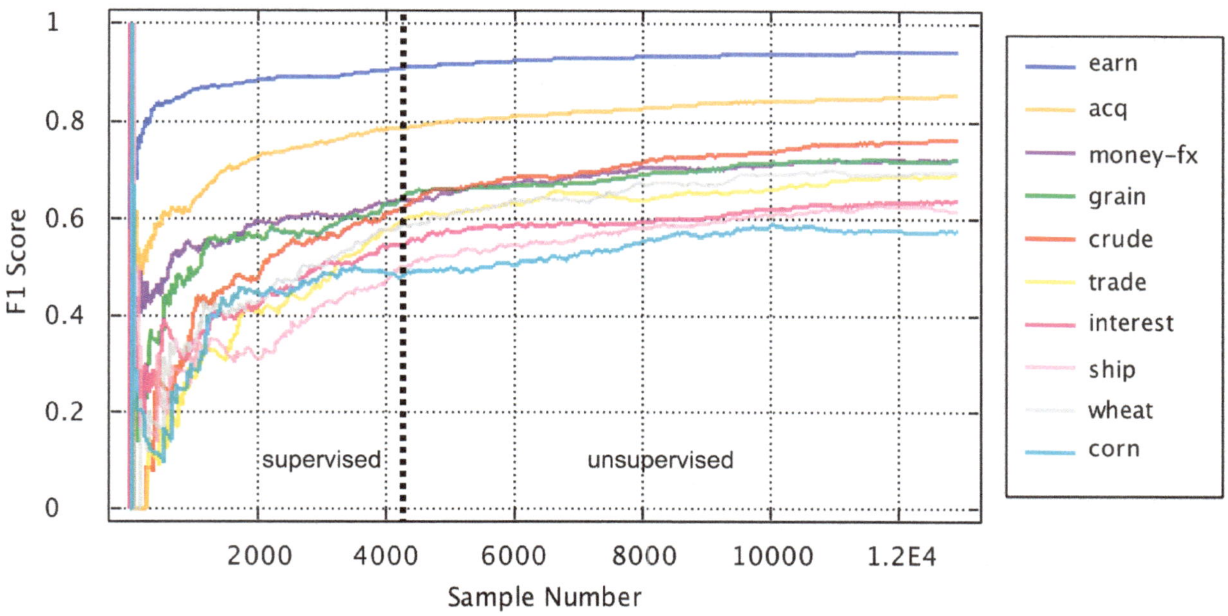

Figure 17: Semi-supervised operation of the AHaH classifier.

For the first 30% of samples from the Reuters-21578 data set, the AHaH classifier was operated in supervised mode followed by operation in unsupervised mode for the remaining samples. A confidence threshold of 1.0 was set for unsupervised application of a learn signal. The F1 score for the top ten most frequently occurring labels in the Reuters-21578 data set were tracked. These results show that the AHaH classifier is capable of continuously improving its performance without supervised feedback.

http://dx.doi.org/10.1371/journal.pone.0085175.g017

Results to date indicate that the AHaH classifier is an efficient incremental optimal linear classifier. The AHaH classifier displays a range of desirable classifier characteristics hinting that it may be an ideal general classifier capable of handling a wide range of classification applications. The classifier can learn online in a feed-forward manner. This is important for large data sets and applications that require constant adaptation such as prediction, anomaly detection and motor control. The classifier can associate an unlimited number of labels to a pattern, where the addition of a label is simply the addition of another AHaH node. By allowing the classifier to process unlabeled data it can improve over time. This has practical implications in any situation where substantial quantities of unlabeled data exist. Through the use of spike encoders, the classifier can handle mixed data types such as discrete or continuous numbers and strings. The classifier tolerates missing values, noise, and irrelevant attributes and is computationally efficient. The most significant advantage, however, is that the circuit can be mapped to physically adaptive hardware. Optimal incremental classification can now become a hardware resource.

AHaH Signal Prediction

The results of the temporal signal prediction experiment are shown in Figure 18. The solid line drawn on top of the true signal represents the predictor's accurate prediction of the true complex waveform after a period of supervised learning (mostly not shown). One advantage of the recursive prediction is that the forward-looking time window can be dynamically chosen. Although the predictor was trained to predict only the next time step, the recursive prediction can be carried forward to the desired point in the future for which the prediction should be made, which was 300 time steps in this example. At some point forward the prediction will degrade if the signal is not deterministic and cyclical. Not all applications require the recursive prediction, and a simpler statically set forward-looking time window could be set.

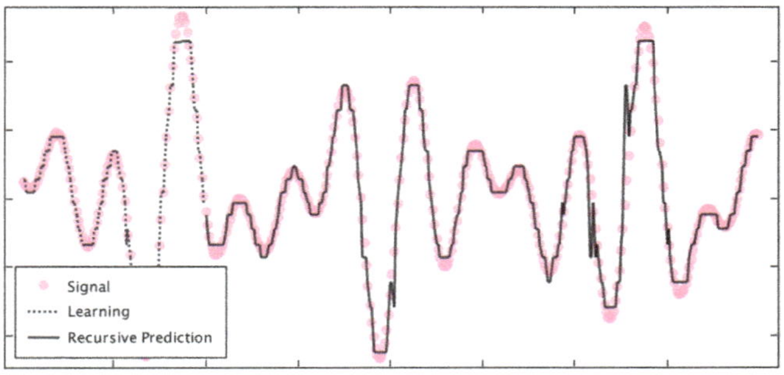

Figure 18: Complex signal prediction with the AHaH classifier.

By posing prediction as a multi-label classification problem, the AHaH classifier can learn complex temporal waveforms and make extended predictions via recursion. Here, the temporal signal (dots) is a summation of five sinusoidal signals with randomly chosen amplitudes, periods, and phases. The classifier is trained for 10,000 time steps (last 100 steps shown, dotted line) and then tested for 300 time steps (solid line).

http://dx.doi.org/10.1371/journal.pone.0085175.g018

While this temporal signal prediction demonstration is not by any means an exhaustive comparison of AHaH signal prediction to other forecasting algorithms, it demonstrates the utility and flexibility of the AHaH classifier and provides the first glimpse of using AHaH nodes in the large application space of signal forecasting. These results also shed light on how AHaH node supervisory signals could be generated in a completely self-organizing system with zero human intervention. Time is the supervisor and prediction is the Hebbian reward. From the practical perspective, prediction provides the ability to prepare or optimize for the future. It also provides the ability to detect when a system is changing. If a prediction fails to meet with reality, an anomaly has occurred.

AHaH Motor Control

The results of the motorized robotic arm experiment are shown in Figure 19. The performance of the AHaH-guided robotic arm is compared with a random-guided robotic arm by measuring the average total joint actuation needed to capture 100 moving targets. The results demonstrate that the collective of AHaH nodes are performing a gradient descent of the value function and can rapidly guide the arm to its target, independent of the number of joints.

The average total joint actuation required for the robot arm to capture the target remains constant as the number of arm joints increases for actuation using the AHaH motor controller. For random actuation, the required actuation grows exponentially.

http://dx.doi.org/10.1371/journal.pone.0085175.g019

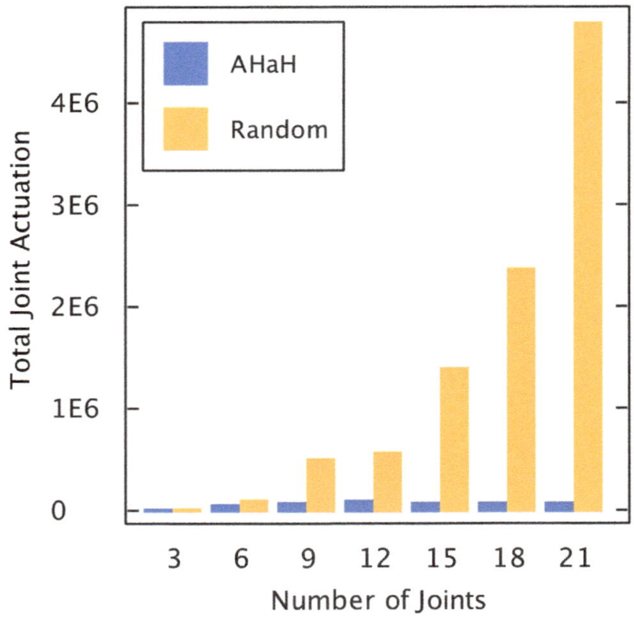

Figure 19: Unsupervised robotic arm challenge.

Our results show that populations of independent AHaH nodes can effectively control multiple degrees of freedom so as to ascend (or descend) a value function. This process is spontaneous and results from the emergent behavior of many AHaH nodes acting as *self configuring classifiers* competing for Hebbian reward. Real world applications of this effect could of course include actuation of robotic appendages as well as autonomous controllers. Our results are significant primarily because the controller can be reduced to physically adaptive circuits and hence can be made to consume very little power and space, an important consideration in mobile platforms.

AHaH Combinatorial Optimization

The results of the traveling salesman problem experiment are shown in Figure 20. Our experiment demonstrates that an AHaH combinatorial optimizer performing a strike search can outperform a strike search backed by a random path chooser (Figure 20A). This result shows that the strike is performing a directed search as expected. Trials with higher convergence times resulted from cases where the optimizer was given a relatively lower learning rate. Recall, a lower learning rate allows for a finer-grained search resulting in the longer convergence times. Figure 20B shows the relationship between the learning rate and the solution value (distance), while Figure 20C shows the relationship between the learning rate and the convergence time. Lowering the learning rate causes more evidence to be accumulated before positive-feedback forces selection and the solution proceeds from the trunk to leaf node, one bit at a time.

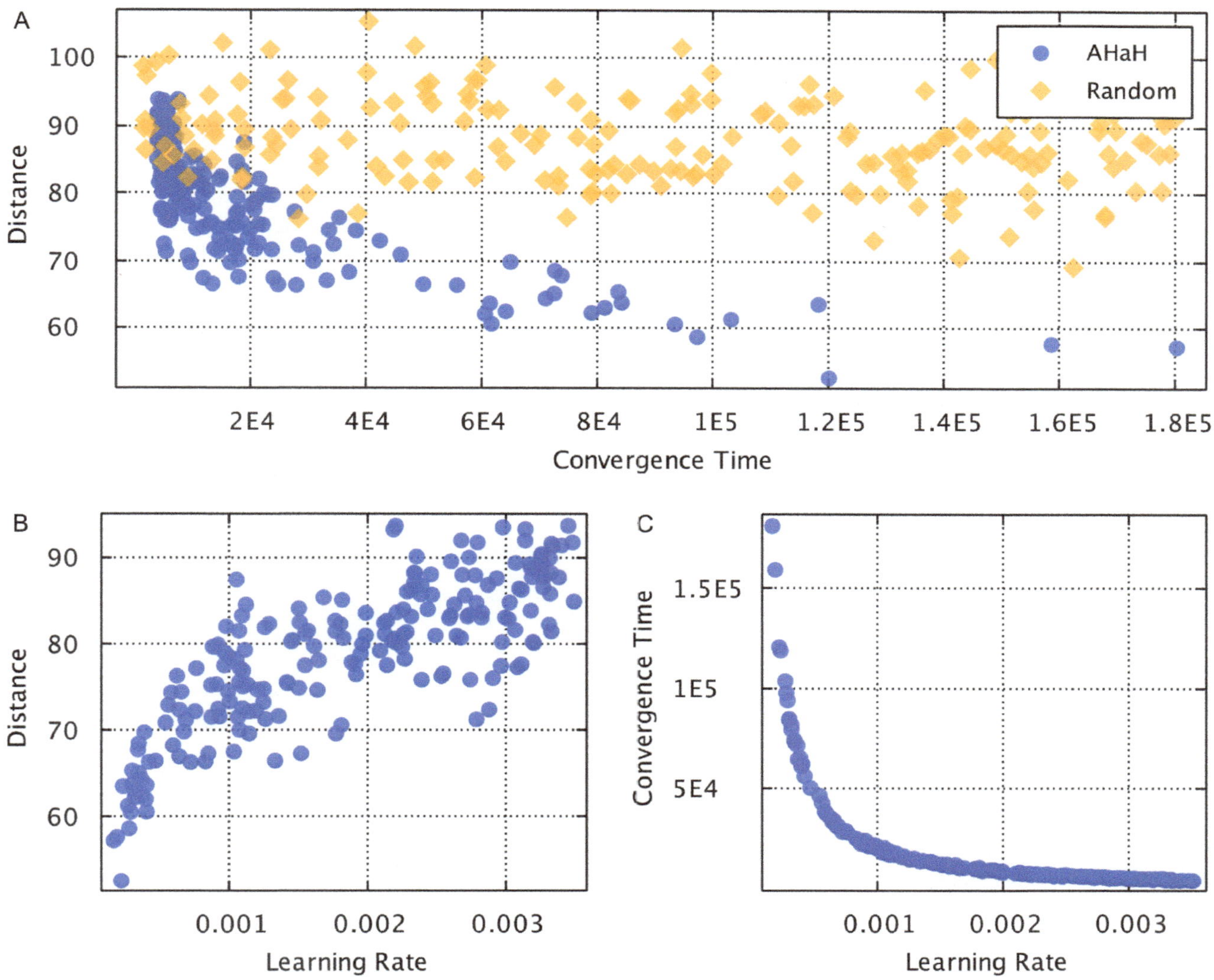

Figure 20: 64-city traveling salesman experiment.

By using single-input AHaH nodes as nodes in a routing tree to perform a strike search, combinatorial optimization problems such as the traveling salesman problem can be solved. Adjusting the learning rate can control the speed and quality of the solution. A) The distance between the 64 cities versus the convergences time for the AHaH-based and random-based strike search. B) Lower learning rates lead to better solutions. C) Higher learning rates decrease convergence time.

http://dx.doi.org/10.1371/journal.pone.0085175.g020

A strike evolves in time as bits are sequentially locked in via the positive feedback selection mechanism after a period of evidence accumulation. The lower the learning rate, the more evidence is accumulated before a path is locked in. In this way, a strike search appears to be a relatively generic method to accelerate the search for a procedure. Using the traveling salesman problem as an example, we could just as easily encode the strike path as a relative procedure for re-ordering a list of cities rather than an absolute ordering. For example, we could swap the cities at indices A and B, then swap the cities at indices C and D, and so on. Furthermore, we could utilize the strike procedure in a recursive manner. In the case of the traveling salesman problem we could assign lower-level strikes to find optimal sub-paths and higher-order strikes to assemble larger paths from the sub-paths. Most generally, if (1) a problem can be represented as a bit configuration and (2) the configuration can be assigned a value in an efficient manner, then a strike can be used as an adaptive learning hardware resource for optimization tasks. The ability to change the convergence times allows dynamic choices to be made in the time available.

Synaptic Power Consumption

Both static and dynamic power consumption pathways must be considered when calculating the energy dissipation of neuromorphic chips containing AHaH circuit architecture. The static power component is dominated by the current flowing through the AHaH node synapse arrays during the read and write phases. The dynamic power component is dominated by the charging and discharging of the capacitive components of the circuitry. This capacitance includes parasitics from circuit elements and interconnect wires. Industry best practices can optimize dynamic power consumption. Here we focus on an estimation of static power consumption. Note that by not including the dynamic power consumption in this estimation, these values represent only a lower bounds on the synaptic power consumption of a neuromorphic chip. Dynamic power consumption, which is heavily dependent on chip design and architecture may have a significant power contribution. Recall that one of the major motivations of AHaH computing is the elimination of the von Neumann bottleneck for machine learning applications. Considering static and dynamic power consumption together with the elimination of this bottleneck, the net gain in power efficiency compared to modern digital electronics will most likely increase.

Static power dissipation of a single AHaH node is equal to $P = V^2 G_{eq}$, where V is the voltage drop across the memristor pairs and G_{eq} is the equivalent conductance of the combined active input and bias memristor pairs in Figure 5. Since each synapse only dissipates energy when it is active (it remains floating otherwise), and since only a small number of synapses are active at any given time (given the sparse spike encoding), the current flowing through the AHaH node during the read and write phases is very low. The total dissipative energy per synaptic event is.

$$E = E_r + E_w = \Delta T V^2 \left(\frac{G_a G_b}{G_a + G_b} + (G_a | G_b | 0) \right), \quad (37)$$

where E_r and E_w are the energy of the read and write phases, respectively, ΔT is the pulse width of the read and write phases and $(G_a|G_b|0)$ denotes the three possible outcomes G_a, G_b or 0. That is, the synapse could select the *A Path*, the *B Path* or else feedback is withheld. Utilizing Equation 1, it is straightforward to show the conductance under maximum power dissipation for each condition, as shown in Table 7.

Table 7: Maximum power and corresponding synaptic weights

http://dx.doi.org/10.1371/journal.pone.0085175.t007

Condition	G_a	G_b	Maximum Power
Path A Selected	k	0	$\frac{1}{2}kV^2$
Path B Selected	0	k	$\frac{1}{2}kV^2$
No Feedback	$k/2$	$k/2$	$\frac{1}{8}kV^2$

The maximum power dissipation of a differential synaptic weight changes depending on whether feedback is present or not. In the absence of feedback, the power is maximized when the conductance of each path is the same and the output descends into randomness. When feedback is present the synapse may converge to one of two possible configurations, and the power dissipation increases by a factor of four.
doi:10.1371/journal.pone.0085175.t007

From Equation 37 it is clear that lower operating voltages, shorter pulse widths, and lower conductance memristors will reduce static power consumption. Developing an ideal candidate memristor for AHaH computing will play an important role in static power reduction. Given an operating resistance of the Ag-chalcogenide memristor of 250Ω, a pulse with of 10 μs and a voltage of 1 V, the estimated per-synapse static energy consumption is 40 nJ. Several synapse-like memristors achieving ultra-low power pulsed synaptic updates have recently been reported [68], [70], [113]. The energy consumption per synapse using roughly average voltage, pulse width, and resistance values taken from the pulse-driven memristive behavior plots in the three references results in a calculated energy consumption per synapse per update of ~12 pJ, ~67.5 pJ, and ~56 pJ respectively. Kuzum et al. [113] claim that scaling trends project energy consumption for electronic synapses down to 2 pJ. As an example, a 100 kΩ device driven with 1 V, 100 ns pulse widths would consume 1.5 pJ of static energy per synapse. Such devices will play a significant role in reaching biological efficiency.

In all applications, the spike encoding plays an important role in reducing the number of spikes and hence the power consumption. Table 8 tabulates the coactive spikes, the spike space, and the number of AHaH nodes used for most of the demonstration applications and benchmarks in this paper. Different applications require different configurations; some will have few AHaH nodes and a large spike space while others may have many AHaH nodes and few inputs.

Table 8: Application spike sparsity and AHaH node count

http://dx.doi.org/10.1371/journal.pone.0085175.t008

Application	Coactive Spikes	Spike Space	Sparsity	AHaH Node Count
Breast Cancer	31	70	0.44	2
Census Income	63	~1800	~0.035	2
MNIST	~1000	~27,500	~0.036	10
Reuters 21578	~100	~46,000	~0.002	119
Robotic Arm	92	341	0.27	345
Comb. Opt.	1	1	n/a	~600,000
Clusterer	16	256	0.0625	20
Prediction	300	9600	0.031	32

The applications and benchmarks presented in this paper to demonstrate various machine learning tasks using AHaH plasticity require different AHaH node configurations depending on the type of data being processed and what the desired result is. The sparsity is a function of the incoming data and is defined as the number of coactive spikes divided by the total spike space.
doi:10.1371/journal.pone.0085175.t008

LIMITATIONS OF THE STUDY, OPEN QUESTIONS, AND FUTURE WORK

We have attempted to connect a low-level general statistical model of collections of metastable switches with dissipative attractor-based computation and machine learning in a physically realizable circuit. Our aim is to provide a road map for others to follow so that we may all explore and exploit this interesting and potentially useful form of computing. Our ultimate goal is to provide a physical adaptive learning hardware resource (the AHaH circuit) in much the same way as modern RAM memory provides a memory resource to computing systems. However, only when we have investigated the circuit and functional models and have demonstrated real world utility is it necessary to move toward simulation of nonideal circuits effects, such as parasitic impedances, signal delays, settling times and variations in memristor properties. These details are certainly required for the eventual construction of a neural processing unit (NPU) but to include them in this paper would obfuscate our core message that "a new type of computing is possible that appears to offer a solution of general machine learning".

Our demonstrations of utility include results across the field of machine learning, from clustering and classification to prediction, control and combinatorial optimization. Given the intended broad scope of this paper it was not possible to provide much elaboration on some of our results, comparison with many other methods, nor discuss the implications. For this reason we have open-sourced all code used to generate the results of this paper. We encourage the reader to investigate our methods carefully and come to their own conclusions.

Although it was important to develop specific techniques to address the broad capabilities we have demonstrated, we wish to convey the idea that the AHaH node is a building block from which many higher-order adaptive algorithms may be built including many we have not yet conceived of. As an example consider our results with the AHaH motor controller and AHaH classifier. By using the classifier's confidence estimation as the value function for the AHaH motor controller, which in turn controls the viewing position, angle and rotation of an "eye", it should be possible to spontaneously control the gaze of a vision system to find and center previously trained objects. Alternately, by pairing the AHaH signal prediction with the AHaH combinatorial optimizer, it should be possible learn to predict a reward signal while simultaneously optimizing actions to attain reward. We can infer from our results that other capabilities are possible. Anomaly detection, for example, is the inverse of prediction. If a prediction can be made about a temporally dynamic signal, then an anomaly signal can be generated should predictions fail to match with reality. Tracking of non-stationary statistics is also a natural by-product of the attractor nature of the AHaH rule, and was slightly touched upon in the 2D clustering videos. Attractor points of the AHaH rule are created by the data structure. It follows logically that these same states will shift as the structure of the information changes. It also follows that a system built of components locked in attractor states will spontaneously heal if damaged [43], [44]. This property could provide new developments in fault-tolerant electronics.

CONCLUSIONS

We have introduced the concept of AHaH computing. We have shown how the simple process of particles dissipating into containers through adaptive channels competing for conduction resources leads to AHaH plasticity. We have shown that memristive devices can arise from metastable switches, how differential synaptic weights may be built of two or more memristors, and how an AHaH node may be built of arrays of synapses. A simple read and write cycle driving an AHaH circuit results in physical devices implementing AHaH plasticity. We have demonstrated that the attractor states of the AHaH rule can configure computationally complete logic functions, and have shown their use in supervised and unsupervised classification, clustering, complex signal prediction, unsupervised robotic arm actuation and combinatorial optimization. We have demonstrated unsupervised clustering and supervised classification in circuit simulations, and have further shown a correspondence between our functional and circuit forms of the AHaH node.

The AHaH node may offer us a building block for a new type of computing with likely application in the field of machine learning. Indeed, we hope that our work demonstrates that functions needed to enable perception (clustering, classification), planning (combinatorial optimization, prediction), control (robotic actuation) and generic computation (universal logic) are possible with a simple circuit that, technologically speaking, may be very close at hand.

ACKNOWLEDGMENTS

The Breast Cancer Wisconsin, Reuters-21578 Distribution 1.0, and Census Income classification benchmark datasets were obtained from the UCI Machine Learning Repository[114].

The authors would like to thank Kristy A. Campbell from Boise State University for graciously providing us with memristor device data.

Special thanks to Air Force Research Laboratory's (AFRL) Information Directorate.

Alex Nugent would like to personally thank Hillary Riggs, Kermit Lopez, Luis Ortiz, and Todd Hylton for their support over the years. This work would definitely not have existed without them.

This manuscript has been approved for public release; distribution unlimited. Case Number: 88ABW-2014-0103.

AUTHOR CONTRIBUTIONS

Conceived and designed the experiments: MAN TWM. Performed the experiments: MAN TWM. Analyzed the data: MAN TWM. Contributed reagents/materials/analysis tools: MAN TWM. Wrote the paper: MAN TWM.

REFERENCES

1. Grime JP, Crick JC, Rincon JE (1985) The ecological significance of plasticity. In: Proc. 1985 Symposia of the Society for Experimental Biology. volume 40, 5–29.
2. Desbiez MO, Kergosien Y, Champagnat P, Thellier M (1984) Memorization and delayed expression of regulatory messages in plants. Planta 160: 392–399. doi: 10. 1007/bf00429754
3. Aphalo PJ, Ballaré CL (1995) On the importance of information-acquiring systems in plant–plant interactions. Functional Ecology 9: 5–14. doi: 10. 2307/2390084
4. Falik O, Reides P, Gersani M, Novoplansky A (2003) Self/non-self discrimination in roots. Journal of Ecology 91: 525–531. doi: 10. 1046/j. 1365-2745. 2003. 00795. x
5. Scialdone A, Mugford ST, Feike D, Skeffington A, Borrill P, et al. . (2013) Arabidopsis plants perform arithmetic division to prevent starvation at night. eLife 2 doi: 10. 7554/eLife. 00669.
6. von Bodman SB, Bauer WD, Coplin DL (2003) Quorum sensing in plant-pathogenic bacteria. Annual Review of Phytopathology 41: 455–482. doi: 10. 1146/annurev. phyto. 41. 052002. 095652
7. Nakagaki T, Yamada H, Tóth Á (2000) Intelligence: Maze-solving by an amoeboid organism. Nature 407: 470–470. doi: 10. 1038/35035159
8. Bonabeau E, Dorigo M, Theraulaz G (1999) Swarm Intelligence: from Natural to 1172 Artificial Systems, volume 4. Oxford University press New York.
9. Choudhary S, Sloan S, Fok S, Neckar A, Trautmann E, et al. . (2012) Silicon neurons that compute. In: Artificial Neural Networks and Machine Learning – ICANN 2012, Springer Berlin Heidelberg, volume 7552 of Lecture Notes in Computer Science. 121–128.

10. Izhikevich EM (2003) Simple model of spiking neurons. IEEE Transactions on Neural Networks 14: 1569–1572. doi: 10. 1109/tnn. 2003. 820440

11. Eliasmith C, Stewart TC, Choo X, Bekolay T, DeWolf T, et al. (2012) A large scale model of the functioning brain. Science 338: 1202–1205. doi: 10. 1126/science. 1225266

12. Boser BE, Guyon IM, Vapnik VN (1992) A training algorithm for optimal margin classifiers. In: Proc. 1992 ACM 5th Annual Workshop on Computational Learning Theory. 144–152.

13. MacQueen J (1967) Some methods for classification and analysis of multivariate observations. In: Proc. 1967 5th Berkeley Symposium on Mathematical Statistics and Probability. 281–297, p. 14.

14. Breiman L (2001) Random forests. Machine Learning 45: 5–32. doi: 10. 1023/a:1010933404324

15. Serrano-Gotarredona R, Oster M, Lichtsteiner P, Linares-Barranco A, Paz-Vicente R, et al. (2009) CAVIAR: A 45k neuron, 5M synapse, 12G connects/s AER hardware sensory–processing–learning–actuating system for high-speed visual object recognition and tracking. IEEE Transactions on Neural Networks 20: 1417–1438. doi: 10. 1109/tnn. 2009. 2023653

16. Sardar S, Tewari G, Babu KA (2011) A hardware/software co-design model for face recognition using cognimem neural network chip. In: Proc. 2011 IEEE International Conference on Image Information Processing (ICIIP). 1–6.

17. Arthur JV, Merolla PA, Akopyan F, Alvarez R, Cassidy A, et al. . (2012) Building block of a programmable neuromorphic substrate: A digital neurosynaptic core. In: Proc. 2012 IEEE International Joint Conference on Neural Networks (IJCNN). 1–8.

18. Ananthanarayanan R, Esser SK, Simon HD, Modha DS (2009) The cat is out of the bag: cortical simulations with 109 neurons, 1013 synapses. In: Proc. 2009 IEEE Conference on High Performance Computing Networking, Storage and Analysis. 1–12.

19. Le QV, Ranzato M, Monga R, Devin M, Chen K, et al. . (2011) Building high-level features using large scale unsupervised learning. preprint arXiv cs. LG/1112. 6209.

20. Pakkenberg B, Pelvig D, Marner L, Bundgaard MJ, Gundersen HJG, et al. (2003) Aging and the human neocortex. Experimental Gerontology 38: 95–99. doi: 10. 1016/s0531-5565(02)00151-1

21. Turing AM (1936) On computable numbers, with an application to the entscheidungsproblem. Proceedings of the London Mathematical Society 42: 230–265. doi: 10. 1017/s002248120003958x

22. Turing A (1948) Intelligent machinery. Report, National Physical Laboratory.

23. Schrödinger E (1992) What is Life?: With Mind and Matter and Autobiographical Sketches. Cambridge University Press.

24. Hebb DO (2002) The Organization of Behavior: A Neuropsychological Theory. Psychology Press.

25. Rosenblatt F (1958) The perceptron: a probabilistic model for information storage and organization in the brain. Psychological Review 65: 386. doi: 10. 1037/h0042519

26. Barlow HB (1953) Summation and inhibition in the frog's retina. The Journal of Physiology 119: 69–88.

27. Hubel DH, Wiesel TN (1959) Receptive fields of single neurones in the cat's striate cortex. The Journal of Physiology 148: 574–591. doi: 10. 1113/jphysiol. 2009. 174151

28. Widrow B (1987) The original adaptive neural net broom-balancer. In: Proc. 1987 IEEE International Symposium on Circuits and Systems. volume 2, 351–357.

29. Minsky M, Seymour P (1969) Perceptrons. MIT press.

30. Hopfield JJ (1982) Neural networks and physical systems with emergent collective computational abilities. Proceedings of the National Academy of Sciences of the United States of America 79: 2554–2558. doi: 10. 1073/pnas. 79. 8. 2554

31. Rumelhart DE, Hinton GE, Williams RJ (1986) Learning representations by back-propagating errors. Nature 323: 533–536. doi: 10. 1038/323533a0

32. Grossberg S (1987) Competitive learning: from interactive activation to adaptive resonance. Cognitive Science 11: 23–63. doi: 10. 1016/s0364-0213(87)80025-3

33. Chua L (1971) Memristor–the missing circuit element. IEEE Transactions on Circuit Theory 18: 507–519. doi: 10. 1109/tct. 1971. 1083337

34. Chua LO, Kang SM (1976) Memristive devices and systems. Proceedings of the IEEE 64: 209–223. doi: 10. 1109/proc. 1976. 10092

35. Mead C, Conway L (1980) Introduction to VLSI Systems. Boston, MA, USA: Addison-Wesley Longman Publishing Co. , Inc.

36. Mead C, Ismail M (1989) Analog VLSI Implementation of Neural Systems. Springer.

37. Bienenstock EL, Cooper LN, Munro PW (1982) Theory for the development of neuron selectivity: orientation specificity and binocular interaction in visual cortex. The Journal of Neuroscience 2: 32–48. doi: 10. 1142/9789812795885_0006

38. Barlow HB (1989) Unsupervised learning. Neural Computation 1: 295–311. doi: 10. 1162/neco. 1989. 1. 3. 295

39. Bell AJ, Sejnowski TJ (1997) The independent components of natural scenes are edge filters. Vision Research 37: 3327–3338. doi: 10. 1016/s0042-6989(97)00121-1

40. Hyvärinen A, Oja E (1997) A fast fixed-point algorithm for independent component analysis. Neural Computation 9: 1483–1492. doi: 10. 1162/neco. 1997. 9. 7. 1483

41. Comon P (1994) Independent component analysis, a new concept? Signal Processing 36: 287–314. doi: 10. 1016/0165-1684(94)90029-9

42. Schölkopf, Simard P, Vapnik V, Smola AJ (1997) Improving the accuracy and speed of support vector

machines. Advances in Neural Information Processing Systems 9: 375–381.

43. Nugent A, Kenyon G, Porter R (2004) Unsupervised adaptation to improve fault tolerance of neural network classifiers. In: Proc. 2004 IEEE NASA/DoD Conference on Evolvable Hardware. 146–149.
44. Nugent MA, Porter R, Kenyon GT (2008) Reliable computing with unreliable components: using separable environments to stabilize long-term information storage. Physica D: Nonlinear Phenomena 237: 1196–1206. doi: 10. 1016/j. physd. 2008. 03. 043
45. Nugent A (2008). Plasticity-induced self organizing nanotechnology for the extraction of independent components from a data stream. US Patent 7,409,375.
46. Nugent A (2008). Universal logic gate utilizing nanotechnology. US Patent 7,420,396.
47. Nugent A (2009). Methodology for the configuration and repair of unreliable switching elements. US Patent 7,599,895.
48. Yang JJ, Pickett MD, Li X, Ohlberg DA, Stewart DR, et al. (2008) Memristive switching mechanism for metal/oxide/metal nanodevices. Nature Nanotechnology 3: 429–433. doi: 10. 1038/nnano. 2008. 160
49. Snider GS (2008) Spike-timing-dependent learning in memristive nanodevices. In: Proc. 2008 IEEE International Symposium on Nanoscale Architectures (NANOARCH). 85–92.
50. Stewart DR, Ohlberg DAA, Beck PA, Chen Y, Williams RS, et al. (2004) Molecule-independent electrical switching in Pt/organic monolayer/Ti devices. Nano Letters 4: 133–136. doi: 10. 1021/nl034795u
51. Kozicki MN, Gopalan C, Balakrishnan M, Mitkova M (2006) A low-power nonvolatile switching element based on copper-tungsten oxide solid electrolyte. IEEE Transactions on Nanotechnology 5: 535–544. doi: 10. 1109/tnano. 2006. 880407
52. Szot K, Speier W, Bihlmayer G, Waser R (2006) Switching the electrical resistance of individual dislocations in single-crystalline SrTiO3. Nature Materials 5: 312–320. doi: 10. 1038/nmat1614
53. Dong R, Lee D, Xiang W, Oh S, Seong D, et al. (2007) Reproducible hysteresis and resistive switching in metal-Cu_xO-metal heterostructures. Applied Physics Letters 90: 042107. doi: 10. 1063/1. 2436720
54. Tsubouchi K, Ohkubo I, Kumigashira H, Oshima M, Matsumoto Y, et al. (2007) High-throughput characterization of metal electrode performance for electric-field-induced resistance switching in metal/$Pr_{0.7}Ca_{0.3}M_nO_3$/metal structures. Advanced Materials 19: 1711–1713. doi: 10. 1002/adma. 200601957
55. Oblea AS, Timilsina A, Moore D, Campbell KA (2010) Silver chalcogenide based memristor devices. In: Proc. 2010 IEEE International Joint Conference on Neural Networks (IJCNN). 1–3.
56. Yang Y, Sheridan P, Lu W (2012) Complementary resistive switching in tantalum oxide-based resistive memory devices. Applied Physics Letters 100: 203112. doi: 10. 1063/1. 4719198
57. Valov I, Kozicki MN (2013) Cation-based resistance change memory. Journal of Physics D: Applied Physics 46: 074005. doi: 10. 1088/0022-3727/46/7/074005
58. Hasegawa T, Nayak A, Ohno T, Terabe K, Tsuruoka T, et al. (2011) Memristive operations demonstrated by gap-type atomic switches. Applied Physics A 102: 811–815. doi: 10. 1007/s00339-011-6317-0
59. Jackson BL, Rajendran B, Corrado GS, Breitwisch M, Burr GW, et al. (2013) Nanoscale electronic synapses using phase change devices. ACM Journal on Emerging Technologies in Computing Systems (JETC) 9: 12. doi: 10. 1145/2463585. 2463588
60. Choi S, Ambrogio S, Balatti S, Nardi F, Ielmini D (2012) Resistance drift model for conductivebridge (CB) RAM by filament surface relaxation. In: Proc. 2012 IEEE 4th International Memory Workshop (IMW). 1–4.
61. Pino RE, Bohl JW, McDonald N, Wysocki B, Rozwood P, et al. . (2010) Compact method for modeling and simulation of memristor devices: ion conductor chalcogenide-based memristor devices. In: Proc. 2010 IEEE/ACM International Symposium on Nanoscale Architectures (NANOARCH). 1–4.
62. Menzel S, Bottger U, Waser R (2012) Simulation of multilevel switching in electrochemical metallization memory cells. Journal of Applied Physics 111: 014501. doi: 10. 1063/1. 3673239
63. Chang T, Jo SH, Kim KH, Sheridan P, Gaba S, et al. (2011) Synaptic behaviors and modeling of a metal oxide memristive device. Applied Physics A 102: 857–863. doi: 10. 1007/s00339-011-6296-1
64. Sheridan P, Kim KH, Gaba S, Chang T, Chen L, et al. (2011) Device and SPICE modeling of RRAM devices. Nanoscale 3: 3833–3840. doi: 10. 1039/c1nr10557d
65. Biolek D, Biolek Z, Biolkova V (2009) SPICE modeling of memristive, memcapacitative and meminductive systems. In: Proc. 2009 IEEE European Conference on Circuit Theory and Design (ECCTD). 249–252.
66. Chang T, Jo SH, Lu W (2011) Short-term memory to long-term memory transition in a nanoscale memristor. ACS Nano 5: 7669–7676. doi: 10. 1021/nn202983n
67. Merrikh-Bayat F, Shouraki SB, Afrakoti IEP (2010) Bottleneck of using single memristor as a synapse and its solution. preprint arXiv: cs. NE/1008. 3450.
68. Jo SH, Chang T, Ebong I, Bhadviya BB, Mazumder P, et al. (2010) Nanoscale memristor device as synapse in neuromorphic systems. Nano Letters 10: 1297–1301. doi: 10. 1021/nl904092h
69. Hasegawa T, Ohno T, Terabe K, Tsuruoka T, Nakayama T, et al. (2010) Learning abilities achieved by a single solid-state atomic switch. Advanced Materials 22: 1831–1834. doi: 10. 1002/adma. 200903680
70. Li Y, Zhong Y, Xu L, Zhang J, Xu X, et al. . (2013) Ultrafast synaptic events in a chalcogenide memristor. Scientific Reports 3 doi:10. 1038/srep01619.
71. Merrikh-Bayat F, Shouraki SB (2011) Memristor-based circuits for performing basic arithmetic operations. Procedia Computer Science 3: 128–132. doi: 10. 1016/j. procs. 2010. 12. 022

72. Merrikh-Bayat F, Shouraki SB (2013) Memristive neuro-fuzzy system. IEEE Transactions on Cybernetics 43: 269–285. doi: 10. 1109/tsmcb. 2012. 2205676

73. Morabito FC, Andreou AG, Chicca E (2013) Neuromorphic engineering: from neural systems to brain-like engineered systems. Neural Networks 45: 1–3. doi: 10. 1016/j. neunet. 2013. 07. 001

74. Klimo M, Such O (2011) Memristors can implement fuzzy logic. preprint arXiv cs. ET/1110. 2074.

75. Klimo M, Such O (2012) Fuzzy computer architecture based on memristor circuits. In: Proc. 2012 4th International Conference on Future Computational Technologies and Applications. 84–87.

76. Kavehei O, Al-Sarawi S, Cho KR, Eshraghian K, Abbott D (2012) An analytical approach for memristive nanoarchitectures. IEEE Transactions on Nanotechnology 11: 374–385. doi: 10. 1109/tnano. 2011. 2174802

77. Rosezin R, Linn E, Nielen L, Kugeler C, Bruchhaus R, et al. (2011) Integrated complementary resistive switches for passive high-density nanocrossbar arrays. IEEE Electron Device Letters 32: 191–193. doi: 10. 1109/led. 2010. 2090127

78. Kim KH, Gaba S, Wheeler D, Cruz-Albrecht JM, Hussain T, et al. (2011) A functional hybrid memristor crossbar-array/CMOS system for data storage and neuromorphic applications. Nano Letters 12: 389–395. doi: 10. 1021/nl203687n

79. Jo SH, Kim KH, Lu W (2009) High-density crossbar arrays based on a Si memristive system. Nano Letters 9: 870–874. doi: 10. 1021/nl8037689

80. Xia Q, Robinett W, Cumbie MW, Banerjee N, Cardinali TJ, et al. (2009) Memristor-1327 tor-CMOS hybrid integrated circuits for reconfigurable logic. Nano Letters 9: 3640–3645. doi: 10. 1021/nl901874j

81. Strukov DB, Stewart DR, Borghetti J, Li X, Pickett M, et al. . (2010) Hybrid CMOS/memristor circuits. In: Proc. 2010 IEEE International Symposium on Circuits and Systems (ISCAS). 1967–1970.

82. Snider G (2011) Instar and outstar learning with memristive nanodevices. Nanotechnology 22: 015201. doi: 10. 1088/0957-4484/22/1/015201

83. Thomas A (2013) Memristor-based neural networks. Journal of Physics D: Applied Physics 46: 093001. doi: 10. 1088/0022-3727/46/9/093001

84. Indiveri G, Linares-Barranco B, Legenstein R, Deligeorgis G, Prodromakis T (2013) Integration of nanoscale memristor synapses in neuromorphic computing architectures. preprint arXiv cs. ET/1302. 7007.

85. Turing AM (1952) The chemical basis of morphogenesis. Philosophical Transactions of the Royal Society of London Series B, Biological Sciences 237: 37–72. doi: 10. 1098/rstb. 1952. 0012

86. Getling AV (1998) Rayleigh-Bénard Convection: Structures and Dynamics. World Scientific.

87. Athelogou M, Merté B, Deisz P, Hübler A, Lüscher E (1989) Extremal properties of dendritic patterns: biological applications. Helvetica Physica Acta 62: 250–253.

88. Swenson R (1989) Emergent attractors and the law of maximum entropy production: foundations to a theory of general evolution. Systems Research 6: 187–197. doi: 10. 1002/sres. 3850060302

89. Bejan A (1997) Constructal-theory network of conducting paths for cooling a heat generating volume. International Journal of Heat and Mass Transfer 40: 799–816. doi: 10. 1016/0017-9310(96)00175-5

90. Jorgensen SE, Svirezhev YM (2004) Towards a Thermodynamic Theory for Ecological Systems. Elsevier.

91. Schneider ED, Sagan D (2005) Into the Cool: Energy Flow, Thermodynamics, and Life. University of Chicago Press.

92. Lotka AJ (1922) Contribution to the energetics of evolution. Proceedings of the National Academy of Sciences of the United States of America 8: 147–151. doi: 10. 1073/pnas. 8. 6. 147

93. Hatsopoulos GN, Keenan JH (1981) Principles of General Thermodynamics. RE Krieger Publishing Company.

94. Shang DS, Shi L, Sun JR, Shen BG, Zhuge F, et al. (2010) Improvement of reproducible resistance switching in polycrystalline tungsten oxide films by in situ oxygen annealing. Applied Physics Letters 96: 072103. doi: 10. 1063/1. 3300637

95. Kvatinsky S, Friedman EG, Kolodny A, Weiser UC (2013) TEAM: ThrEshold Adaptive Memristor model. IEEE Transactions on Circuits and Systems I: Regular Papers 60: 211–221. doi: 10. 1109/tcsi. 2012. 2215714

96. Strukov DB, Williams RS (2009) Exponential ionic drift: fast switching and low volatility of thin-film memristors. Applied Physics A 94: 515–519. doi: 10. 1007/s00339-008-4975-3

97. Pickett MD, Strukov DB, Borghetti JL, Yang JJ, Snider GS, et al. (2009) Switching dynamics in titanium dioxide memristive devices. Journal of Applied Physics 106: 074508. doi: 10. 1063/1. 3236506

98. Williams RS, Pickett MD, Strachan JP (2013) Physics–based memristor models. In: 1365 Proc. 2013 IEEE International Symposium on Circuits and Systems (ISCAS). 217–220.

99. Zhang JJ, Sun HJ, Li Y, Wang Q, Xu XH, et al. (2013) AgInSbTe memristor with gradual resistance tuning. Applied Physics Letters 102: 183513. doi: 10. 1063/1. 4804983

100. Mladenov VM, Kirilov SM (2012) Analysis of a serial circuit with two memristors and voltage source at sine and impulse regime. In: Proc. 2012 IEEE 13th International Workshop on Cellular Nanoscale Networks and Their Applications (CNNA). 1–6.

101. Jain AK, Murty MN, Flynn PJ (1999) Data clustering: a review. ACM Computing Surveys (CSUR) 31: 264–323. doi: 10. 1145/331499. 331504

102. Lloyd S (1982) Least squares quantization in PCM. IEEE Transactions on Information Theory 28: 129–137.

doi: 10. 1109/tit. 1982. 1056489

103. Kriegel HP, Kröger P, Sander J, Zimek A (2011) Density-based clustering. Wiley Interdisciplinary Reviews: Data Mining and Knowledge Discovery 1: 231–240. doi: 10. 1002/widm. 30
104. Ankerst M, Breunig MM, Kriegel HP, Sander J (1999) OPTICS: ordering points to identify the clustering structure. ACM SIGMOD Record 28: 49–60. doi: 10. 1145/304181. 304187
105. Dempster AP, Laird NM, Rubin DB (1977) Maximum likelihood from incomplete data via the EM algorithm. Journal of the Royal Statistical Society: Series B 39: 1–38.
106. Kotsiantis SB (2007) Supervised Machine Learning: a Review of Classification Techniques. IOS Press. 3–24 pp.
107. Yu HF, Hsieh CJ, Chang KW, Lin CJ (2012) Large linear classification when data cannot fit in memory. ACM Transactions on Knowledge Discovery from Data (TKDD) 5: 23. doi: 10. 1145/2086737. 2086743
108. Ndong J, Salamatian K (2011) Signal processing-based anomaly detection techniques: a comparative analysis. In: Proc. 2011 3rd International Conference on Evolving Internet. 32–39.
109. Zhang G, Patuwo BE, Hu MY (1998) Forecasting with artificial neural networks: The state of the art. International Journal of Forecasting 14: 35–62. doi: 10. 1016/s0169-2070(97)00044-7
110. Hoeting JA, Madigan D, Raftery AE, Volinsky CT (1999) Bayesian model averaging: a tutorial. Statistical Science: 382–401.
111. Cireşan DC, Meier U, Gambardella LM, Schmidhuber J (2012) Deep big multilayer perceptrons for digit recognition. In: Neural Networks: Tricks of the Trade, Springer Berlin Heidelberg, volume 7700 of Lecture Notes in Computer Science. 581–598.
112. Chaaban I, Scheessele MR (2007) Human performance on the USPS database. Report, Indiana University South Bend.
113. Kuzum D, Jeyasingh RGD, Lee B, Wong HSP (2011) Nanoelectronic programmable synapses based on phase change materials for brain-inspired computing. Nano Letters 12: 2179–2186. doi: 10. 1021/nl201040y
114. Bache K, Lichman M (2013). UCI machine learning repository. Available:http://archive. ics. uci. edu/ml.
115. Chen HL, Yang B, Liu J, Liu DY (2011) A support vector machine classifier with rough set-based feature selection for breast cancer diagnosis. Expert Systems with Applications 38: 9014–9022. doi: 10. 1016/j. eswa. 2011. 01. 120
116. Kohavi R (1996) Scaling up the accuracy of naïve-Bayes classifiers: a decision-tree hybrid. In: Proc. 1996 2nd International Conference on Knowledge Discovery and Data Mining. 202–207.
117. Deng L, Yu D (2011) Deep convex net: A scalable architecture for speech pattern classification. In: Proc. 2011 Interspeech. 2285–2288.
118. Joachims T (1998) Text Categorization with Support Vector Machines: Learning with Many Relevant Features. Springer.
119. Bennett KP, Blue JA (1998) A support vector machine approach to decision trees. In: Proc. 1998 IEEE International Joint Conference on Neural Networks Proceedings. The 1998 IEEE World Congress on Computational Intelligence. volume 3, 2396–2401.
120. Ranzato M, Huang FJ, Boureau YL, Lecun Y (2007) Unsupervised learning of invariant feature hierarchies with applications to object recognition. In: Proc. 2007 IEEE Conference on Computer Vision and Pattern Recognition (CVPR). 1–8.
121. Quinlan JR (1996) Improved use of continuous attributes in C4. 5. preprint arXiv cs. AI/9603103.

CHAPTER 3 BALANCING THE LIFETIME AND STORAGE OVERHEAD ON ERROR CORRECTION FOR PHASE CHANGE MEMORY

Ning An[1], Rui Wang[1], Yuan Gao[2], Hailong Yang[2], Depei Qian[2]

[1]State Key Laboratory of Software Development Environment, Beihang University, Beijing, China

[2]School of Computer Science and Engineering, Beihang University, Beijing, China

ABSTRACT

As DRAM is facing the scaling difficulty in terms of energy cost and reliability, some nonvolatile storage materials were proposed to be the substitute or supplement of main memory. Phase Change Memory (PCM) is one of the most promising nonvolatile memory that could be put into use in the near future. However, before becoming a qualified main memory technology, PCM should be designed reliably so that it can ensure the computer system's stable running even when errors occur. The typical wear-out errors in PCM have been well studied, but the transient errors, that caused by high-energy particles striking on the complementary metal-oxide semiconductor (CMOS) circuit of PCM chips or by resistance drifting in multi-level cell PCM, have attracted little focus. In this paper, we propose an innovative mechanism, Local-ECC-Global-ECPs (LEGE), which addresses both soft errors and hard errors (wear-out errors) in PCM memory systems. Our idea is to deploy a local error correction code (ECC) section to every data line, which can detect and correct one-bit errors immediately, and a global error correction pointers (ECPs) buffer for the whole memory chip, which can be reloaded to correct more hard error bits. The local ECC is used to detect and correct the unknown one-bit errors, and the global ECPs buffer is used to store the corrected value of hard errors. In comparison to ECP-6, our method provides almost identical lifetimes, but reduces approximately 50% storage overhead. Moreover, our structure reduces approximately 3.55% access latency overhead by increasing 1.61% storage overhead compared to PAYG, a hard error only solution.

INTRODUCTION

DRAM-based main memory is facing severe challenges due to its high leakage, limited scaling and increasingly refresh cost. In addition, in future 64Gb devices, almost 50% of DRAM power will be consumed by refresh operations [1]. Emerging non-volatile memory (NVRAM) is a promising technology to be the substitute or supplement of the DRAM main memory system. The phase change memory (PCM) is one of the most promising non-volatile memory for its good scalability, high density and compatibility with complementary metal-oxide semiconductor (CMOS) process.

PCM stores information by setting the phase change materials to different resistance states, which are called crystal and amorphous states respectively. The state of the material can be switched for a certain time after a strong current has passed through. The required strengths of current are different if we switch the state to different directions. Adding a tiny

Citation: An N, Wang R, Gao Y, Yang H, Qian D (2015) Balancing the Lifetime and Storage Overhead on Error Correction for Phase Change Memory. PLoS ONE 10(7): e0131964. doi:10.1371/journal.pone.0131964.

current to the material can detect its resistance, and then read out the stored information. This operation will not change the resistance of this PCM cell. In other words, the information is permanently stored in the PCM cells even when the electric is turned off. The tiny electric charges from the high-energy particle are far from being able to trigger the PCM memory cells to switch the state. Therefore, PCM cells have barely normal transient errors that often appear in DRAM cells.

Like other memory technologies, PCM has its own disadvantages. The average write endurance that phase change memories usually have is close to 10^8, which is far smaller than the DRAM endurance (approximately 10^{15}). If the number of writes in one PCM cell has exceeded the write endurance, this cell would be permanently stuck at fault, or worn out. With the process variation and unevenly writing, the worn out cells may appear more earlier [2]. These worn out faults are normally referred to as "hard errors". A majority of prior works have focused on this type of errors and try to employ methods to prolong the PCM lifetime [3][4][5].

As the phase change materials have immunity to the high energy particles, conventional researches including ECP [5] and PAYG [3] seldom focus the soft errors in PCM. However, PCM also suffers soft errors. Soft errors are generally refer to the transient, undestroyed state inverses caused by the striking of high-energy particles in the storage cells or transmission lines. When transient errors occur, the device can still work. This issue has been concerned in DRAM memory, which is mainly composed by capacitors. ECC [6], and Chipkill [7] are the normal solutions. But in PCM chips, CMOS circuit also takes a big part, which also introduces the possibility of state inversion caused by high energy particles. Another issue comes from the resistance drifting in multi-level cell PCM [4] [8]. We will discuss it further in section 2.

In this paper, we propose a novel method, Local-ECC-Global-ECPs (LEGE), to solve not only hard errors, but also soft errors by deploying a small local ECC, a local ECP-1 and a global ECPs buffer for each cache line size memory block(normally 512 bits). The local ECC is a single-error-correction-double-error-detection (SEC-DED) code, which can detect two errors and correct one error normally. Therefore, it can detect and correct one-bit soft error effectively in PCM, just as it is used in DRAM. As the number of soft errors in PCM is far lower than that in DRAM [4][9][10][11][12], we allocate a ECC fielded in a more large data block(512 bits) than that normally used in DRAM(64 bit). Also, note that the SEC-DED ECC can correct a one-bit hard error without having to reload the hard error data, which is one key idea in our work. The rest errors caused by worn-out cells will be corrected by a local ECP-1 field and a global ECP buffer.

Besides error corrections, LEGE also has a good tradeoff between performance and storage overhead. The benefit from our design is as follows:

As the normally used ECC, this method can solve soft errors in the PCM circuit;

- The ECPs section in a global buffer can store corrected errors caused by worn out cells (hard errors) and decrease storage overhead;
- As the ECC is associated with the data array, it can correct one-bit error whatever the kind of the error is;
- The storage overhead of error correction is 7.1%, which is much lower than the normally used hard error correction method, such as ECP-6; this design trivially affects memory read performance.

The rest of the paper is organized as follows: Section 2 provides a background on PCM technology. Section 3 discusses the related works and motivation. Section 4 describes the framework of LEGE and the error correction methods in our work. Section 5 interprets our experimental methodology and evaluation. Section 6 concludes this paper.

BACKGROUND

PCM Technology

A typical PCM cell consists of an NMOS access transistor and a phase change material ($Ge_2Sb_2\ Te_5$). Each PCM cell has two material states with different properties: amorphous and crystalline states. When the cell is heated to its melting point and then cools down, it is in an amorphous state. In the amorphous phase, which is also called the RESET state, the cell's resistivity is high. However, when the material is heated above its crystallization point and below its melting point, it will transform to the crystalline phase (the SET state) and have low resistivity. The PCM technology utilizes the reversible phenomena to store values. When the cell is at a low resistance state, it logically stores '1'. Conversely, it stores '0' at a high resistance state.

Moreover, the multi-level cell (MLC) PCM has more than two states because there is a wide range of varying resistance levels in it. The MLC PCM relies on a finer-grain control in order to write accurately to expected states in a cell; it stores more data per cell by using intermediate states that are located between the crystalline and amorphous states of the phase material. Although this type of PCM can increase density by storing more information, it has a problem caused by the resistance drifting, which has been presented in some prior papers [8][9][13] in recent years. The resistance drift phenomenon is attributable to the resistance values that may increase over time. When one or more intermediate states' resistances cross their state boundaries and the cells' states change, PCM soft errors emerge.

Errors in PCM

The temperature changes frequently for the storage of values. As the states change between amorphous and crystalline, the phase-change materials in PCM cells become invalid as time passes. Therefore, the lifetime of PCM decreases distinctly. Due to the process variation, the worn out phenomena in each cell occur at different times. In some cells these phenomena may emerge earlier than usual. The wear-out error problem is the topic that has received the most focus in previous research on PCM error correction.

Although phase change material is immune to the high-energy particles, the PCM memory still suffers soft errors. For example, In PCM chips, up to 40% of the whole area consists of CMOS circuits, such as row buffers, latches, and registers [14]. The data may be changed during the transmission, which caused by the high energy particles striking on CMOS. Additionally, the room temperature may cause of crystallization of phase change materials, lowering cells' resistances beyond normal values, which is due to the unstable nature of amorphous phase [4]. Particularly, the multi-level cells may suffer from resistance drifting. They store data in different resistance levels; for example, a two-bit cell has four resistance levels (00, 01, 10 and 11). Experimental results showed that the four resistance levels of the two-bit cell drift after 400 hours at room temperature and 12 hours at 130°C [8]. While, prior work [9] has shown that four levels PCM cannot be used as main memory because of the great effect of the resistance drift.

RELATED WORKS AND MOTIVATION

As time goes on, all PCM cells' lifetimes decay as the frequent changes between RESET and SET states in the phase change materials. Wear-out error solutions are used to discover and correct the cells only when error cells emerge far earlier than average.

The majority of prior works have focused on cells' wear-out problems; far fewer works pay attention to solving soft errors. Before this paper, the work paying close attention to both kinds of errors is FREE-p [4]. Besides that, the related works in the area which focus on wear-out error corrections are PAYG [3] and ECP-6 [5].

The ECC is an outstanding method used to detect and correct errors in a DRAM-based system. However, it seems less-effectively used to discover and correct wear-out errors in non-volatile memory, because these kinds of errors are dominant, irreversible errors and can accumulate.

The ECP's aim is only focused on the PCM wear-out errors. It includes a pointer and a correction bit; the pointer can indicate the error location, and the correction bit can record the corrected data. For example, at a 512-bit memory line, ECP-n uses n entries; each entry is comprised of a 9-bit ($\log_2(512)$) pointer and a one-bit correction bit for one error. Therefore, ECP-n can correct n bits errors for this line. Beyond that, there is a global bit to show whether n entries are all in use. When they are all in use, the value in this bit is set to '1'. If they are not, it is set to '0'. In particular, ECP-1 uses 9 bits for storing the error location and 1 bit for correcting the error. When the system reads or writes at one line comprised of several wear-out errors, it will read or write the corrected data in ECPs instead of the wrong values in the worn out cells. This method may be effective in preventing the occurrence of a large number of wear-out errors. When hard errors in one line accumulate over time, the storage overhead of ECPs would gradually increase.

As the ECP means high overhead for memory manufactures, prior work PAYG [3] has found that ECP-6 is inefficient, and most lines do not need many error correction bits before large areas of cells become invalid. PAYG [3] proposed an attractive method for efficiently using the ECP-6 and optimizing its storage overhead, but it can only solve detectable hard errors. Their emphasis, demonstrated by their experiment, showed that more than 75% of PCM lines required no more than an ECP-1 throughout the age of ECP-6. Therefore, PAYG presented a dynamic mechanism to improve the organization of ECP: setting an ECP-1 in the local line and many ECP entries for global error correction. The global entries are used when one line appears more than one wear-out errors. This mechanism can reduce storage overhead effectively relative to the ECP-6.

Besides the wear-out errors, soft errors cannot be ignored. Both ECP and PAYG can only detect and correct hard errors. The perfect memory system must be strong in dealing with both types of errors.

FREE-p [4] is one of a few pieces of literature that addresses both soft errors and hard errors of the PCM. It uses a staged 61-bit 6-bit error correcting 7-bit error detecting (6EC-7ED) BCH code for solving both soft errors and hard errors. If even one error occurs in one line, the information in this line should be remapped and stored to a new line, but storage overhead is too expensive. When the hard errors accumulate in a data block, the BCH decoder is consequently slower. In addition, Free-p needs the help of the OS to remap the address of the disabled data block to a new area. However, this process may lead to access traffic and latency, limiting its performance. Moreover, when the page is swapped out to the disk, the OS needs to know which line is remapped. Because FREE-p lacks of a fine-grained remapping table between memory and disk.

As the PCM also faces the threat of soft errors, the existing hard error solutions cannot be put into applications unless they are as friendly as soft error solutions in normal DRAM systems. This paper provides a novel, low-cost mechanism to protect PCM-based memory against both soft errors and hard errors, while improving system reliability and prolonging cells' lifetimes with the premise of not losing too much performance.

LOCAL-ECC-GLOBAL-ECPS (LEGE)

Comparing with 6EC-7ED BCH [4], our method is even simpler and faster. We use a SEC-DED ECC to solve both soft and hard errors, and present a novel process that can use ECC more efficiently, since ECC was only used in DRAM in the past. We present the architecture of our work at first in this section, and then discuss the memory access process.

Architecture of LEGE

We designed the PCM-based memory with a global ECPs buffer and a detection and correction matrix that includes valid-bits, an ECP-1 and a SEC-DED ECC.

Fig 1 shows the simple architecture of our scheme, which consists memory lines and an associated detection and correction matrix. Below that, there is a global ECPs buffer. We assume that one memory line can store 512-bit data, which is the same size as the last level cache line. It is just as suitable to use eight bursts incurred by a memory controller for transferring such 512-bit data.

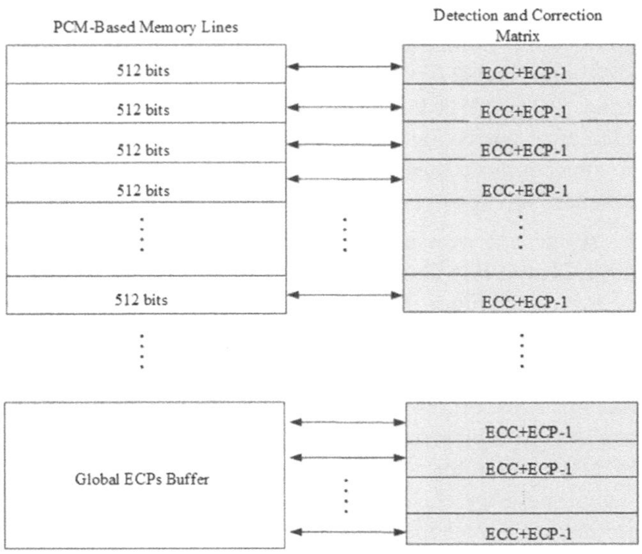

Figure 1: The structure of LEGE.

We allocated one detection and correction matrix for one memory line. Due to the global ECPs buffer was set in memory region, each global buffer line also had a matrix for detection and correction.

http://dx.doi.org/10.1371/journal.pone.0131964.g001

Fig 2 presents the structure of the detection and correction matrix. One detection and correction matrix line corresponds to one memory line, which is comprised of a SEC-DED ECC, an ECP-1 and valid-bits. By setting a local ECP-1, the data lines that have one bit error don't have to reload the correction information from global buffer. We tentatively set 24 bits for one matrix line, 11 bits suited a SEC-DED ECC.

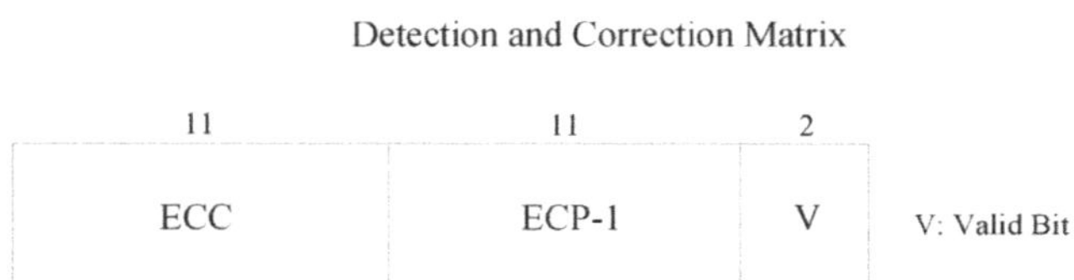

Figure 2: The structure of a detection and correction matrix line.

Each detection and correction matrix had an 11-bit ECC, an 11-bit ECP-1 and a 2-bit valid bits.

http://dx.doi.org/10.1371/journal.pone.0131964.g002

Soft error is dominant in all kinds of errors of DRAM cells, while it is trivial in that of DRAM peripheral circuit [15]. Phase change material is almost immune to the high-energy particles which produce the DRAM cell's soft errors [4][12]. The reliability concern of PCM is mainly caused by the resistance drift. Resistance drift was not a problem in single-level-cell (SLC) PCM which is mentioned in [4][9][10][11][12]. Because the rate of resistance drift is proportional to the initial resistance of the cell and is nearly-zero for the crystalline state [9]. The drift occurred in amorphous state does not cause error because resistance always tends to drift to a high values and all values above the boundary resistance threshold represent the amorphous state [8][11][12]. Prior work [16] has presented that the thermal interference between PCM cells at 85°C is very low even over a 10-year retention period. Moreover, as prior work [12] demonstrated that tri-level-cell (3LC) PCM has a lower soft error rate (SER) than DRAM, the SER in SLC PCM is lower than that in DRAM since boundary resistance thresholds are more closely spaced in 3LC PCM and the resistance can cross the state boundary easier than that in SLC PCM, which can lead to soft error due to state changes. So, One ECC is sufficient for a 512-bit line. We allocated 11 bits for ECP-1, which is comprised of one full bit, a 9-bit pointer and one corrected value bit. Moreover, we allocated 2 bits for valid-bit (symbol 'V' in Fig 2).

In valid-bit, the value '00' means this memory line has no hard error so far. '01' means only one hard error has occurred, so we read the corrected value from the local ECP-1. '10' means there are two hard errors. '11' means there are more than two hard errors. Both such situations show that the corrected data have been stored to the global ECPs buffer.

We can use three more data bits for transmitting the value in 24-bit associated detection and correction matrix during an eight-burst period; there is no need to add any more peripheral circuitry. Current server memories provide a 72-bit data path and this mechanism uses only 67 bits (64 bits for data and 3 bits for detection and correction matrix) of them.

As each local matrix can store one error, we allocated five more ECPs in global ECPs buffer for memory lines having two or more errors, which can prolong its lifetime so that it is equivalent to ECP-6. The corrected value will be put into ECP-1 or a global ECPs buffer that depends on whether this error is the first hard error. The global ECPs buffer is used to store other stuck-at-fault data bits that should be stored in worn out areas. The reloading makes the memory controller issue another memory access. These five ECPs are located in the global buffer but not in the corresponding local detection and correction matrix of each memory line. In the global buffer, there is not a one-to-one correspondence between a memory line and a five-ECPs set, which we will explain further in this section. This structure can result in a remarkable reduction of memory storage overhead.

In Fig 3 we have provided a draft about the global mapping line in the global ECPs buffer. The architecture can provide fast access by using a Hash table structure between the memory lines and the global ECPs buffer. We assumed a global ECPs buffer line size would be equal to the memory line granularity, so it can be set into the memory region. A global mapping line includes a valid bit, six G entries and two pointers. The valid-bit indicates whether the six G entries are all occupied and whether the access operation needs one extra memory access. The value '1' in the valid-bit means that all the six G entries have been allocated, and new memory line address must be mapped to a new buffer line by the pointer in this mapping line. '0' means there are several G entries that are not allocated yet, and new memory line address can be mapped into this line.

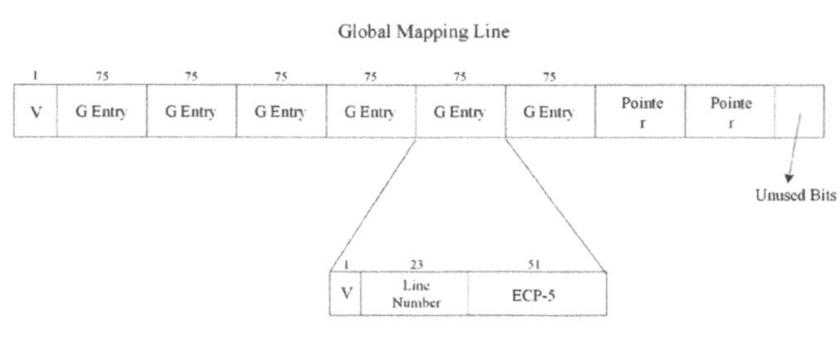

Figure 3: The structure of a global mapping line.

Each global mapping line used a valid bit to indicate whether these six entries were all occupied. One G entry can be allocated to one memory line which had more than one error.

http://dx.doi.org/10.1371/journal.pone.0131964.g003

Each G entry is an ECP-5 with a valid bit and a 23-bit line number. Unless this G entry is occupied by one memory line, the valid-bit in the G entry is always '0'. The line number section stores the corresponding memory line number, indicating in which memory line the errors occurred. The space of the line number occupied depends on the memory's addressable space, so it is allocated with 23 bits that depends upon the memory size assumed in our work. In order to insure the pointer bits do not wear out and can function well, we used one pointer and a duplicate to protect against getting a wrong address.

Fig 4 shows the simple architecture of a global ECPs buffer. The global buffer contains two parts: global mapping lines and collision pool lines. We assumed that the structure of the collision pool line is the same as the global mapping line, but they have different access patterns: the addresses of the global mapping lines must be mapped by the Hash table and the collision pool lines must be directed by the pointer in the global mapping line. The collision pool part can be flexibly deployed and only contact with the global mapping part, but not memory lines. When a global mapping line is full (the valid-bit in front of this mapping line is '1'), a new memory line that appears with the second error (the first error has been stored in associated local ECP-1) and is mapped into this full line should be remapped to a collision pool line by the pointer. If the valid-bit in the pointed collision pool line is '1', meaning this line is already full with six entries and errors in a new memory line must be stored to another collision pool line. The pointer in this collision pool line can address a new collision line for it. The traversal continues until the valid-bit of one collision pool line is '0', at which point the data can be stored in it. In the following section, we will discuss the global ECPs buffer size through experimentation.

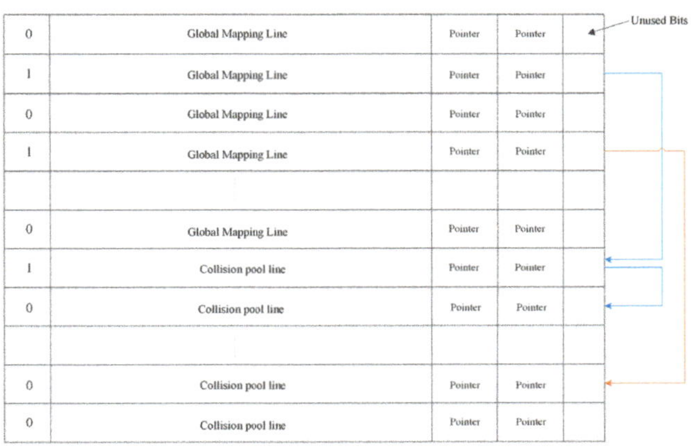

Figure 4: The structure of global ECPs buffer.

When valid bit displayed '1', it means these six G entries have already been occupied. we must use the address in pointer area to map to a collision pool line.

http://dx.doi.org/10.1371/journal.pone.0131964.g004

We also deploy a detection and correction matrix to this area, like it in the memory line. When data will be written to the worn out cells in memory line, they will be written to the corresponding global buffer line. If cells in this buffer line have been written many times, they may wear out. Although the probability of occurring hard errors in corresponding global buffer line is less than it in memory line (the address bits will not be changed by rewriting and the hard error occurrence in buffer line is later than it in memory line), the reliability of global buffer cannot be ignored.

Access Process

If the number of writes in one PCM cell has exceeded the write endurance, this cell would be permanently stuck at fault, or worn out. This phenomenon is called hard error. Verify-after-write method is widely used in write process to detect the hard error and put the corrected value into ECP.

In our work, we model the detection and correction process with considering not only the hard error but also the soft error in PCM. We use an ECC for detecting and correcting in read process. ECC can detect and correct soft error, besides that, ECC also can rapidly report the correct value of the data in hard error, but not really change the value in memory cell. The right value of data in hard error is put into ECP when it is found by verify-after-write in write process.

We explain the ECC and verify-after-write further as follow:

In the write process, we apply a widely used detection mechanism, verify-after-write. This method is successful in detecting hard errors. It writes data to one memory line and immediately reads the stored data, and then compares the written data and read data. If some bits in stored data are different from written data, it indicates that several invalid worn out bits have occurred in this line. If hard error is detected, we use one or more ECPs to record these errors (the first error will be stored in the local ECP, the others will be stored in the global ECPs buffer), and set the value of valid-bits in the detection and correction matrix. The write process can be extended by using some PCM lifetime-prolonging mechanisms, such as Flip-N-Write [17] and Coset coding [18].

In the read process, we first use SEC-DED ECC for checking memory lines; if the ECC reveals there is one error, no matter whether it is a hard error or a soft error, the ECC can immediately detect and correct it. That is the key difference in our work from prior papers. This method can largely improve performance because ECC can even correct one hard error without refetching correction information from main memory. As a majority of the memory lines have no more than a one-bit hard error in the ECP-6 service lifetime [3], our work will be effective and can maintain low storage overhead. When the error is a soft error, the local ECC corrects it directly. When the error is a hard error, the local ECC rapidly report the right data. As ECC can only report the correct values of the data in error but not really correct data by changing the memory cell values, the right value should be stored in the ECP. The hard errors can be found and corrected by the verify-after-write method in the write operation. If it is not the first one that has been found, the corrected value will be stored in the global ECPs buffer. After storing the data into the ECP, the valid-bit in the detection and correction matrix will be set automatically.

As lifetimes of PCM cells are controlled by the number of writes, the wear-out errors always occur in write operations. Our work is based on using the verify-after-write method for detecting and correcting hard errors in write operations, and using ECC for discovering and correcting soft errors in read operations. Besides that, ECC can be also used to correct wear-out errors to improve memory access performance.

Because the SEC-DED ECC can only detect two errors, when errors in one memory line exceed one, the ECC can detect but not correct it. When the errors exceed two, the ECC is unable to work. This is not suitable for PCM-based memory, because it can allow for the accumulation of errors. Combined with the use of valid-bits, the SEC-DED ECC will be available if we make small changes in the ECC circuitry. Using the ECPs' corrected data merged ECC results (using Exclusive-OR Logic) in the ECC circuitry and letting ECC regard them as already corrected values, then the ECC can work to detect and correct soft errors. As two soft errors in a DRAM memory line scarcely occur simultaneously, the probability of two soft errors occurring at the same time can be negligible in PCM. SEC-DED ECC can be well applied in a PCM like in DRAM.

In read operations, the first step is to read the valid bits in the detection and correction matrix, and fetch the corrected data from the ECPs based on the display of valid bits. The second step is to change the ECC circuit by merging the corrected data in the ECPs, making occurred hard errors transparent to the ECC. The third step is to use the ECC for detecting and correcting the soft error.

In first step, there are four states in valid-bit:

- If valid bits display '00', there is no hard error in this line. If the ECC finds a soft error, it will correct it immediately.
- If valid bits display '01', this line has emerged only one hard error and its real value has been stored in local ECP-1. We fetch the value, and then use the ECC for detecting.
- Valid bits display '10', meaning that there are two hard errors in this line; one of the real values has been stored in the local ECP-1 and another has been stored in the global ECPs buffer. This situation is a special one. We only utilize one corrected data in the local ECP-1 rather than using both errors to change the ECC circuit. Then, if the ECC displays one error, it means that no soft error occurs in this line, and the error displayed in the ECC is another hard error for which the corrected value has been stored in the global ECPs buffer. The ECC can send the corrected value immediately without fetching it from the global buffer, which can reduce extra memory access effectively. On the other hand, if the ECC displays two errors, it means a soft error has occurred in this line. We must fetch the value in the global buffer and change the ECC circuit again (using Exclusive-OR Logic is fast), then the ECC can work to correct this soft error. The access to the global buffer in this process leads to memory access latency, but there are very few because soft errors in PCM are rare.
- If the valid bits display '11', it means that more than two hard errors have occurred. This situation indicates that we must take one extra memory access process. We need to fetch data from the local ECP-1 and the global ECPs buffer.

The changes to ECC must be ordered especially in the third state. To begin with, we must use the value in local ECP-1 to change ECC's circuit instead of using the values from local ECP-1 and global ECPs buffer. This is a key point in our method. Because using the data from both local ECP-1 and global ECPs buffer seems easy enough, factually fetching data from global buffer can lead to at least one more memory access. Due to PCM has few soft errors, more than one memory access in the above third state rarely appears. From the above four conditions, we can see that the only conditions that will lead to taking an additional memory access are when there are two hard errors with a soft error or more than two hard errors.

In the global ECPs buffer, we allocated an ECP-5 to an associated memory line which has more than one error. Adding the local ECP-1, each memory line has six places to store corrected data. Theoretically, it can fix 7-bit hard errors with an ECC. But, we assume that it can only correct 6-bit hard errors because the ECC must be available to correct soft errors at any moment. If the ECC is used to correct the seventh hard error, and the first six hard errors have already been stored in the ECPs, the seventh right value must be sent by ECC with no place for it to be stored. At this moment, a soft error has occurred in this line; the ECC will display that there are two errors in this line and it is unable to correct any further errors. So, we assume that when the read-verify-write find the seventh hard error on it, the operating system can invalidate the physical page on the software level.

EXPERIMENTS

Using the Distribution of Hard Errors in PAYG to Set LEGE Configuration

In order to compare with PAYG, we first use the distribution of hard errors in PAYG to set LEGE configuration. The distribution of lines with different number of hard errors tested by PAYG [3] is shown in the first line of Table 1. The distribution indicates 96.06% of lines need only one ECP, and 99.61% of lines need no more than two ECPs. Our analysis of the LEGE architecture in the previous section shows that only when there is more than one error occurring in one line will lead to extra memory access to write, so our global buffer size must be set to accommodate the hard errors in this approximately 4% of memory lines. We set the global ECPs buffer size to 64K lines, 55K lines for the global mapping portion that can store the corrected values of approximately 4% of memory lines, and 9K for the collision pool portion. We used the simple Hash function: H (k) = k% P for handling the mapping relationships, where the value of P was the largest prime number in 0 ~ 55K-1. Our simulation generate the addresses of these memory lines by using the data of PAYG, which have more than one hard error and need access to global ECPs buffer to store corrected data at the end lifetime of ECP-6. Our experimental results show that less than 4% of memory lines need one more memory access to store data to the global mapping lines; only approximately 46000 lines need store to the collision pool once, resulting in a remaining 0.55% of all memory accesses if using our architecture. The possibility of two or more accesses to the collision pool seems negligible.

Table 1: The distribution of lines that using different number of ECP at the end age of ECP-6 in PAYG and our experiment

http://dx.doi.org/10.1371/journal.pone.0131964.t001

Num ECP per line	ECP-1	ECP-2	ECP-3 ~ ECP-6	ECP-0	ECP0 + ECP1	ECP0 + ECP1 + ECP2
lines/2^{23}(PAYG) [3]	22.82%	3.55%	0.40%	73.24%	96.06%	99.61%
lines/2^{23}(LEGE)	30%	9.2%	2.12%	58.68%	88.68%	97.88%

doi:10.1371/journal.pone.0131964.t001

Based on the above analysis of ECC part in section 4, we have to go to the global buffer to fetch data only when there are more than two hard errors or two hard errors with a soft error. Due to soft errors are rare in PCM, our method can effectively reduce the access latency.

LEGE's latency overhead (0.39%) is 3.55% less than PAYG (3.94%) throughout the lifetime of ECP-6. The storage overhead of LEGE using the data derived from PAYG is 5.5%, which is only 1.61% higher than PAYG (3.89%). LEGE used data derived from PAYG that could reduce storage overhead by 7%, compared to ECP-6(12.5%).

Fig 5 shows that one extra memory access by using PAYG architecture covers 3.94% of all memory accesses, by contrast, only incurs 0.4% in all memory accesses by using LEGE architecture at the end age of ECP-6 based PCM. The first additional access by using PAYG architecture happens at 10% of the ECP-6 lifetime, by contrast, happens at 43% of the ECP-6 lifetime by using LEGE architecture. In the first 40% of write operations, LEGE architecture will not cause any latency overhead. Moreover, LEGE only results in 0.4% latency overhead at its end age (LEGE's lifetime is equal to ECP-6's).

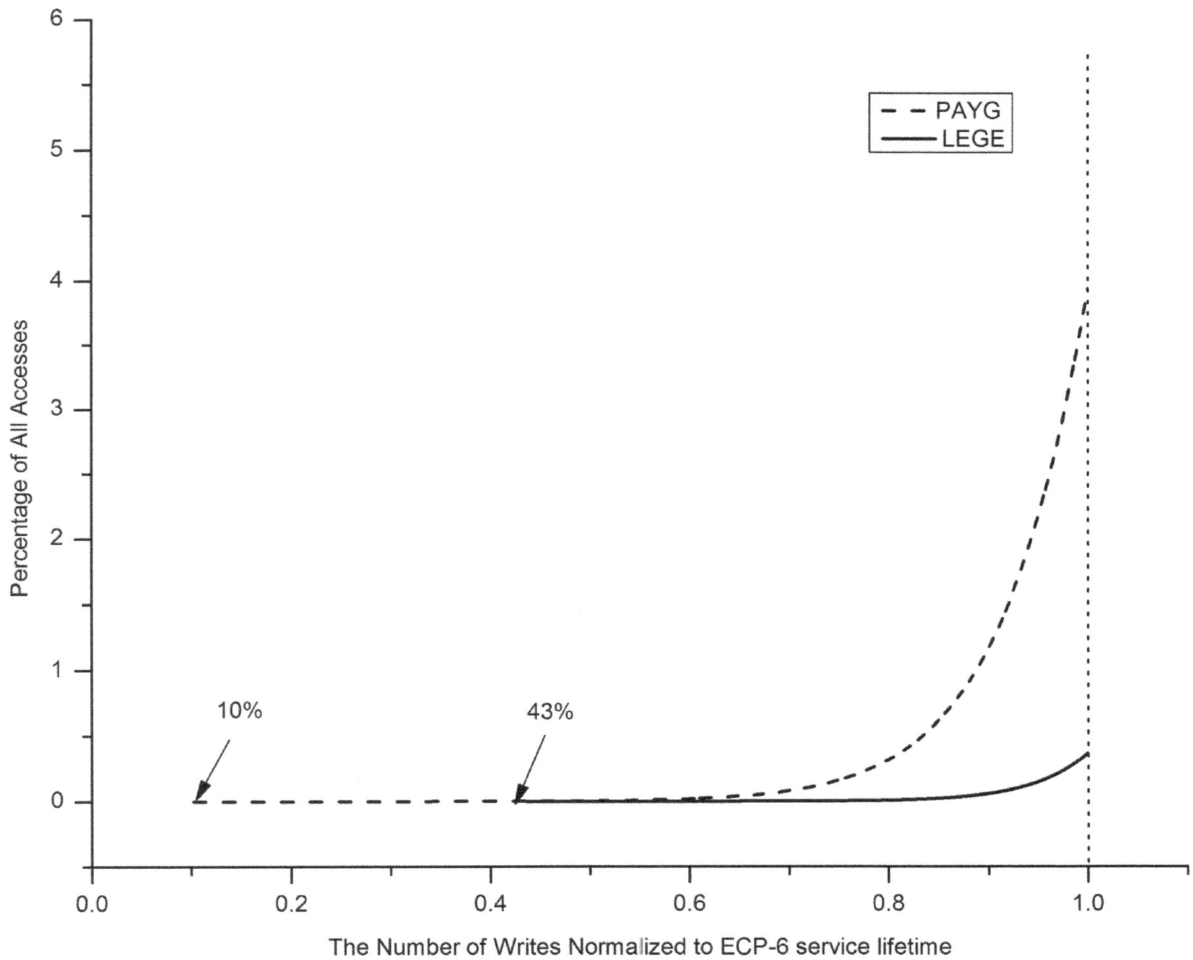

Figure 5: The proportion of one additional memory access among all accesses.

This figure indicated that the proportion of LEGE and PAYG fetch data from global buffer among all accesses.

http://dx.doi.org/10.1371/journal.pone.0131964.g005

Using the Distribution of Hard Errors in Our Experiment to Set LEGE Configuration

In order to verify the view that only a few lines require high levels of hard error correction [3]. We take an experiment similar to PAYG.

Firstly, in order to prove the accuracy of our simulation, we set the ECP-6 lifetime to 35% of theoretical maximum lifetime (each cell can be written 10^8 and PCM can maintain 18 years in such ideal condition) with a COV of 20%, which was tested out by PAYG, to simulate the number of hard errors. After repeating the experiment several times, we found the proportion of different numbers of hard errors is almost the same as PAYG.

Secondly, we try to detect the distribution of hard errors using our own tested ECP-6 lifetime.

Memory lifetime simulation is a big challenge since it is impractical to simulate all wear patterns in a long term (PCM lifetime is more than 10 years). We use a Monte Carlo simulation to test the lifetime of ECP-6, and gather the numbers of lines using different types of ECPs. Both above two experimental configurations and assumptions are as follow:

1. We assume that the experimental memory size is 512MB with a contiguous address space. The size of the memory line is equal to the last level cache line size of 512 bits, so there are 2^{23} lines in this memory.
2. We assume that the lifetime in each cell of one PCM memory line obeyed the random Gaussian distribution which is used in prior works [3][4][5][19][20][21]. The failure bit distribution is from [21]. The mean is the average writes 10^8[4][5] and the coefficient of variation (COV) is 20% [3].
3. We utilizes an ideal wear-leveling, which means all the memory lines will have the same number of writes, to allow our experimental results to trend normal and conservative and only focus on the error correction method impact. According to the calculation in [11] that an ECC-8 (73 bits) can reduce 7% lifetime and ECC-4 (37 bits) can reduce 4% lifetime with a wear-leveling optimization, our ECC-1 (11 bits) may only approximately reduce 1% of PCM lifetime and it is trivial.
4. The end lifetime of ECP-6 is the time that the first line fails out of all memory lines. In ECP-6, the first line fails implies that there is a data line which occurs the first uncorrectable error by deploying an ECP-6. We sort the lifetime of cells for each memory line and collect the seventh shortest cell endurance for each memory line. Then, finding out the shortest one from collection.
5. The simulation is aimed to verify the view that only a few lines require high levels of hard error correction and generate the configuration of each part according to the experimental evaluation. The soft error is rare in PCM and cannot affect the lifetime as it can be corrected by ECC immediately when it occurs. Because soft error in DRAM cell is the dominant, and it in peripheral circuit is trivial in DRAM system [15]. The soft error in PCM cell is always caused by resistance drift. The resistance drift is not a problem in single-level-cell (SLC) PCM [4][12]. So, the soft error rate in SLC PCM is far less than it in DRAM. So that, the soft error problem is not concerned in this simulation.

After repeating this experiment several times, we find the distribution of the number of failure errors is not ideal compared with it in PAYG. The detailed data in our experimental result is shown in the second line of Table 1. Table 1 shows that the distribution of lines that needs different number of ECPs according to PAYG and our experimental results. 88.68% of lines have one or no hard errors and 97.88% of lines have no more than two wear-out errors in our result. The number of no more than one error memory lines is 88.68% in our result, which is much less than 96.06% tested by PAYG. These mean that the proportion of one more access per memory access used LEGE architecture and PAYG architecture are 2.12% and 11.32% respectively, if using our experimental data. Because local area in LEGE can correct two hard errors and in PAYG can only correct one hard error. Due to the first error is stored in local ECP-1 and the excess part should be stored in global buffer, PAYG should have one more accesses to write the corrected data to corresponding global buffer. So, it gets 11.32% memory access latency if using our data.

The architecture of our work can balance storage overhead and lifetime by deploying an ECC and a local ECP-1 to reduce the number of memory accesses, and save storage overhead by setting a global ECPs buffer to store the remaining data, located in the relatively small number of lines, to decrease the associated matrix storage overhead. According to the data from the second line of Table 1, the global ECPs buffer size must be suitable in order to place lines that have more than one hard error under ECP-6 service lifetime. The size of the global mapping section we set was 192K lines, which is large enough to support 11.32% of memory lines to be mapped. We used the Hash function: $H(k) = k\%P$, which has been proven to work well; in it, P is the largest prime number 167917 in 0 ~ 164K-1. The result shows that less than 1.3% needs access to the collision pool portion. The size of the collision pool section we set is 28K, which is sufficient for storing data from the extra accesses. So, the pointer in the global ECPs buffer can be allocated to 15 bits.

The structure of our LEGE is presented in Fig 6. Except for the main memory, there are two parts in our framework. One is the detection and correction matrix and another is global ECPs buffer.

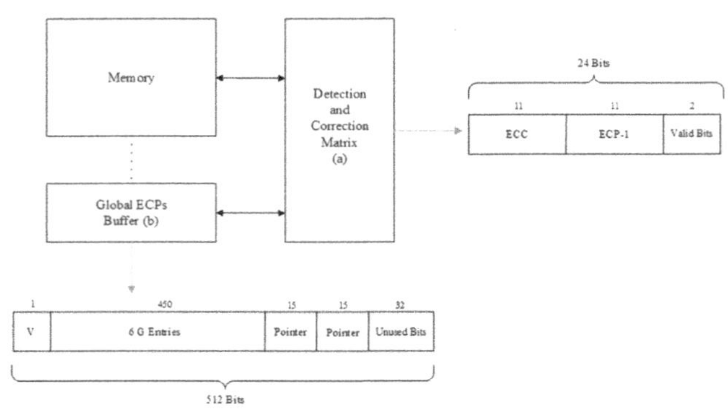

Figure 6: The structure of LEGE.

Bits allocation for different parts in LEGE architecture.

http://dx.doi.org/10.1371/journal.pone.0131964.g006

The calculation of detection and correction matrix's storage overhead is as follows:

The number of detection and correction lines is 2^{23} + 192K. Where 192K is the number of these detection and correction lines that correspond to the global ECPs buffer. The total overhead of the detection and correction matrix is 24bit × (2^{23} + 192K) = 24.563MB.

The number of global ECPs buffer lines is 164K + 28K, where 164K lines is the global mapping section and 28K lines is the collision pool portion. The total overhead of global ECPs buffer is 512bit × (164K + 28K) = 12MB.

We can see the major overhead is derived from the detection and correction matrix. The total overhead of both the matrix and global buffer is 36.563MB, 7.1% of memory capacitance.

Albeit FREE-p also can correct both hard errors and soft errors, it differs from LEGE in architecture and implement method, so we cannot have a valid comparison. In FREE-p, when an error occurs in one 64B block, this block is no longer used to store data; the stored data must be remapped to a new memory line. Although FREE-p uses the other bits in this dead memory block to store the remapping pointer, the time cost in the process of remapping data cannot be ignored. Especially in Table 1, our experimental result shows that more than 40% memory lines have hard errors in the end of ECP-6 service lifetime, it means more than 40% of memory line cannot be used, we must have one more remap when these lines appear only one error by using FREE-p. However, FREE-p takes advantage in providing longer lifetime compared with LEGE. If we focus on the capacity of soft error correction, it is worth noting that comparing with FREE-p's 6EC-7ED BCH, LEGE uses even simple and fast method to detect and correct PCM's soft errors.

CONCLUSION

In this paper, we presented a simple method for PCM-based memory reliability. This mechanism protects PCM memory against both soft errors and hard errors, provides equal lifetime for PCM chips comparing to ECP-6, and incurs only 7.1% storage overhead, which is 5.4% less than ECP-6. Comparing with PAYG, our structure can reduce approximately 3.55% latency overhead by increasing 1.61% storage overhead. Our proposal shows almost equal performance and lifetime with the hard-error-only design. This architecture is promising to handle the reliable challenges posed by PCM.

AUTHOR CONTRIBUTIONS

Conceived and designed the experiments: NA RW YG HY DQ. Performed the experiments: NA YG. Analyzed the data: NA RW HY. Contributed reagents/materials/analysis tools: RW HY. Wrote the paper: NA RW.

REFERENCES

1. Liu J, Jaiyen B, Veras R, Mutlu O (2012) Raidr: Retention-aware intelligent dram refresh. In: Computer Architecture (ISCA), 2012 39th Annual International Symposium on. IEEE, pp. 1–12.
2. Yoon DH, Erez M (2010) Virtualized and flexible ecc for main memory. In: ACM SIGARCH Computer Architecture News. ACM, volume 38, pp. 397–408.
3. Qureshi MK (2011) Pay-as-you-go: low-overhead hard-error correction for phase change memories. In: Proceedings of the 44th Annual IEEE/ACM International Symposium on Microarchitecture. ACM, pp. 318–328.
4. Yoon DH, Muralimanohar N, Chang J, Ranganathan P, Jouppi NP, et al. (2011) Free-p: Protecting non-volatile memory against both hard and soft errors. In: High Performance Computer Architecture (HPCA), 2011 IEEE 17th International Symposium on. IEEE, pp. 466–477.
5. Schechter S, Loh GH, Straus K, Burger D (2010) Use ecp, not ecc, for hard failures in resistive memories. In: ACM SIGARCH Computer Architecture News. ACM, volume 38, pp. 141–152.
6. Chen CL, Hsiao M (1984) Error-correcting codes for semiconductor memory applications: A state-of-the-art review. IBM Journal of Research and Development 28: 124–134. doi: 10. 1147/rd. 282. 0124.
7. BIOS A kernel developer's guide for amd npt family 0fh processors, jul. 2007. URLhttp://support amd com/us/ Processor_TechDocs/32559 pdf.
8. Kang D, Lee JH, Kong J, Ha D, Yu J, et al. (2008) Two-bit cell operation in diode-switch phase change memory cells with 90nm technology. In: VLSI Technology, 2008 Symposium on. IEEE, pp. 98–99.
9. Yeo S, Seong NH, Lee HHS (2012) Can multi-level cell pcm be reliable and usable? analyzing the impact of resistance drift. In: the 10th Ann. Workshop on Duplicating, Deconstructing and Debunking (WDDD).
10. Azevedo R, Davis JD, Strauss K, Gopalan P, Manasse M, et al. (2013) Zombie memory: Extending memory lifetime by reviving dead blocks. In: Proceedings of the 40th Annual International Symposium on Computer Architecture. ACM, pp. 452–463.
11. Awasthi M, Shevgoor M, Sudan K, Rajendran B, Balasubramonian R, et al. (2012) Efficient scrub mechanisms for error-prone emerging memories. In: High Performance Computer Architecture (HPCA), 2012 IEEE 18th International Symposium on. IEEE, pp. 1–12.
12. Seong NH, Yeo S, Lee HHS (2013) Tri-level-cell phase change memory: Toward an efficient and reliable memory system. In: Proceedings of the 40th Annual International Symposium on Computer Architecture. ACM, pp. 440–451.
13. Zhang W, Li T (2011) Helmet: A resistance drift resilient architecture for multi-level cell phase change memory system. In: Dependable Systems & Networks (DSN), 2011 IEEE/IFIP 41st International Conference on. IEEE, pp. 197–208.
14. Yokoyama Y, Itoh N, Hasegawa M, Katayama M, Akasaki H, et al. (2001) A 1. 8-v embedded 18-mb dram macro with a 9-ns ras access time and memory-cell area efficiency of 33%. Solid-State Circuits, IEEE Journal of 36: 503–509. doi: 10. 1109/4. 910489.
15. Baumann R (2005) Soft errors in advanced computer systems. Design & Test of Computers, IEEE 22: 258–266. doi: 10. 1109/MDT. 2005. 69.

16. Burr GW, Breitwisch MJ, Franceschini M, Garetto D, Gopalakrishnan K, et al. (2010) Phase change memory technology. Journal of Vacuum Science & Technology B 28: 223–262. doi: 10. 1116/1. 3301579.

17. Cho S, Lee H (2009) Flip-n-write: a simple deterministic technique to improve pram write performance, energy and endurance. In: Microarchitecture, 2009. MICRO-42. 42nd Annual IEEE/ACM International Symposium on. IEEE, pp. 347–357.

18. Jacobvitz AN, Calderbank R, Sorin DJ (2013) Coset coding to extend the lifetime of memory. In: High Performance Computer Architecture (HPCA2013), 2013 IEEE 19th International Symposium on. IEEE, pp. 222–233.

19. Ipek E, Condit J, Nightingale EB, Burger D, Moscibroda T (2010) Dynamically replicated memory: building reliable systems from nanoscale resistive memories. In: ACM SIGARCH Computer Architecture News. ACM, volume 38, pp. 3–14.

20. Seong NH, Woo DH, Srinivasan V, Rivers JA, Lee HHS (2010) Safer: Stuck-at-fault error recovery for memories. In: Proceedings of the 2010 43rd Annual IEEE/ACM International Symposium on Microarchitecture. IEEE Computer Society, pp. 115–124.

21. Kim K, Ahn SJ (2005) Reliability investigations for manufacturable high density pram. In: Reliability Physics Symposium, 2005. Proceedings. 43rd Annual. 2005 IEEE International. IEEE, pp. 157–162.

CHAPTER 4

FROM BOOLEAN NETWORK MODEL TO CONTINUOUS MODEL HELPS IN DESIGN OF FUNCTIONAL CIRCUITS

Bin Shao[1, 2], Xiang Liu[2], Dongliang Zhang[1], Jiayi Wu[1], and Qi Ouyang[1, 2]

[1]The State Key Laboratory for Artificial Microstructures and Mesoscopic Physics, School of Physics, Peking University, Beijing, China

[2]The Center for Quantitative Biology and Peking-Tsinghua Center for Life Sciences, Peking University, Beijing, China

ABSTRACT

Computational circuit design with desired functions in a living cell is a challenging task in synthetic biology. To achieve this task, numerous methods that either focus on small scale networks or use evolutionary algorithms have been developed. Here, we propose a two-step approach to facilitate the design of functional circuits. In the first step, the search space of possible topologies for target functions is reduced by reverse engineering using a Boolean network model. In the second step, continuous simulation is applied to evaluate the performance of these topologies. We demonstrate the usefulness of this method by designing an example biological function: the SOS response of E. coli. Our numerical results show that the desired function can be faithfully reproduced by candidate networks with different parameters and initial conditions. Possible circuits are ranked according to their robustness against perturbations in parameter and gene expressions. The biological network is among the candidate networks, yet novel designs can be generated. Our method provides a scalable way to design robust circuits that can achieve complex functions, and makes it possible to uncover design principles of biological networks.

INTRODUCTION

Synthetic biology is an emerging field focusing on understanding the behaviors of biological systems through designing and constructing of synthetic gene circuits. Inspired by electric circuits and cellular systems, various synthetic genetic circuits have been created, including toggle switch [1], oscillator [2, 3], counter [4], and logic gates [5]. In addition, multiple well-characterized parts are combined together to achieve more complex functions, such as biosensing [6], edge detection [7] and Pavlovian-like conditioning [8]. However, designing circuits with complex functions remains a challenge. In previous work, methods based on continuous simulation have been developed to select out networks that are capable of executing different functions. A very useful method is enumeration of network structures [9–11]. Through computer simulation, the full range of possible network architecture was explored and the interacting network of biological components was converted into a set of differential equations. Solutions to these ordinary differential equations (ODEs) provided the dynamics of each component of the network. Functional topologies with fewer parameter constraints can thus be selected. A list of these functions includes adaptation [9], dose-

Citation: Shao B, Liu X, Zhang D, Wu J, Ouyang Q (2015) From Boolean Network Model to Continuous Model Helps in Design of Functional Circuits. PLoS ONE 10(6): e0128630. doi:10.1371/journal.pone.0128630.

response alignment [10], and segment polarity in development [11]. However, enumeration of network structures usually leads to a dramatic rise in computational cost with increasing number of genes, thus making it difficult to scale-up. In other approaches, network structures are evolutionarily optimized from a finite set of independent circuits. However, this method may not result in an optimal design, as only a limited space of topology is sampled [12–14].

On the other end of the spectrum lies the class of discrete models ranging from simple logic circuits to finite state machines and Boolean networks. In these models, each component in the system has only two states: ON (1) and OFF (0). The regulation rule for each component is defined by Boolean functions, such as two-input AND gate and OR gate. In electronic circuit design, constructing digital circuits with a given truth table and logic gates is a standard procedure called combinational logic design. One of the basic methods is "Karnaugh map", which is also applied in the design of biological digital circuits [15]. However, it is difficult to construct biological circuits with sequential logic behaviors using standard methods of electronics, because these methods usually involve flips-flops that are quite rare in standard biological parts [16]. Boolean network model has been used to illustrate the dynamic behaviors of biological systems and to reconstruct biological networks underlying specific functions, thus contributing to a natural method of constructing sequential logic circuits in biology [17, 18]. Although these discrete models are quite efficient in computation, they suffer from several inherent limitations: It is unclear whether the circuits designed using discrete framework can execute desired functions robustly in a wet-lab implementation. In addition, it is difficult to rank the possible circuits based on their ability to tolerate perturbations in parameters and gene expression levels.

In this paper, we present a two-step method that combines the discrete model and the continuous model to generate a novel design of functional circuits. In our approach, first, a Boolean network model is applied to generate candidate networks that are better capable of executing the target functions. Then, continuous simulation is used to quantitatively assess the robustness of these candidate networks. Here, we focus on one critical biological behavior, the SOS response in E. coli., wherein DNA repair is induced in response to the existence of a single stranded DNA (ssDNA). The desired function of this network is as follows: Upon accumulation of ssDNA, RecA is recruited to the single stranded regions of DNA and becomes activated. Activation of RecA releases the inhibition of SOS genes by facilitating self-cleavage of their repressor LexA. The main activator of SOS gene is σ^{70}, which belongs to a family of transcription initiation factors responsible for stress response. Its downstream genes can be simplified into two genes, SSB and UmuDC, which are responsible for the repair of DNA damage and inhibition of RecA. When the DNA repair is completed, LexA is activated and the expression of SOS genes is down-regulated [19, 20]. The natural network performing this function is presented in Fig 1A. This DNA damage response may represent a large class of response pathways and we use our approach to design functional circuits underpinning this function. The analysis of functional circuits obtained by our approach allows us to discover core motifs responsible for robust response.

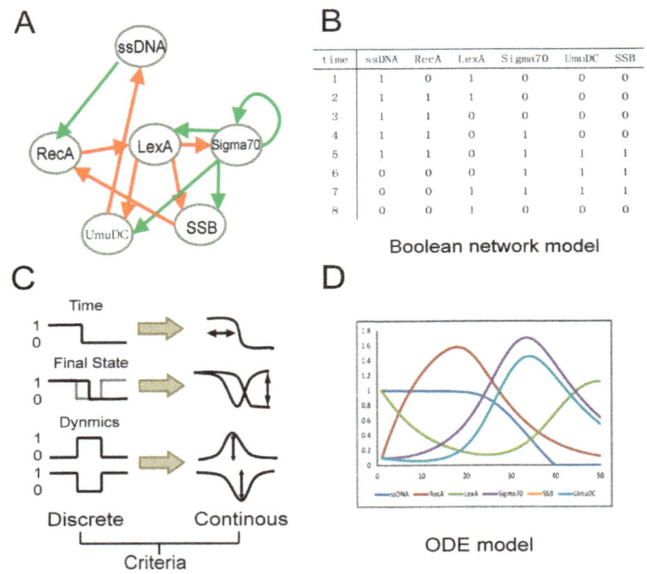

Figure 1: The SOS network and its dynamics.

(A) Regulatory network of the SOS response of E. coli. The nodes represent the signal and the essential proteins. The green lines represent activation and the red lines represent inhibition. (B) Dynamics of the SOS response in the Boolean network model. (C) Three criteria and their representations in the discrete and continuous model. The first criteria addresses degradation time of ssDNA, i.e., ssDNA should be down-regulated to zero at the end of the simulation. In the second criteria, the final state of the system should return to the initial state, except for the degradation of ssDNA. The third criteria requires that the dynamics of each node in the ODE model should be in accordance with those in Boolean trajectory. (D) Example of a successful response in the ODE model.

http://dx.doi.org/10.1371/journal.pone.0128630.g001

METHODS

Boolean Network Model and Reverse Engineering

In the Boolean network model, each node represents a biological species. $S_i(t) \in \{0,1\}$ represents the state of node i at time t. The network topology can be described by its connection matrix A, in which a_{ij} indicates regulation from node i to node j. a_{ij} is positive for activation and negative for inhibition [17]. As changes in weight of inhibition regulation render the Boolean trajectory nearly unchanged [21], we take a dominant inhibition form of regulation in our Boolean network model, which means that inhibition has a much larger weight than activation, i.e., $|a_{inh}| \gg |a_{act}|$. If the state of node j is 1 (On) at time t and one of its inhibitor is activated, then $S_i(t + 1) = 0$ regardless of all the activation terms. This captures the combinatorial mechanism of transcription regulation, in which existence of an inhibitor can block transcription of the target gene. Although some limitations persist, our method is able to recover the regulatory network from different types of data in an efficient way [22]. The level of genes in the next time step is determined by the level of genes in the current time step by the following rule:

$$\begin{cases} S_i(t+1) = \theta(\sum_j S_j(t)a_{ji}), \sum_j S_j(t)a_{ji} \neq 0 \\ S_i(t+1) = S_i(t), \sum_j S_j(t)a_{ji} = 0 \end{cases} \quad (1)$$

Where $\theta(x)$ is a Heaviside step function with $\theta(x) = 1$ for $x > 0$ and $\theta(x) = 0$ for $x < 0$. In our example of SOS response pathway, LexA is constitutively expressed to bind to the SOS box of the target genes, whereas σ^{70} is not stimulated in the normal state. RecA is expressed in normal conditions, yet it only becomes functional when forming filaments around single stranded DNA. Thus only LexA and ssDNA are 1 in the initial state and the dynamics (trajectory) of the network can be generated according to Fig 1A and using Eq 1 until the system returns to its normal state, as illustrated in Fig 1B. In the first step of our circuit design method, the purpose is to generate a set of networks that can perform this trajectory of the SOS network.

We apply reverse engineering methods [18] in the first step to limit the number of possible topologies. Reverse engineering presents a class of methods aiming to uncover biological regulatory networks based on experimental data. In our inhibition dominant Boolean network model, the constraint of network topology by the Boolean trajectory can be represented in an analytical manner [21].

$$s_i(t+1) = \left(\sum_{j \neq i} (s_j(t) \cdot g_{ji}) + s_i(t) \cdot \overline{r_{ii}} + \overline{s_i(t)} \cdot g_{ii} \right) \cdot \prod_{j \neq i} (\overline{s_j(t) \cdot r_{ji}}) \quad (2)$$

In the equation, g_{ij} and r_{ij} are the Boolean variables corresponding to activation and inhibition from node i to node j, respectively. The OR gate is indicated by addition ('+' and '$\sum$'), and used to combine all activation terms in the first bracket, whereas inhibition terms are linked by AND gate, which is represented by multiplication ('•' and '$\prod$'). The bar in Eq 2 denotes the NOT logic gate. For each node in the pathway, the logical constraints (Eq 2) of different time steps are combined together to get all possible regulations for that node. This reverse engineering method generates 7.1×10^6 possible networks using the biological trajectory of SOS response.

Minimal Network Constraint

When the number of nodes is fixed, the complexity of practical implementation of our design depends largely on the number of edges. Circuits with fewer edges are more favored in a wet-lab implementation. Minimal networks have been proposed to contribute to the core motif responsible for the main functional response [21]. Moreover, our previous work indicates that biological networks may prefer to use minimal networks to fulfill their functions, providing evidence that the edge number is limited in the evolutionary process [22]. To apply minimal network constraint in our approach, we enumerate all possible regulations of each node and obtain the ones with the fewest edges. These regulations are then combined together to obtain 48 minimal networks. By definition of our target function, we exclude all networks with self-loop on the input node, i.e., ssDNA.

Continuous Model Simulation

To investigate the continuous dynamics of our designed circuits, we modeled the selected networks via ordinary differential equations (ODEs) in the second step. In the simulation, we limited ourselves to transcriptional regulatory networks, which are quiet common in synthetic biological networks. Input node is ssDNA, and all other nodes are assumed to be transcriptional factors (TFs) that can bind to upstream sequences of the target genes. In our model, Hill functions with Hill coefficient 2 are used to model the activation terms. The inhibition terms of nodes take a multiplication form, except for the input node (ssDNA), implying that the inhibitors are independent of one another and can independently block the transcription of the target gene. Repair of ssDNA is modeled by sum of the Hill function terms in order to take the cooperative nature of the functional proteins into account [23]. Basal production and degradation is also introduced in the equations. The ordinary equations are as follows:

$$\begin{cases} \frac{dx_i}{dt} = \left(\sum_{positive} W_p a_j \frac{x_j^N + \delta_{ij}}{b_j + x_j^N} \right) \prod_{negative} \frac{1}{c_j + W_N d_j x_j^N} + 0.1 - x_i, i \neq 1 \\ \frac{dx_1}{dt} = - \sum_{negative} \frac{x_j^N}{c_j + x_j^N} \end{cases} \quad (3)$$

In this equation, b denotes the dissociation constant and a denotes the kinetic constant. δrepresents leakage of the promoter. Wp and Wn are the two parameters determining the relative weight of the inhibition and activation terms. All nodes are numbered as in Fig 1 to avoid confusion in the following analysis, i.e., ssDNA as node 1, RecA as node 2, LexA as node 3, σ^{70} as node 4, UmuDC and SSB as nodes 5 and 6. The initial value of node 1 (ssDNA) and node 3 (LexA) is set to be 1.0, and the rest of the nodes are set to be 0.1.

To evaluate the performance of candidate networks, quantitative criteria should be employed. Although we start from a Boolean trajectory, it is meaningless to reproduce the same trajectory in ODE simulations, partly because of the lack of time scale in the Boolean network model and arbitrary ways of discretization. We assume that the core dynamics of the DNA damage response lies behind the exact Boolean trajectory and can be abstracted to several criteria. Three conditions are assigned to capture the main characteristic of a successful response:

i. Level of ssDNA (node 1) should be down-regulated to zero at the end of simulation, which is exactly the function we need.

ii. The final state of the system will be same as the initial state, except for the decrease of node 1, i.e., the system relaxes to its normal state after a transient response to input signal (node 1). At the initial state, LexA (node 3) is ON in order to repress the downstream SOS genes, so that the level of node 3 should be larger than any other nodes in the end.

iii. The dynamics patterns of nodes in continuous simulation must be the same as those in the Boolean trajectory; the correspondence is presented in Fig 1C. We identify the dynamic patterns of each node, which are classified into three categories, namely, 'peak', 'valley' and 'decline' according to the 'shape' of their time trajectories. The dynamics of 'peak' should have a maximum level of expression and 'valley' should have a minimum level.

In criteria ii), we restricted the final state of the system to be the normal state of the cell, in which only LexA is activated. Many response pathways share similar characteristics, yet it is not always true. For example, activation of the mating pathway of budding yeast drives the cell to a different phenotype called 'shmoo'. Then, different criteria for the final state should be chosen to take into consideration the Boolean trajectory and the biological interpretations of the biological pathway. In addition, we add a constraint that maximum expression of each gene should be larger than 0.1. These criteria correspond to the main characteristics of the Boolean trajectory, leaving details, such as specific time of activation, unconstrained.

Performance of Circuit

The evaluation process of circuit performance is presented in Fig 2, in which two scores that can reflect robustness of networks are employed. The first score is Q value that estimates the volume of the functional parameter space [9]. For each network, we randomly choose 1000 sets of parameters using Latin hypercube sampling. Starting from the same initial state, the network dynamics with each parameter set is obtained by solving the ODEs. Time series of all nodes are then checked to see if they present a successful response, i.e., if they meet three criteria above. For a particular network structure, the proportion of parameter sets that can achieve our desired function is defined as the Q value of that structure. A larger Q value implies that the network can generate successful response to DNA damage over a wide range of parameters and, hence, is less sensitive to parameter variation. The second score is relevant to robustness in state space. For the Boolean network model, the state of the system is updated using Eq 1 until it reaches a fixed point, which can also be called an attractor. The number of initial states that will flow into an attractor is defined as the basin size of that attractor. It is proposed that a biological state should have a large basin size in order to generate stability [17]. We held similar assumption in our design procedure that normal state with node3 activated should be a big attractor and stable against fluctuations in gene expression. In the Boolean network model, all possible initial conditions (2^N for network with N nodes) are enumerated to calculate the basin size, which is not practical in continuous simulation. To sample the space of initial states in the ODE model, we employ a similar approach in which the state of the nodes is treated as a continuous variable instead of a Boolean variable, e.g., the initial state (1.0, 0.1, 1.0, 0.1, 0.1, 1.0) is used instead of (1, 0, 1, 0, 0, 1). Criteria (i) and (ii) are used to see if the state of system flows into the normal state. Criteria (iii) is abandoned, as changing of the initial state may affect the dynamic patterns of the nodes. ssDNA is set to be 1.0, leaving 32 initial states in total.

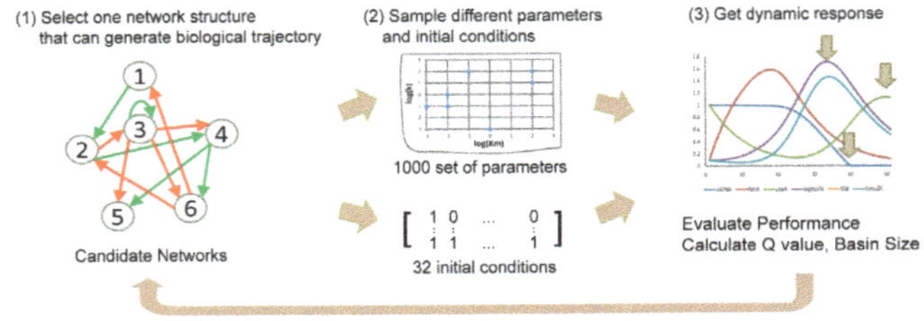

Figure 2: Workflow of continuous simulation.

For networks selected by Boolean network model, 1000 random parameter sets and 32 initial conditions are used to quantitatively assess their robustness. The ordinary differential equations describing these transcriptional networks are solved numerically to generate the dynamics of each network component. The proportion of parameter sets and number of initial conditions that can achieve a successful response is termed as the Q value and basin size, respectively, for the corresponding network.

http://dx.doi.org/10.1371/journal.pone.0128630.g002

RESULTS

Performance of Minimal Networks and Candidate Networks

In addition to minimal networks, we randomly selected 100 candidate networks that can generate the Boolean trajectory to compare the performance of both types of networks. The Q value distributions of minimal and candidate networks are illustrated in Fig 3A. Most of the first-step-selected networks have Q values larger than zero. On the contrary, only ~15% random networks have a positive Q value (data not shown), suggesting that filtering networks by the Boolean network model is efficient in obtaining the functional topologies. Former work has shown that the transform from differential equations to Boolean networks is possible under several assumptions [24]; our results reveal that the Boolean dynamics can also be transformed to continuous ones. Application of Boolean network model can largely reduce the search space of topologies in which most networks have a non-zero Q value, thus helping us focus on networks that are better capable of achieving target functions.

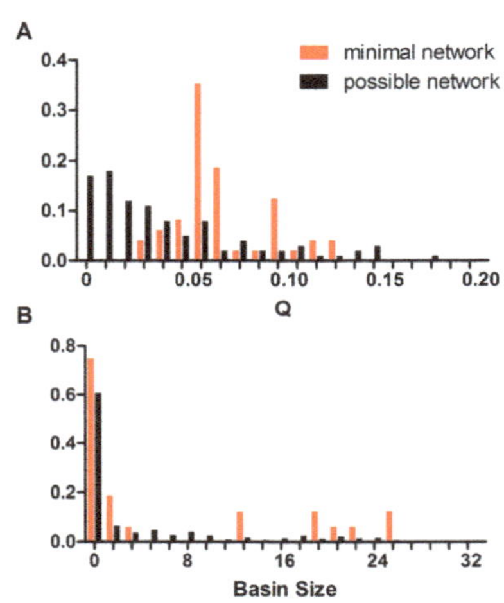

Figure 3: Distribution of Q value and basin size for minimal networks and candidate networks.

(A) Distribution of Q value for minimal networks and candidate networks. Minimal networks are represented in red and the candidate networks in black. (B) Distribution of basin size for minimal networks and candidate networks.

http://dx.doi.org/10.1371/journal.pone.0128630.g003

The distribution of Q value and basin size of minimal and candidate networks differs dramatically. For candidate networks, the proportion of networks decreases quickly as the Q value increases. Only a small fraction of networks can robustly execute the response function. Approximately 15% of the networks have a zero Q value, indicating that among all candidate networks filtered out in the first step, a small fraction is not functional in a more realistic circumstance. Although our criteria have relaxed the requirement for time sequence of events, they may be violated by some networks due to incoherence of regulations or mismatch of time scales. All minimal networks have a Q value larger than zero and the distribution peaks in 0.05, providing evidence that minimal networks are more robust against parameter perturbations. To discriminate between the two distributions in a quantitative way, we employed a measure of Jensen-Shannon Divergence (JSD), which describes the similarity of the two distributions and used statistical tests to examine the significance of difference. Our null hypothesis is that the distribution of minimal network and candidate networks have no significant difference, i.e., H_0:JSD(p_m,p_c) = 0. The JSD of the Q value distribution for minimal networks and candidate networks is 0.31 with a p-value < 0.0001, indicating that the two distributions are significantly different. Similar results hold for the basin size distribution (Fig 3B), where the JSD of the two distributions is 0.13 with a p-value < 0.0001. Minimal networks are more likely to have a large basin size compared to candidate networks. Interestingly, although most networks we sampled have a full basin in the Boolean network model, only one of them has a full basin attractor in the continuous model simulation.

Our results suggest that minimal networks may be more robust to parameter and expression perturbations than candidate networks, which is apparent in Fig 4. Minimal networks tend to outperform the candidate networks, and mainly distribute in the upper right corner of the figure. In our previous work [22], we have proposed that least number of regulatory edges is preferred to implement biological functions. Our results here provide further evidence and suggest that minimal network constraints may also be useful in the design of functional circuits. As shown in Fig 4, the natural SOS network has a larger Q value and basin size than most networks, indicating that the biological design may be optimized in the course of evolution. Nonetheless, there are a small number of minimal networks that outperform the biological network; these networks may provide the basis for rational design of functional networks that can outperform the natural network.

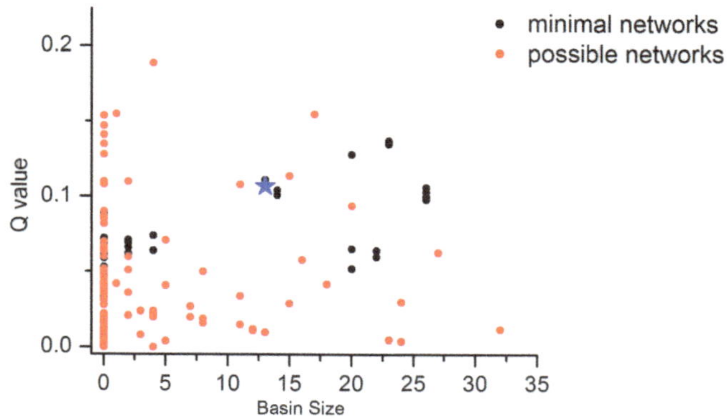

Figure 4: Performances of minimal networks and candidate networks.

The distribution of Q value and basin size for minimal networks and candidate networks. Minimal networks are represented in red and candidate networks in black. The large blue star represents the biological network.

http://dx.doi.org/10.1371/journal.pone.0128630.g004

Topology Analysis

We next analyze the backbone motifs (minimal networks) responsible for robust response. Among all 48 minimal networks, the main difference is the activation mode of node 3 (LexA) and node 4 (σ^{70}). In the biological network (Fig 1A), node 3 is activated by node 4, and node 4 has a self-activation loop, whereas 4 minimal networks that have a better performance than the biological network employ another regulation pattern (Fig 5). These networks all include self-activation of node 3, and activation of node 4 by node 1. The only difference is the different combination of inhibition of node 1 and node 2 by node 5 and node 6, which is not significantly different, as node 5 and node 6 both belong to the genes regulated by node 4. These results suggest that direct activation of node 4 by upstream signals may be more reliable than a self-activation loop in signal transduction. In addition, self-activation of node 3 may help build up its expression level when DNA is repaired.

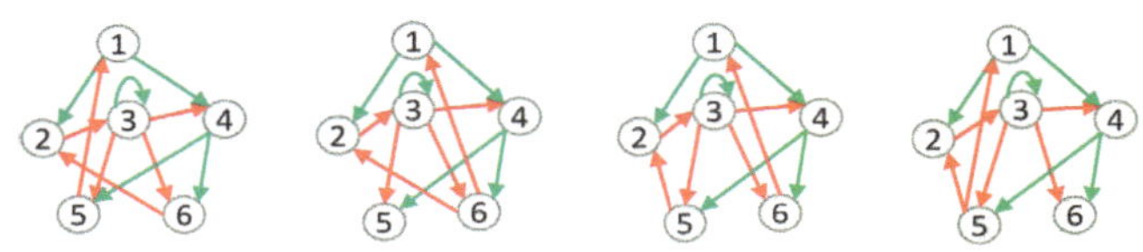

Figure 5: Core motifs responsible for successful response.

Four minimal networks have larger Q values and basin sizes than the biological network. Aside from the edges shared by all of the candidate networks, all these topologies employed self-activation of node 3 and direct activation of node 4 by node 1, which contributes to the robustness of the network.

http://dx.doi.org/10.1371/journal.pone.0128630.g005

DISCUSSION

By limiting the search space of networks through reverse engineering of the Boolean network model, we have developed an efficient method to generate networks that can achieve robust response to the DNA damage signal. Starting from a discretized time trajectory, reduction of candidate networks using Boolean network model can make the following simulation more practical. In a network of 6 interacting nodes, as in our case, there are 1.5×10^{17} different topologies, making it impossible to select out networks that are capable to perform the desired function by enumerating all topologies. Even when enumeration is possible, computational resources are largely wasted because only a small fraction of these networks are able to execute a successful response. Application of Boolean network model can reduce the number of possible networks by 10^{10} fold, and there only exist 48 minimal networks.

Networks found by the discrete model are better capable to reproduce the desired dynamics in continuous model simulation than random networks, which may suggest a good correspondence between the ODE and Boolean models in our problem. However, these networks are not equivalent in their ability to achieve robust response in more realistic implementation. Network performance is characterized by continuous model simulation under different parameters and initial conditions. The Q value and Basin size are used to quantitatively assess the ability of networks to robustly respond to the damage signal. Among all the candidate networks derived by the Boolean network model, those with the minimal number of edges are more robust than the rest. This implies that adding edges to minimal networks can barely improve the network performance. Biological networks usually not only have more edges, but also show better performance than minimal networks, providing evidence of the efficiency of network evolution.

The main feature of SOS response is the transit down-regulation of the inhibitor and the activation of functional proteins. Another similar response pathway in E coli. is the heat shock response pathway, in which the cell expresses chaperones to refold unfolded proteins induced by high temperature or chemicals such as ethanol. In the response system, the heat shock genes are activated by another sigma factor named σ^{32}. σ^{32} is inhibited by chaperone DnaK, which can also bind to unfolded proteins [25]. Upon temperature up-shift, DnaK is driven to form a complex with unfolded proteins, thus releasing σ^{32} from inhibition. Translation of σ^{32} is increased at the same time [26] and heat shock genes are expressed [27], leading to a process as damage is repaired [28]. We found that the heat shock pathway shares a similar structure with networks that are designed to response to DNA damage (Fig 6). Although the two biological networks function in different biological contexts, our finding suggests that response pathways may employ a limited number of strategies to achieve reliable response, and investigating the design principle of one function may hint to the control logic of another.

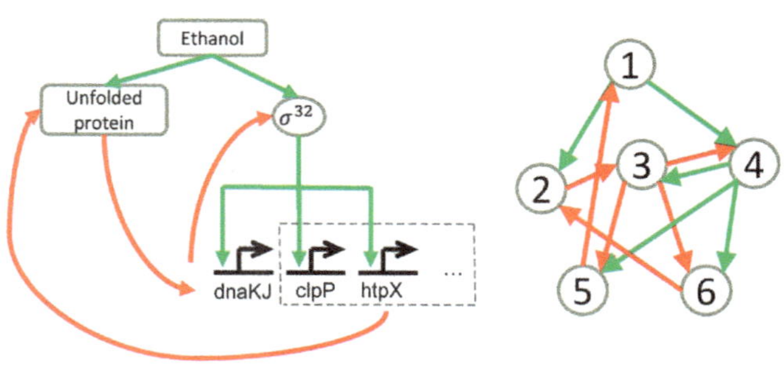

Figure 6: The network for heat shock response in E. coli. is similar to one of the minimal networks.

Left: heat shock response network in E. coli. Right: one of the 48 minimal networks identified by our algorithm.

http://dx.doi.org/10.1371/journal.pone.0128630.g006

Our proposed method can largely facilitate design of synthetic circuits that can achieve sequential logic behaviors and can be easily scaled up. Design of functional circuits opens up the possibility of broad application, ranging from biological sensors to medical care [29, 30]. Moreover, illustration of function circuits underlying known biological functions helps us understand the design principles of biological networks [9]. Nevertheless, our method is not without limitations: the performance of our approach may be deteriorated when the dynamics of the network components do not have a significant ON and OFF state. Another issue should be called is that our method does not address the optimization of parameters. In our approach, the networks are ranked according to the volume of the functional parameter space, which is estimated by random sampling of the parameters. However, if possible, the functional parameters should be determined in a wet-lab design. Some of the parameters are not easily adjusted in the experiment, which imposes constraints on the parameter space. In addition, perturbation of several sensitive parameters can dramatically affect the network performance. Thus, further work should be performed to introduce limitations of parameters to contribute to a better design procedure.

AUTHOR CONTRIBUTIONS

Conceived and designed the experiments: BS QO. Performed the experiments: XL BS DLZ JYW. Analyzed the data: XL BS DLZ QO. Wrote the paper: BS QO.

REFERENCES

1. Gardner TS, Cantor CR, Collins JJ. Construction of a genetic toggle switch in Escherichia coli. Nature. 2000; 403(6767):339–42. pmid: ISI:000084899700060. doi: 10. 1038/35002131
2. Elowitz MB, Leibler S. A synthetic oscillatory network of transcriptional regulators. Nature. 2000;403(6767):335–8. doi: 10. 1038/35002125 pmid:ISI: 000084899700059.
3. Danino T, Mondragon-Palomino O, Tsimring L, Hasty J. A synchronized quorum of genetic clocks. Nature. 2010;463(7279):326–30. doi: 10. 1038/Nature08753 pmid:ISI: 000273748100037.
4. Friedland AE, Lu TK, Wang X, Shi D, Church G, Collins JJ. Synthetic Gene Networks That Count. Science. 2009;324(5931):1199–202. doi: 10. 1126/science. 1172005. pmid:19478183
5. Wang BJ, Kitney RI, Joly N, Buck M. Engineering modular and orthogonal genetic logic gates for robust digital-like synthetic biology. Nature Communications. 2011;2. Artn 508 doi: 10. 1038/Ncomms1516 pmid: ISI: 000296787300021.
6. Prindle A, Samayoa P, Razinkov I, Danino T, Tsimring LS, Hasty J. A sensing array of radically coupled genetic

'biopixels'. Nature. 2012;481(7379):39–44. doi: 10. 1038/Nature10722 pmid:ISI:000298836900024.

7. Tabor JJ, Salis HM, Simpson ZB, Chevalier AA, Levskaya A, Marcotte EM, et al. A Synthetic Genetic Edge Detection Program. Cell. 2009;137(7):1272–81. doi: 10. 1016/j. cell. 2009. 04. 048 pmid:ISI:000267373400020.

8. Zhang HQ, Lin M, Shi HD, Ji WY, Huang LW, Zhang XM, et al. Programming a Pavlovian-like conditioning circuit in Escherichia coli. Nature Communications. 2014;5. Artn 3102 doi: 10. 1038/Ncomms4102 pmid:ISI:000331084400011.

9. Ma WZ, Trusina A, El-Samad H, Lim WA, Tang C. Defining Network Topologies that Can Achieve Biochemical Adaptation. Cell. 2009;138(4):760–73. doi: 10. 1016/j. cell. 2009. 06. 013 pmid:ISI:000269156100017.

10. Long Y, Qi O, Wang HL. Dose-Response Aligned Circuits in Signaling Systems. PLoS ONE. 2012;7(4). ARTN e34727 doi: 10. 1371/journal. pone. 0034727 pmid:ISI:000305010500026.

11. Ma WZ, Lai LH, Qi OY, Tang C. Robustness and modular design of the Drosophila segment polarity network. Molecular Systems Biology. 2006;2. Artn 70 doi: 10. 1038/Msb4100111 pmid:ISI:000243245400068.

12. Francois P, Hakim V. Design of genetic networks with specified functions by evolution in silico. Proceedings of the National Academy of Sciences of the United States of America. 2004;101(2):580–5. doi: 10. 1073/pnas. 0304532101 pmid:ISI:000188210400031.

13. Rodrigo G, Carrera J, Jaramillo A. Genetdes: automatic design of transcriptional networks. Bioinformatics. 2007;23(14):1857–8. doi: 10. 1093/bioinformatics/btm237 pmid:ISI:000249248300026.

14. Dasika MS, Maranas CD. OptCircuit: An optimization based method for computational design of genetic circuits. Bmc Systems Biology. 2008;2. Artn 24 doi: 10. 1186/1752-0509-2-24 pmid:ISI:000255178200001.

15. Marchisio MA, Stelling J. Automatic Design of Digital Synthetic Gene Circuits. Plos Computational Biology. 2011;7(2). ARTN e1001083 doi: 10. 1371/journal. pcbi. 1001083 pmid:ISI:000287698700018.

16. Canton B, Labno A, Endy D. Refinement and standardization of synthetic biological parts and devices. Nature Biotechnology. 2008;26(7):787–93. doi: 10. 1038/Nbt1413 pmid:ISI:000257491600023.

17. Li FT, Long T, Lu Y, Ouyang Q, Tang C. The yeast cell-cycle network is robustly designed. Proceedings of the National Academy of Sciences of the United States of America. 2004;101(14):4781–6. doi: 10. 1073/pnas. 0305937101 pmid:ISI:000220761200013.

18. Zhang XM, Shao B, Wu YL, Qi O. A Reverse Engineering Approach to Optimize Experiments for the Construction of Biological Regulatory Networks. PLoS ONE. 2013;8(9). ARTN e75931 doi: 10. 1371/journal. pone. 0075931 pmid:ISI:000324777300100.

19. Sassanfar M, Roberts JW. Nature of the Sos-Inducing Signal in Escherichia-Coli—the Involvement of DNA-Replication. Journal of Molecular Biology. 1990;212(1):79–96. doi: 10. 1016/0022-2836(90)90306-7 pmid:ISI:A1990CV28600011.

20. Gardner TS, di Bernardo D, Lorenz D, Collins JJ. Inferring genetic networks and identifying compound mode of action via expression profiling. Science. 2003;301(5629):102–5. doi: 10. 1126/science. 1081900 pmid:ISI:000183914700043.

21. Wang GY, Du CH, Chen H, Simha R, Rong YW, Xiao Y, et al. Process-based network decomposition reveals backbone motif structure. Proceedings of the National Academy of Sciences of the United States of America. 2010;107(23):10478–83. doi: 10. 1073/pnas. 0914180107 pmid:ISI:000278549300029.

22. Shao B, Wu JY, Tian BH, Ouyang Q. Minimum network constraint on reverse engineering to develop biological regulatory networks. Journal of Theoretical Biology. 2014, unpublished. doi: 10. 1016/j. jtbi. 2015. 05. 005

23. Fisher RJ, Fivash M, Casafinet J, Bladen S, Mcnitt KL. Real-Time Biacore Measurements of Escherichia-Coli Single-Stranded-DNA Binding-Protein to Polydeoxythymidylic Acid Reveal Single State Kinetics with Steric Cooperativity. Faseb J. 1994;8(7):A1271-A. pmid:ISI:A1994NH51600288. doi: 10. 1006/meth. 1994. 1015

24. Davidich M, Bornholdt S. The transition from differential equations to Boolean networks: A case study in simplifying a regulatory network model. Journal of Theoretical Biology. 2008;255(3):269–77. doi: 10. 1016/j. jtbi. 2008. 07. 020 pmid:ISI:000261430200001.

25. Straus D, Walter W, Gross CA. Dnak, Dnaj, and Grpe Heat-Shock Proteins Negatively Regulate Heat-Shock Gene-Expression by Controlling the Synthesis and Stability of Sigma-32. Gene Dev. 1990;4(12A):2202–9. doi: 10. 1101/gad. 4. 12a. 2202 pmid:ISI:A1990EN02600015.

26. Yuzawa H, Nagai H, Mori H, Yura T. Heat Induction of Sigma(32) Synthesis Mediated by Messenger-Rna Secondary Structure—a Primary Step of the Heat-Shock Response in Escherichia-Coli. Nucleic Acids Research. 1993;21(23):5449–55. doi: 10. 1093/nar/21. 23. 5449 pmid:ISI:A1993MK39700025.

27. Kanemori M, Mori H, Yura T. Induction of Heat-Shock Proteins by Abnormal Proteins Results from Stabilization and Not Increased Synthesis of Sigma(32) in Escherichia-Coli. Journal of Bacteriology. 1994;176(18):5648–53. pmid:ISI:A1994PF22200008.

28. Morimoto RI, Kline MP, Bimston DN, Cotto JJ. The heat-shock response: regulation and function of heat-shock proteins and molecular chaperones. Essays in Biochemistry, Vol 32, 1997. 1997;32:17–29. pmid:ISI:000072180100002.

29. Xiang SL, Fruehauf J, Li CJ. Short hairpin RNA-expressing bacteria elicit RNA interference in mammals. Nature Biotechnology. 2006;24(6):697–702. doi: 10. 1038/Nbt. 1211 pmid:ISI: 000238187300032.

30. Liu YC, Zeng YY, Liu L, Zhuang CL, Fu X, Huang WR, et al. Synthesizing AND gate genetic circuits based on CRISPR-Cas9 for identification of bladder cancer cells. Nature Communications. 2014;5. Artn 5393 doi: 10. 1038/Ncomms6393 pmid:ISI: 000345624800014.

CHAPTER 5

ULTRA-FAST DATA-MINING HARDWARE ARCHITECTURE BASED ON STOCHASTIC COMPUTING

Antoni Morro, Vincent Canals, Antoni Oliver, Miquel L. Alomar, and Josep L. Rossello

Electronic Engineering Group, Physics Department, Universitat de les Illes Balears, Palma de Mallorca, Balears, Spain

ABSTRACT

Minimal hardware implementations able to cope with the processing of large amounts of data in reasonable times are highly desired in our information-driven society. In this work we review the application of stochastic computing to probabilistic-based pattern-recognition analysis of huge database sets. The proposed technique consists in the hardware implementation of a parallel architecture implementing a similarity search of data with respect to different pre-stored categories. We design pulse-based stochastic-logic blocks to obtain an efficient pattern recognition system. The proposed architecture speeds up the screening process of huge databases by a factor of 7 when compared to a conventional digital implementation using the same hardware area.

INTRODUCTION

Data explosion is the capability of current technologies to generate large amounts of data at different scientific disciplines. Data volumes are doubling every year in most areas of modern science [1], and its proper analysis is becoming more and more complex. As a matter of fact, data explosion has not led to an information explosion since current data analysis techniques are unable to handle billions of data records in a reasonable period of time. Large scientific databases containing several terabytes of information need to be continuously screened by scientists, and current processor-based techniques are unable to provide an efficient response to this problem. To solve this, different solutions have been developed based on artificial neural networks [2–5], the use of simple metrics [6] or the extraction of simplified datasets from the original data [7].

An alternative to traditional deterministic computational methodologies is the use of stochastic logic, introduced more than 40 years ago [8,9]. Stochastic computing is the result of applying probabilistic laws to logic cells where variables are represented by random pulse streams, thus providing a natural way of representing analog quantities with digital systems [10]. Pulses can be converted to binary numbers by using digital counters (P2B converters) while binary numbers can be translated to stochastic signals by combining a random (or a pseudo-random) number generator and a comparator (B2P converters). Stochastic computing makes use of digital technology to perform complex arithmetic operations with a reduced number of gates. When performing those types of operations, stochastic signals must be completely uncorrelated for a proper behavior.

Stochastic computing is useful for those applications requiring parallel-processing techniques [11–14]. Traditional

Citation: Morro A, Canals V, Oliver A, Alomar ML, Rossello JL (2015) Ultra-Fast Data-Mining Hardware Architecture Based on Stochastic Computing. PLoS ONE 10(5): e0124176. doi:10.1371/journal.pone.0124176.

parallel processing architectures have the shortcoming of requiring a large amount of hardware resources. Therefore, the number of tasks that can be executed in parallel within a single chip is relatively small. Stochastic computing could represent a solution to this problem since the hardware used to solve each task is reduced in size if compared to traditional digital implementations. As a result, more complex tasks can be executed in parallel when using stochastic computing elements. The greatest advantage of stochastic computing is the potential of implementing hundreds of smart computing elements in one single integrated circuit, thus obtaining a highly parallelized processing chip with a computing capacity that can be several orders of magnitude higher than traditional binary-logic-based microprocessors.

In this work we generalize the basic stochastic computing principles, thus creating a more general *probabilistic processing*. The primary basic principle of traditional stochastic computing is that the stochastic bit streams must be completely uncorrelated in time and space to obtain the desired arithmetic operations when using simple logic gates. Our extended *probabilistic processing* approach will also use correlated signals in order to implement a set of non-linear operations. This set of non-linear operations can be used for an efficient and fast comparison between signals as we will show later.

Accordingly, *probabilistic computing* is characterized by the combination of two different types of operations:

- Arithmetic operations implemented by uncorrelated signals (such as the multiplication, the division or the addition). These ones are taken from the traditional stochastic computing concepts [8].
- Non-linear operations performed by correlated signals evaluated through logic gates.

The introduction of the possibility of coherence between the signal phases enhances the capacity of stochastic computing by including those non-linear operations that can be applied to perform fast similarity searches. Consequently, complex pattern recognition tasks can be executed in parallel when using probabilistic processing operations.

In this work we present a highly efficient methodology for data mining based on probabilistic processing. High dimensional data is inherently complex in clustering, classification and similarity search [15]. The proposed approach is evaluated showing its application to a similarity search over a huge database. Most data mining algorithms use similarity search as a subroutine core [16–18], and thus the time taken for this task is the bottleneck of virtually all data mining algorithms [19]. Similarity search plays a fundamental role in many data mining and machine learning problems, e.g. text categorization [20], collaborative filtering [21], time-series analysis [22,23], protein sequencing [24] or any application-specific task as petroglyphs comparison [25]. At the same time, the mining of huge datasets implies the use of large computer clusters [26,27]. The proposed approach based on the use of probabilistic processing shows large improvements in terms of hardware resources when compared with conventional solutions.

BASIC PRINCIPLES OF PROBABILISTIC PROCESSING

Stochastic Computing Principles

In stochastic-based computations a global clock provides the time interval during which all stochastic signals are stable (settled to 0 or 1). For each clock cycle, a particular node has a probability p of being in the HIGH state (see Fig 1). Stochastic pulsed signals follow probabilistic laws when evaluated with logic gates. As an example, an AND gate provides at the output the product of its inputs (that is to say, the collision probability between signals) whereas a NOT gate converts a probability p at the input to the complementary $1-p$ at the output. One of the requirements for these stochastic computing blocks is that signals must be un-correlated at different clock cycles and between them. In Fig 2 we show the importance of the temporal de-correlation when implementing arithmetic functions (we use the example of implementing $f(p) = p(1-p)$). The figure illustrates that if the inputs of the AND gate p and $1-p$are correlated the output is always equal to zero. Such correlations can be eliminated using shift registers to delay signals from one arithmetic level to the next one. In the correct case, the AND gate evaluates properly the product between p and a delayed (and therefore uncorrelated) value of $1-p$.

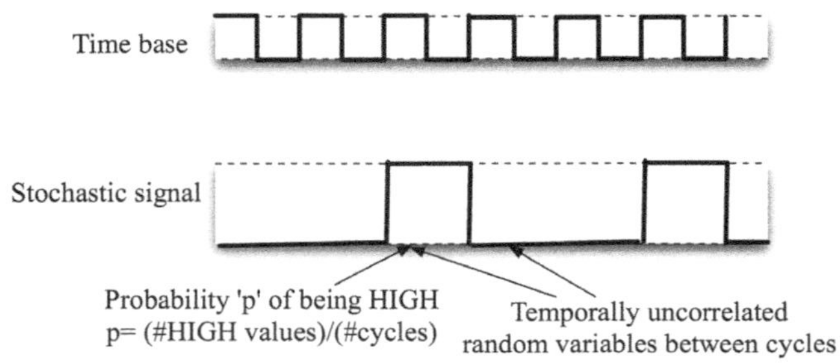

Figure 1: Basic temporal behavior of stochastic signals.

There is no correlation between the signal values at different clock cycles. The number associated with the stochastic signal is the activation probability.

http://dx.doi.org/10.1371/journal.pone.0124176.g001

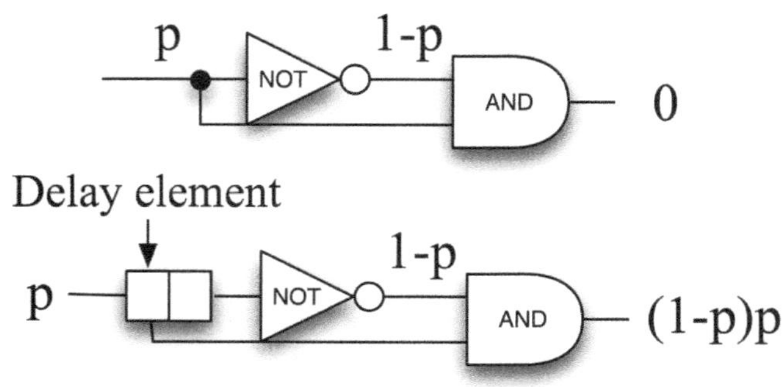

Figure 2: Correlation between signals and mathematical relationship between inputs and outputs.

If we desire to implement the function p(1-p) we must add a delaying element to de-correlate signals at the input of the AND gate.

http://dx.doi.org/10.1371/journal.pone.0124176.g002

Random Number Generation

The generation of pseudorandom sequences is a key issue for the implementation of probabilistic computing systems since stochastic bit streams are required to convert binary magnitudes to their equivalent stochastic signals. In particular, in order to obtain a random variable from Bernoulli sequences with a known generating probability *p*, we have to compare the binary number to convert (*P*) and the generated random number (*R*). If *P*>*R*, the stochastic pulse will be at high level, otherwise it will be low (Fig 3a).

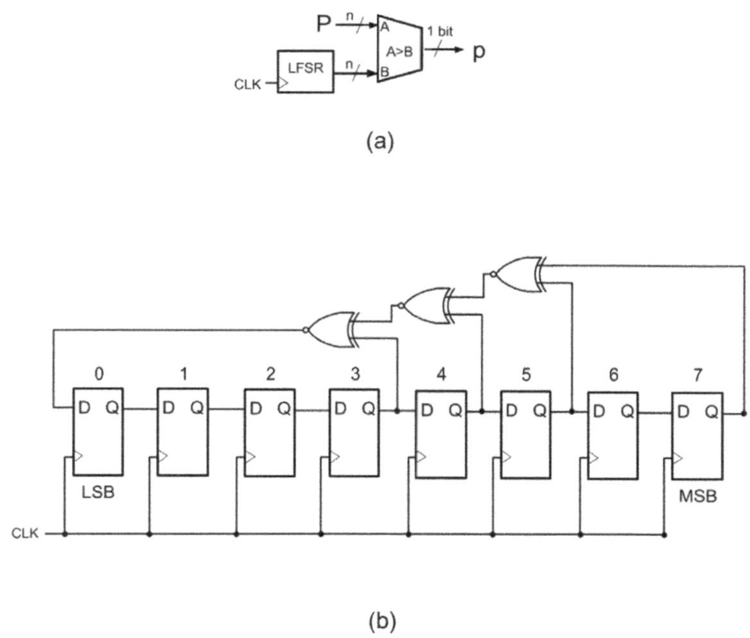

Figure 3: (a) Binary to pulse converter (B2P), an LFSR and a comparator are combined to obtain the pulsed signal. (b) Linear Feedback Shift Register (LFSR) used in the experiments.

http://dx.doi.org/10.1371/journal.pone.0124176.g003

A commonly used [28] source of pseudorandom digital noise is the linear feedback shift register (LFSR), which is an array of interconnected flip-flops with feedback to its input from a combination of the outputs of its various stages gated together in EXCLUSIVE-OR gates (see Fig 3b). This linear feedback structure provides uniformly distributed sequences (which have an autocorrelation delta function), but with a finite period of repetition, which has an exponential dependence with the number of bits. Since the sequences are produced deterministically, uncorrelated sources must be generated using different initial values (seeds) for the registers. The feedback configurations enabling maximal-length generators have been given [29,30]. The configuration applied in the present work, which employs a 8-bit shift register, is shown in Fig 3b. The autocorrelation function of a pseudorandom sequence generated by the LFSR used is shown at Fig 4. This function is a set of repeating delta functions separated by the number of bits in the LFSR's sequence length (2^n-1).

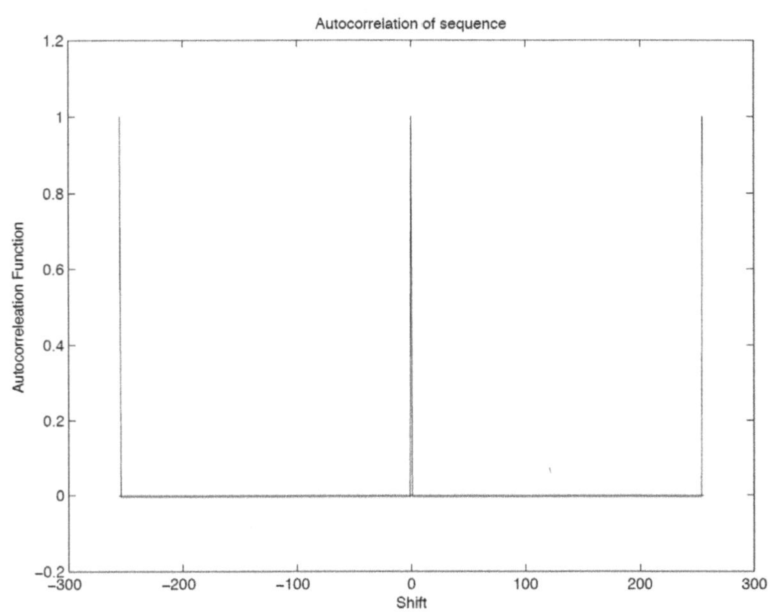

Figure 4: Autocorrelation function.

Autocorrelation function of the pseudo-random number generator used in the experiments.

http://dx.doi.org/10.1371/journal.pone.0124176.g004

For operations that need un-correlated signals, we define different seeds for each LFSR block. On the other hand, for operations requiring correlated signals we employ the same LFSR output for all stochastic variables.

Generalization of Stochastic Computing

Different kinds of stochastic non-linear functions can be reproduced using stochastic correlated signals. In Fig 5a, we show an example of the implementation of the absolute value of a subtraction ($|p_x\text{-}p_y|$). Stochastic signals x and y are derived from binary numbers X and Y when comparing with a random number R, generated using one LFSR. The probability of getting x = '1' or y = '1' (values of p_x and p_y) is proportional to X and Y respectively. Since X and Y are correlated (they share the same LFSR) the probability of getting both signals x and y with a high level (x = y = '1') is $\min(p_x, p_y)$, while the probability of getting both signals with a low level (x = y = '0') is $\min(1\text{-}p_x, 1\text{-}p_y)$. These two situations provide a low signal at the XOR output. On the other hand, the probability of getting different values on x and y is equal to $|p_x\text{-}p_y|$. In this situation, the XOR output (z) provides a high value with probability $p_z = |p_x\text{-}p_y|$.

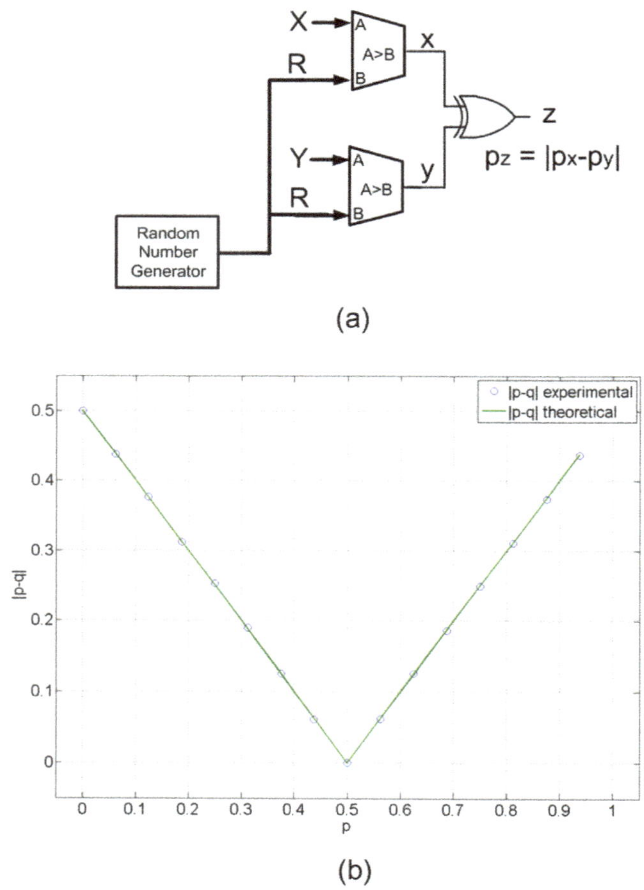

Figure 5: Similarity estimation of X and Y (with correlated signals).

In contrast to the un-correlated scheme (case in which X and Y are compared with different random values), a non-linear function is obtained. The probability of the R signal of being between X and Y is proportional to $|P_x\text{-}P_y|$ that is the switching activity of the output signal. (a) Probabilistic circuitry for the computation of $|P_x\text{-}P_y|$. (b) Experimental results when evaluating $|P_x\text{-}P_y|$.

http://dx.doi.org/10.1371/journal.pone.0124176.g005

The experimental measurements of this circuit confirm this behavior (see Fig 5b). In general, correlated signals evaluated through logic gates would implement max-min algebra functions. These types of functionalities are ideal for pattern recognition.

Data Mining of Huge Databases

The probabilistic nature of stochastic logic is an advantage for the implementation of probabilistic-based pattern recognition methodologies [2] and pattern matching is in the core of many data mining processes. The purpose is to compare parameters (the features) from different objects with reference vectors that represent different categories. All the features define each object, thus configuring an m-dimensional vector (for m different features).

Fig 6 shows the stochastic circuit used to compare two m-dimensional vectors, providing at the output an estimation of the similarity (s_{jr}) of both objects (vector '**r**' in the database to mine, and the vector defining the category '*j*' ($\mathbf{x}_j$)). A total of '*2m*' binary comparators and also '*m*' randomly selected binary numbers (R_k for the k-th descriptor) are configured in parallel to create a total of *2m* switching bits. Each pair of stochastic bits (x_{jk} and r_k) are compared through XNOR gates connected with a block performing the AND function between the *m* signals, thus providing at its output an estimation of the similarity (s_{jr}) between the vector (**r**) and the category j ($\mathbf{x}_j$).

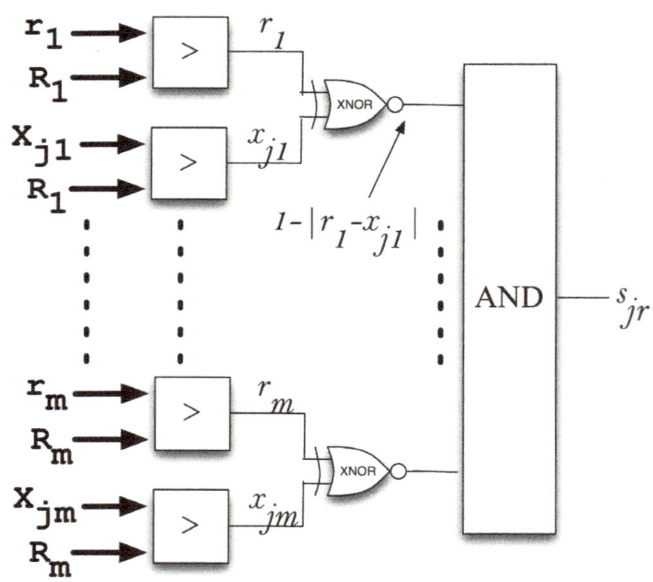

Figure 6: Stochastic architecture for comparing m-dimensional vectors.

A total of 'm' random signals are needed for each comparison.

http://dx.doi.org/10.1371/journal.pone.0124176.g006

Note the combination of both correlated and uncorrelated stochastic signals to obtain s_{jr} (all the AND inputs must be uncorrelated between them so that we need ‹m› random numbers R_k). This block would represent the typical implementation of a probabilistic processing unit combining both types of signaling. Therefore, the probabilistic signal obtained (s_{jr}) can be expressed as:

$$s_{jr} = \prod_{k=1}^{m}\left(1 - |x_{jk} - r_k|\right) \quad (1)$$

The switching activity at the output of the AND block is therefore proportional to the similarity between the object **r** and the category represented by vector $\mathbf{x}_j$. In Fig 7 we show the level curves associated to the Manhattan-based metric used in (1). As it is shown, the selected metric can be considered a good proximity estimator for fast mining of huge databases.

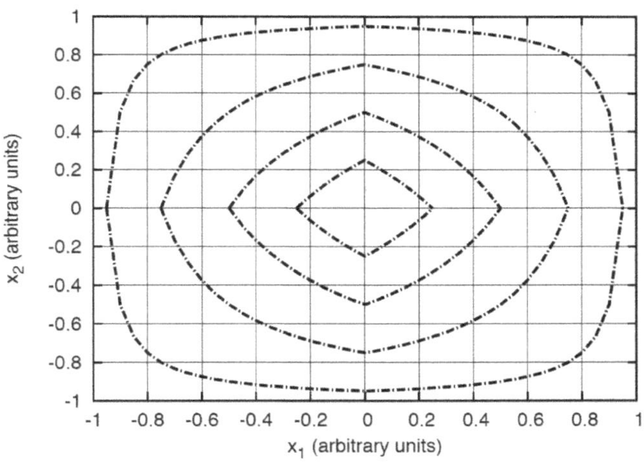

Figure 7: Level curves for the similarity metric used for the case of a two-dimensional space.

http://dx.doi.org/10.1371/journal.pone.0124176.g007

Hundreds of similarity estimators as the one shown in Fig 6 can be configured in parallel in a medium-sized FPGA, thus increasing considerably the mining speed in comparison with traditional processor-based techniques. Different vectors can be compared in parallel by using a Winner-Take All (WTA) architecture (see Fig 8). From ‹n' different vectors of the database, the circuitry provides at its output the closer to reference vector **r**. Then, relative fraction of the area used by the LFSRs with respect to the total circuit area decreases with the number of vectors $\mathbf{x}_j$ that can be compared.

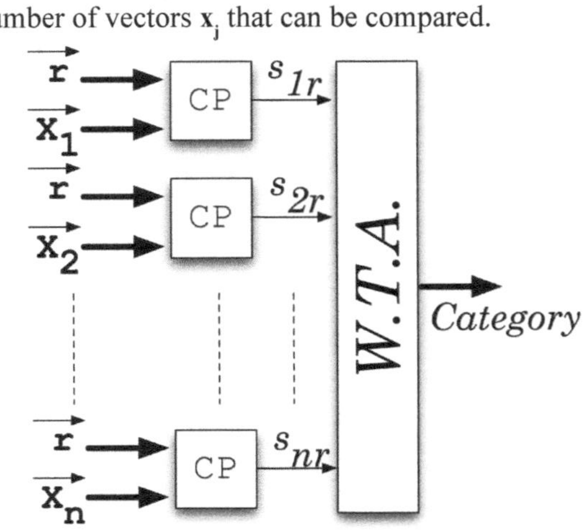

Figure 8: Architecture used to estimate the closest category to the reference vector ‹r›.

Each comparator (CP) provides at its output a switching signal proportional to the similarity of the two vectors connected to it. The Winner-Take All selects the highest frequency signal.

http://dx.doi.org/10.1371/journal.pone.0124176.g008

Thus, at the WTA output the category label *'j'* that is activated is the one with the highest similarity:

$$Category = 'j' | s_{jr} \geq s_{ir} \forall i \in \{1..n\} \quad (2)$$

The WTA can be constructed by using binary counters (module-k) (see Fig 9). Only a maximum number of clock cycles per comparison ($N>k$) are allowed, and therefore the number of cycles needed to overflow fixes a minimum similarity value to be distinguished s_{min} so that$N \cdot s_{min} = k$. If all the similarities at the input of the WTA (s_{jr}) are lower than s_{min}, the most probable scenario is not to obtain any positive result at the output of the WTA.

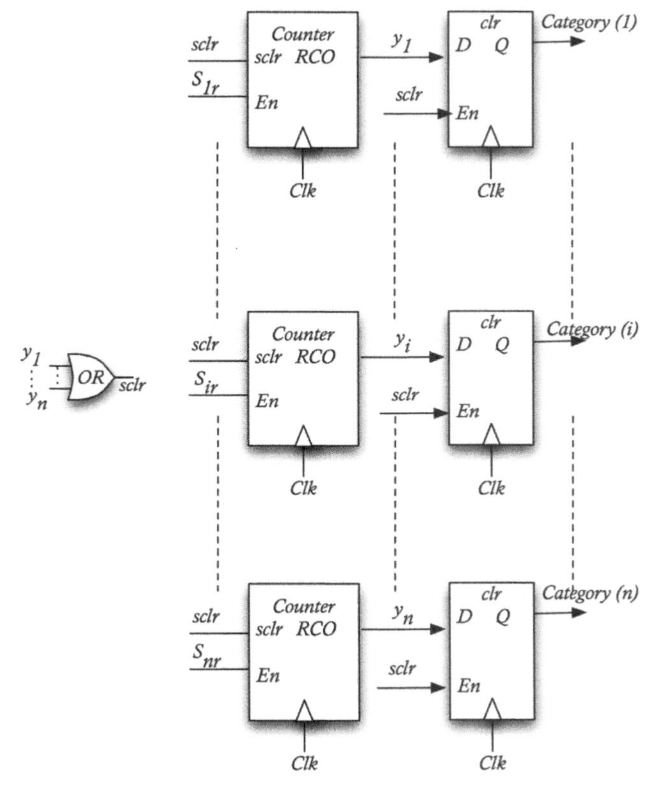

Figure 9: Winner-Take All.

Architecture used to select the category associated to the s_{ir} with the higher activity.

http://dx.doi.org/10.1371/journal.pone.0124176.g009

For any similarity value $s_{jr} > s_{min}$ between two vectors ($\mathbf{x}_j$ and **r**) we have that the probability of identifying vector ***r*** as belonging to class ‹j› is close to 1. In Fig 10 we show the relationship between the distances of two vectors (|*x-y*|) and the number of cycles (N) needed to obtain an overflow of one of the counters of the WTA. We compare the analytical formula of the theoretical behavior (solid line showing the formula $N = k/s$, where the similarity is s = 1-d since we only vary the distance in one dimension) and the measurements obtained with an ALTERA Cyclone III FPGA (dashed lines). As it can be seen, a close relationship is obtained between the expected behavior and the probabilistic implementation.

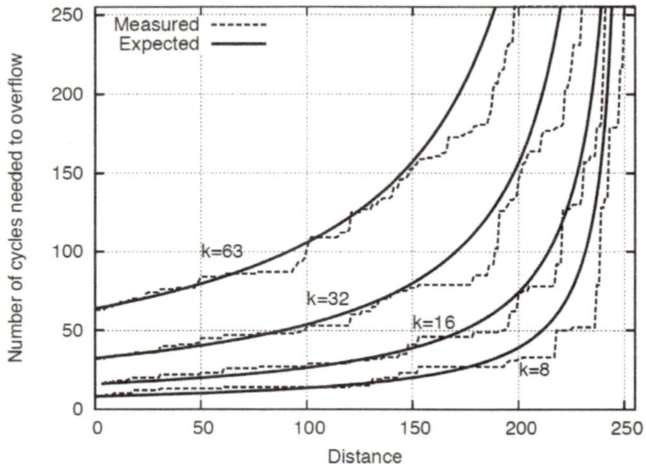

Figure 10: Relationship of distance between vectors (|*x-y*|) and the number of cycles needed to overflow a module-k counter of the WTA.

The switching activity of the similarity stochastic signal (s_{jr}) is dependent on the distance ($s = 1\text{-}d$). As the higher is the distance, lower will be the switching activity of s, and higher the number of cycles needed to overflow the counter.

http://dx.doi.org/10.1371/journal.pone.0124176.g010

The probability of identifying vector $\boldsymbol{r}$ as belonging to class ‹*j*› can be estimated considering the probability of obtaining more than ‹*k*› HIGH values from a signal with switching activity ‹s_{jr}' when waiting a total of N cycles (N>k).

$$P_{jr}\left(s_{jr}\right) = P(r \in j) = \sum_{l=k}^{N} \binom{N}{l} s_{jr}^{l} \left(1 - s_{jr}\right)^{N-l} \tag{3}$$

where N is the total number of cycles used to compare any set of vectors. Function (3) is represented in Fig 11 for the special case in which $k = 16$ and $N = 80$ as a function of the similarity s_{jr}. For this case, $s_{min} = k/N = 0.2$, note that $P_{jr}>0.5$ for $s_{jr}>s_{min}$ in Fig 11.

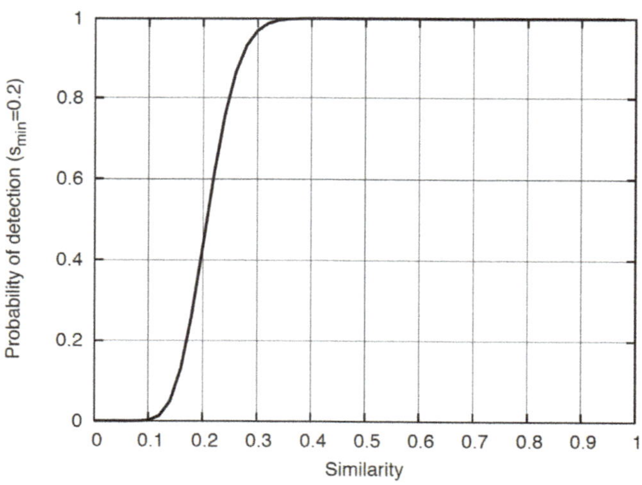

Figure 11: Variation of query identification probability with respect to similarity.

An arbitrary threshold of $s_{min} = 0.2$ is selected.

http://dx.doi.org/10.1371/journal.pone.0124176.g011

The similarity search function (1) has the inherent advantage of its analytical simplicity. In this sense, given two randomly selected m-dimensional vectors $\mathbf{x}_j$ and $\mathbf{r}$, it is possible to compute the cumulative distribution function of obtaining a similarity lower than a given selected value 'z'. From basic probability theory we have that this cumulative distribution function can be estimated as:

$$F_z(z) = \int_{\Omega} dy \tag{4}$$

where $F_z(z)$ is the probability of obtaining a similarity lower than z, Ω is the volume for which$g(y)<z$, being $g(y)$ the function under consideration (1). Vector $\boldsymbol{y}$ is composed by all the parameters for which ‹*g*› is dependent (the components of both $\mathbf{x}_j$ and $\mathbf{r}$ vectors that are stochastic signals with values between 0 and 1).

After some algebra, and considering (1) for the estimation of Ω and that each parameter is bounded between 0 and 1 we have:

$$F_z(z) = z^2 \sum_{j=1}^{m} \frac{[-2log(z)]^{j-1}}{(j-1)!} = \frac{\Gamma(m, -2log(z))}{\Gamma(m)} \tag{5}$$

where ‹*m*› is the dimension of the vectors. If we want to estimate the probability density function $f_z(z)$, we have to derive the cumulative distribution function with respect the selected similarity (z), $f_z(z) = \partial_z F_z(z)$

From expression (5) we can estimate the number of positive identifications from a database with W unknown objects if we select a minimum similarity of s_{min} ($Positives \approx W \bullet (1\text{-}F(s_{min}))$). Expression (5) is also very useful to estimate the minimum database width W_{min} needed to identify an object with a specific property (assuming a random database):

$$W_{min} = \frac{1}{1 - s_{min}^2 \sum_{j=1}^{m} \frac{[-2log(s_{min})]^{j-1}}{(j-1)!}} \tag{6}$$

RESULTS

We implemented the proposed methodology in a FPGA-based PCIe board (ProcStarIV110E-4B) fabricated by GIDEL Ltd (see Fig 12). This board incorporates four ALTERA STRATIX III 110E FPGAs along with 32GB of DDR2 SODIMM memory banks to allocate the database. The communication with the board is done through a PCI-express connector of a PC. The clock frequency of operation of the board is 87.5 MHz and the maximum power dissipation is about 100W.

Figure 12: FPGA-based PCIe board used for the data mining process from GIDEL.

http://dx.doi.org/10.1371/journal.pone.0124176.g012

Inside the four FPGA cores we configured a total of 400 similarity comparators per core (design shown in Fig 6), thus implementing a total of 1600 stochastic comparators in the board that operate in parallel. The dimension of the vectors are selected to be $m = 12$. To test the circuit effectiveness, we created a similarity search in a random database with a total of 2.56•10^6 12-D vectors setting an arbitrary minimum similarity of $s_{min} = 0.2$. The final positive results are shown in Fig 13, where we show, as a function of the similarity 's': the number of vectors of the database with this similarity (dark diamonds), the number of vectors that are finally selected by the circuitry as belonging to class 'c' (circles), the ideal number of vectors to be selected (solid line) and the expected number of vectors to be selected (dashed line$f_z(s_{jr})P_{jr}$). It is visible that the experimental and theoretical results (circles and dashed line) are similar. We also can see that nearly all the vectors with a similarity higher than the selected threshold are identified (circles and solid line). Unfortunately there is a non-negligible fraction of vectors with a similarity lower than s_{min} that are selected (circles with $s<s_{min} = 0.2$). This fact is due to the non-negligible tie of $P_{jr}(s_{jr})$ for $s_{jr}<s_{min}$ and the exponential dependence of$f_z(s_{jr})$with s_{jr}.

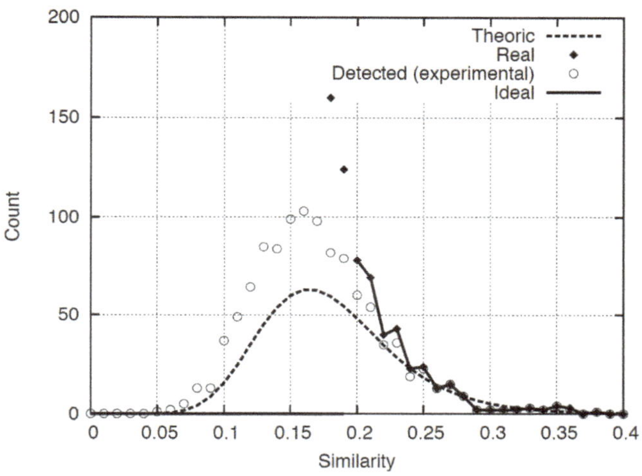

Figure 13: Number of positive identifications from a database with 2.56•10^6 **particles when** s_{min} **= 0.2.**

http://dx.doi.org/10.1371/journal.pone.0124176.g013

The solution to erase such *false positives* is to filter-out the final results using software. The impact of this filtering on the database screening timing is small since only a low volume of vectors must be recomputed (rather than the billions of vectors that the database can contain).

The total number of positives to be selected by the stochastic circuitry can be estimated numerically as:

$$Positives = \int_0^1 P_{jr}(s) f_z(s) ds \quad (7)$$

Expression (7) can be used to select an optimal number for k and N to control the number of positive results within reasonable values. In Fig 14 we show the different types of negatives and positives provided by the system. False positives are those positives with a similarity lower than the threshold s_{min}. False negatives are those vectors that, even presenting a higher similarity than s_{min}, are not found by the system. Finally, the true positives and negatives are those vectors that are correctly classified by the system.

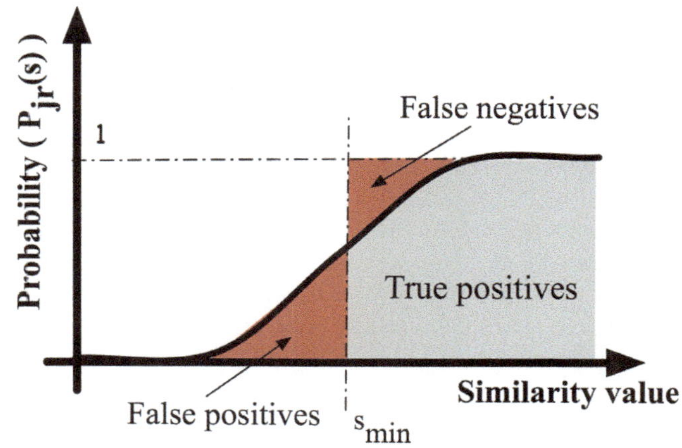

Figure 14: For each computation, a given number of false positives and negatives are provided by the probabilistic system.

False positives can be filtered out by software while false negatives are lost.

http://dx.doi.org/10.1371/journal.pone.0124176.g014

Although the false positives can be filtered-out by software, the vectors with $s>s_{min}$ that are finally not found by the hardware (false negatives) cannot be recovered (in Fig 13, the false negatives are the difference between the solid line and the circles). In this sense, the precision of the system is defined as the ratio between true positives and the total number of possible positives (true positives + false negatives).

$$\eta = \frac{\int_{s_{min}}^1 P_{jr}(s) f_z(s) ds}{\int_{s_{min}}^1 f_z(s) ds} \quad (8)$$

In Table 1 we show the dependence of η with respect to parameter k assuming $f_z = 1$ with $s_{min} = 0.5$. Increasing k increases the efficiency but decreases the screening speed of the circuit. In Table 1 we find the relationship between η and the number of cycles needed to overflow a WTA counter (k value). For most applications, a value of k = 8 provides good results in terms of accuracy and computation time.

Table 1: Relationship between the k value and the precision of the system when fz = 1 and using smin = 0.5 (i.e. N = 2k)

http://dx.doi.org/10.1371/journal.pone.0124176.t001

k	Precision (η)
8	0.93
16	0.94
32	0.96
64	0.97
128	0.97

doi:10.1371/journal.pone.0124176.t001

The circuit speed can be estimated from the time needed to screen a database (t_p) with a total of W vectors:

$$t_p = \frac{W \cdot size}{f_{RAM}} + \frac{NT_{clk}W}{1600} + t_{setup} \quad (9)$$

where W is the number of vectors in the database, f_{RAM} is the data transfer frequency of the DDR-DRAM (of the order of 16GB/s for the PCIe used in the experiment), T_{clk} is the global clock period of the circuit (operating at 87.5MHz), ‹*size*› is the number of bytes per vector (12 bytes in our case), 1600 is the number of stochastic comparators implemented inside the FPGAs, parameter N is the number of clocks used by the WTA to process the inputs (signals s_{jr}), and finally t_{setup} is the setup time needed for board initialization (of the order of 1ms). Finally, the speed of the screening process is estimated as a function of the number of vectors in the database ($f_{proc} = W/t_p$). In Fig 15 we show the screening speed (in millions of comparisons per second) as a function of the database width (in millions of vectors), where the vectors are composed by 12 bytes ($m = 12$). The low speed obtained by the proposed methodology when the database is small is basically due to the fixed setup time of the system.

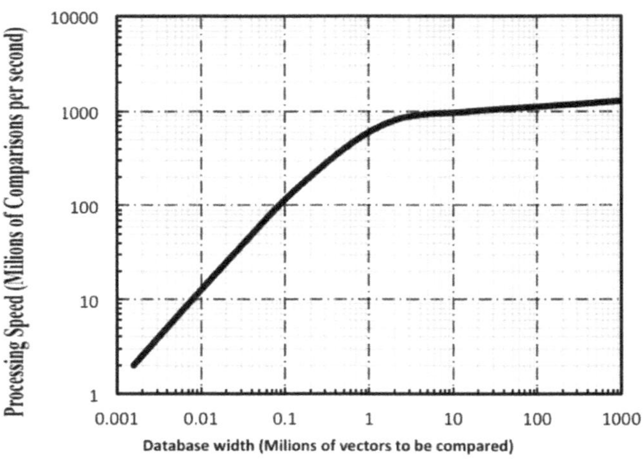

Figure 15: Data mining speed variation with database width when using probabilistic computing.

Each comparison implies the processing of two vectors of 12 Bytes each one.

http://dx.doi.org/10.1371/journal.pone.0124176.g015

We compared a conventional implementation using binary-based digital hardware of a multi-vector comparator implementing the metric proposed in (1) with the proposed probabilistic system. In Fig 16 we show the conventional implementation of expression (1). Finally, in Fig 17 we compare the result of such similarity comparators with the minimum similarity s_{min}. In Table 2 we compare both implementations in terms of FPGA resources when using an ALTERA Cyclone III EP3C25F324C8 device. The vectors to be compared are fixed to 12 dimensions. As it can be seen, the ratio of Logic Elements needed by the conventional and the probabilistic implementation increases as the number of vectors to be compared grows. The area ratio between both systems is of the order of 56 while the ratio in terms of circuit speed is constant and about 1/k (1/8). Therefore, the probabilistic implementation speeds up the screening process of the database by a factor of 7 when compared to a conventional digital implementation using the same hardware area. In other words, the total performance when using the proposed architecture is higher by a factor of 7. It means that, the lack in speed can be compensated by using less hardware area.

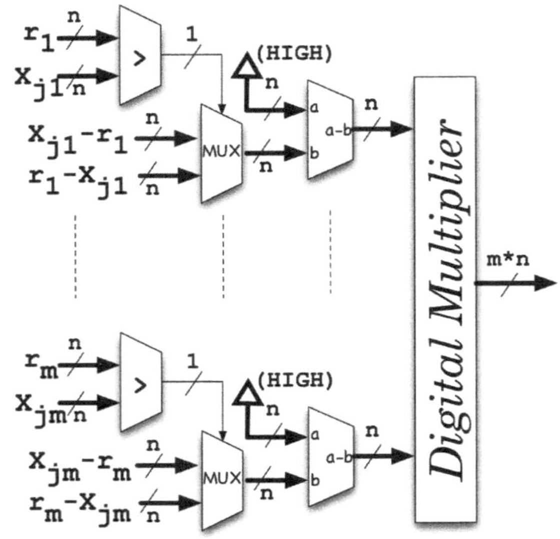

Figure 16: Conventional implementation of a m-dimensional digital comparator where expression (1) is implemented.

The first comparator and the multiplexer implement the absolute value function while the multiplier provide the product for all the dimensions.

http://dx.doi.org/10.1371/journal.pone.0124176.g016

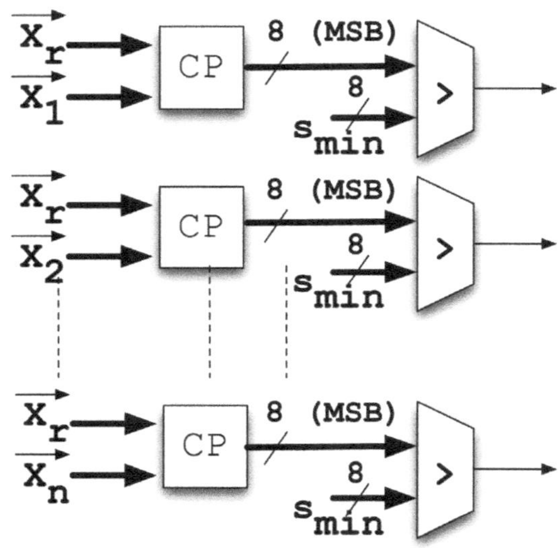

Figure 17: Global comparator between the reference vector and each vector in the database.

Block CP is implementing the circuitry shown in Fig 16. The result is compared with s_{min}.

http://dx.doi.org/10.1371/journal.pone.0124176.g017

Table 2: Relationship between conventional and probabilistic implementation in terms of FPGA resources for an Altera Cyclone III device

http://dx.doi.org/10.1371/journal.pone.0124176.t002

Number of vectors	Classical			Probabilistic		
	Logic Elements (LE)	Multipliers	Delay time (t_D)	Logic Elements	Multipliers	Delay time (t_D)
2	1.955	126	1	321	0	8
4	13.898	132	1	551	0	8
8	39.142	132	1	1.023	0	8
16	91.464	132	1	1.908	0	8
32	196.076	132	1	3.713	0	8
128	823.814	132	1	14.719	0	8

doi:10.1371/journal.pone.0124176.t002

To sum up, the presented application is an example where stochastic computing is advantageous over conventional computing systems. Since the presented implementation takes advantage of the high parallelism of stochastic computing, less circuit area implies a lower power demand to achieve a determined processing speed.

CONCLUSIONS

We have presented a new and unconventional computing technique for ultra-fast mining of huge databases. The methodology is based on the use of probabilistic pulsed signals. We describe how correlated bit streams can be used to implement non-linear functions like the absolute value function, which have been also developed by other research group [31]. In the final architecture we allow the use of both correlated and uncorrelated stochastic bit streams. The combination of both types of switching signals increases the mathematical capacity of original stochastic computing implementations. The data to be mined is translated to spikes and processed by a simple digital circuitry. The simplicity of the circuitry is used to implement hundreds of stochastic comparators inside Field-Programmable Gate Arrays and oriented to screen huge databases. The final implementation uses an FPGA-based PCIe board for the screening. This implementation uses less hardware resources than conventional digital methodologies (based on binary and not probabilistic logic) and is able to process the order of 13GBytes of information per second (in contrast to the estimated 2GBytes/s of speed that could be achieved by the conventional implementation using the same hardware area). With the 12-dimensional space used to allocate each vector in the example shown in this paper we obtain the order of 1 billion of comparisons per second. A patent application has been done for this new mining methodology [32].

AUTHOR CONTRIBUTIONS

Conceived and designed the experiments: AM VC AO MLA JLR. Performed the experiments: AM. Analyzed the data: JLR VC. Wrote the paper: JLR.

REFERENCES

1. Szalay A, Gray J. (2006) 2020 computing: Science in an exponential world. Nature 440: 413–414. pmid:16554783 doi: 10. 1038/440413a
2. Bishop C. M. , editor. (2006) Pattern recognition and machine learning. New York: Springer.
3. Fayos J, Fayos C. (2007) Wind data mining by Kohonen neural networks. PLoS ONE 2. doi: 10. 1371/journal. pone. 0000210
4. Stegmayer G, Gerard M, Milone D. (2012) Data mining over biological datasets: An integrated approach based on computational intelligence. IEEE Computational Intelligence Magazine 7: 22–34. doi: 10. 1109/mci. 2012. 2215122
5. Zou J, Han Y, So S-. (2008) Overview of artificial neural networks. Methods in Molecular Biology 458: 15–23. pmid:19065803 doi: 10. 1007/978-1-60327-101-1_2
6. Ballester PJ, Richards WG. (2007) Ultrafast shape recognition for similarity search in molecular databases. Proceedings of the Royal Society A: Mathematical, Physical and Engineering Sciences 463: 1307–1321. doi: 10. 1098/rspa. 2007. 1823
7. Pan F, Wang W, Tung AKH, Yang J. (2005) Finding representative set from massive data. Proceedings—IEEE International Conference on Data Mining, ICDM: 338–345.
8. Gaines BR. (1975) Stochastic and fuzzy logics. Electron Lett 11: 188–189. doi: 10. 1049/el:19750144
9. Gaines R. (1969) Stochastic Computing Systems. Advances in Information Systems Science. Tou ,ed. , 2:37–172. doi: 10. 1007/978-1-4899-5841-9_2
10. Kim Y, Shanblatt MA. (1995) Random noise effects in pulse-mode digital multilayer neural networks. IEEE Trans Neural Networks 6: 220–229. doi: 10. 1109/72. 363434
11. Canals V, Morro A, Rosselló JL. (2010) Stochastic-based pattern-recognition analysis. Pattern Recog Lett 31: 2353–2356. doi: 10. 1016/j. patrec. 2010. 07. 008
12. Hori M, Ueda M, Iwata A. (2006) Stochastic computing chip for measurement of manhattan distance. Japanese Journal of Applied Physics, Part 1: Regular Papers and Short Notes and Review Papers 45: 3301–3306. doi: 10. 1143/jjap. 45. 3301
13. Rosselló JL, Canals V, Morro A. (2012) Probabilistic-based neural network implementation. Proceedings of the International Joint Conference on Neural Networks.
14. Li P. , Lilja D. J. , Qian W. , Bazargan K. , Riedel M. D. (2014) Computation on stochastic bit streams digital image processing case studies. IEEE Transactions on Very Large Scale Integration (VLSI) Systems, 22: 449–462. doi: 10. 1109/tvlsi. 2013. 2247429
15. Aggarwal CC, Han J, Wang J, Yu PS. (2005) On high dimensional projected clustering of data streams. Data Mining and Knowledge Discovery 10: 251–273. doi: 10. 1007/s10618-005-0645-7
16. Bayardo RJ, Ma Y, Srikant R. (2007) Scaling up all pairs similarity search. 16th International World Wide Web Conference, WWW2007: 131–140. doi: 10. 1145/1242572. 1242591
17. Gullo F, Ponti G, Tagarelli A, Greco S. (2009) A time series representation model for accurate and fast similarity detection. Pattern Recognit 42: 2998–3014. doi: 10. 1016/j. patcog. 2009. 03. 030
18. Kontos D, Megalooikonomou V. (2005) Fast and effective characterization for classification and similarity searches of 2D and 3D spatial region data. Pattern Recognit 38: 1831–1846. pmid:16565747 doi: 10. 1016/j. patcog. 2005. 04. 020
19. Rakthanmanon, T. , Campana, B. , Mueen, A. , Batista, G. , Westover, B. , Zhu et al. (2012) Searching and mining trillions of time series subsequences under dynamic time warping. Proceedings of the ACM SIGKDD International Conference on Knowledge Discovery and Data Mining: 262–270.
20. Zhen Y, Yeung D-. (2013) Active hashing and its application to image and text retrieval. Data Mining and Knowledge Discovery 26: 255–274. doi: 10. 1007/

s10618-012-0249-y

21. Koren Y. (2008) Factorization meets the neighborhood: A multifaceted collaborative filtering model. Proceedings of the ACM SIGKDD International Conference on Knowledge Discovery and Data Mining: 426–434.
22. Wang X, Smith K, Hyndman R. (2006) Characteristic-based clustering for time series data. Data Mining and Knowledge Discovery 13: 335–364. doi: 10. 1007/s10618-005-0039-x
23. Li H, Guo C. (2012) Similarity search for time series based on efficient warping measure. Proceedings of the ACM SIGKDD International Conference on Knowledge Discovery and Data Mining.
24. Kaznadzey A, Alexandrova N, Novichkov V, Kaznadzey D. (2013) PSimScan: Algorithm and utility for fast protein similarity search. PLoS ONE 8. doi: 10. 1371/journal. pone. 0058505
25. Zhu Q, Wang X, Keogh E, Lee S-. (2011) An efficient and effective similarity measure to enable data mining of petroglyphs. Data Mining and Knowledge Discovery 23: 91–127. doi: 10. 1007/s10618-010-0200-z
26. Grossman R, Gu Y. (2008) Data mining using high performance data clouds: Experimental studies using sector and sphere. Proceedings of the ACM SIGKDD International Conference on Knowledge Discovery and Data Mining: 920–927.
27. Zhou P, Lei J, Ye W. (2011) Large-scale data sets clustering based on MapReduce and hadoop. Journal of Computational Information Systems 7: 5956–5963.
28. Kitsos P, Sklavos N, Zewas N, Koufopavlou O. (2001) A reconfigurable linear feedback shift register (LFSR) for the bluetooth system. Proceedings of the IEEE International Conference on Electronics, Circuits, and Systems 2: 991–994. doi: 10. 1109/icecs. 2001. 957640
29. Peterson W. (1961) Error Correcting Codes. Jonh Wiley and Sons.
30. Korn G. (1966) Random-Process Simulation and Measurements. McGraw-Hill Book Co.
31. Alaghi A, Hayes JP. (2013) Exploiting correlation in stochastic circuit design. 2013 IEEE 31st International Conference on Computer Design, ICCD 2013: 39–46. doi: 10. 1109/iccd. 2013. 6657023
32. Rosselló JL, Morro A, Canals V. (2012) Sistema digital probabilístico para la exploración y reconocimiento de formas eficiente en grandes bases de datos.

CHAPTER 6 RECONFIGURABLE BOOLEAN LOGIC USING MAGNETIC SINGLE-ELECTRON TRANSISTORS

M. Fernando Gonzalez-Zalba[1], Chiara Ciccarelli[2], Liviu P. Zarbo[3], Andrew C. Irvine[2], Richard C. Campion[4], Bryan L. Gallagher[4], Tomas Jungwirth[3,4], Andrew J. Ferguson[2], Joerg Wunderlich[1,3]

[1]Hitachi Cambridge Laboratory, Cambridge CB3 0HE, United Kingdom

[2]Cavendsih Laboratory, University of Cambridge, Cambridge, United Kingdom

[3]Institute of Physics, The Academy of Science of the Czech Republic, v.v.i, Cukrovarnick'a 10, 162 53 Praha 6, Czech Republic

[4]School of Physics and Astronomy, University of Nottingham, Nottingham NG7 2RD, United Kingdom

ABSTRACT

We propose a novel hybrid single-electron device for reprogrammable low-power logic operations, the magnetic single-electron transistor (MSET). The device consists of an aluminium single-electron transistor with a GaMnAs magnetic back-gate. Changing between different logic gate functions is realized by reorienting the magnetic moments of the magnetic layer, which induces a voltage shift on the Coulomb blockade oscillations of the MSET. We show that we can arbitrarily reprogram the function of the device from an n-type SET for in-plane magnetization of the GaMnAs layer to p-type SET for out-of-plane magnetization orientation. Moreover, we demonstrate a set of reprogrammable Boolean gates and its logical complement at the single device level. Finally, we propose two sets of reconfigurable binary gates using combinations of two MSETs in a pull-down network.

INTRODUCTION

As the downscaling of conventional CMOS technology is bound to reach its fundamental limit new algorithms will be the answer to achieve increasingly higher performance and reduced power consumption. Reconfigurable digital circuits provide a way to extend the functionalities of conventional CMOS by implementing in the same physical space multiple logic operations and therefore increasing the computational complexity. Reconfiguration of the logic functions at each individual device promises even more compact and flexible circuit design [1–5]. However, the implementation of such reconfigurable logic using single-electron transistors (SETs) [6, 7] is appealing because SETs have good scalability, one of the lowest energy-per-switching-event [8] and the possibility to combine their electrical properties with magnetic elements [9–14]. There have been several proposals to implement programmable SET logic by using the charge degree of freedom such as fixed gate voltages [15], non-volatile charge nodes [16, 17] and the spin degree of freedom [18–20].

In this manuscript, we show a proof of principle for reconfigurable Boolean logic based on magnetically-gated circuit elements and we suggest multi-device reconfigurable logic architectures. More particularly, we report the complementary logic operation of a aluminium MSET. The reconfigurable capability of our MSET stems from the magnetization-dependent work function of GaMnAs back gate. When the back-gate is kept at a constant potential the magnetic-field-induced chemical change causes a charge accumulation in the gate electrode which can be readily sensed by the SET as a change in the Coulomb oscillation phase

Citation: Gonzalez-Zalba MF, Ciccarelli C, Zarbo LP, Irvine AC, Campion RC, Gallagher BL, et al. (2015) Reconfigurable Boolean Logic Using Magnetic Single-Electron Transistors. PLoS ONE 10(4): e0125142. doi:10.1371/journal.pone.0125142.

[21,22]. Although the concept is demonstrated here for MSET devices, the operating principle is general and transferable to any field-effect transistor.

RESULTS

The MSET has two modes of operation: it responds to gate voltage inputs (electric mode) as well as to the orientation of the magnetic moments (magnetic mode). By reorienting the magnetization of the GaMnAs substrate we are able to switch from n-type to p-type MSET. Making use of the magnetic mode we demonstrate two sets of reprogrammable Boolean logic gates implemented at the single device level. Finally, we suggest a strategy to scale the reconfigurable logic operation to the multiple device level.

A schematic cross-section of the proposed aluminium MSET is depicted in Fig 1(a). The key element of this structure is the magnetic gate electrode, an epitaxially grown $Ga_{0.94}$ $Mn_{0.06}$ As layer on GaAs that acts as a back-gate with easy-axis directions [110] and [1–10]. A magnetic field $B = 0.7$ T, larger that the saturation field $B_s \approx 0.3$ T, is applied to rotate the magnetization$M$ with respect to the [001] direction (φ = 0°). The angle θ with respect to the [1–10] direction is kept equal to 90° in all measurements. The results presented here are independent of the magnitude of B for $B > B_s$. A 100 nm alumina layer is used as gate dielectric on top of which the Al-SET is fabricated. The source and drain leads are separated from the micron-sized island by aluminium oxide tunnel junctions, as it can be observed in the SEM image in Fig 1(b). In this letter the side gates were not used but could provide extra-functionality to the structure.

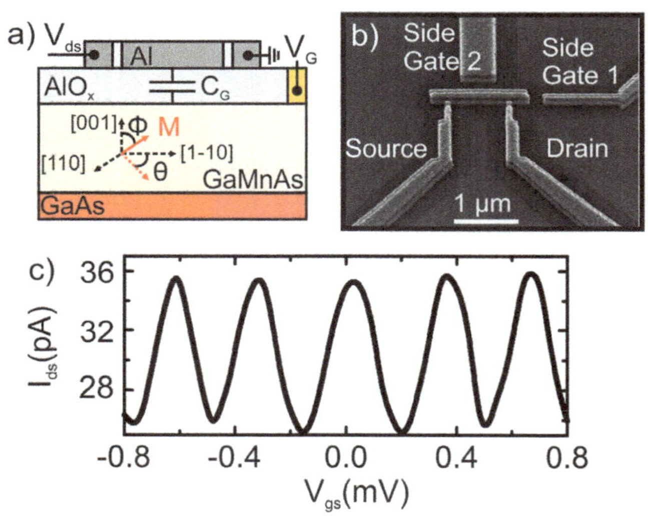

Figure 1: Device structure.

(a) Schematic cross-section of the device sketching the magnetization orientation of the (Ga,Mn)As back-gate layer. (b) SEM image of the device. The aluminium island is separated from the source and drain leads by AlO_x tunnel junctions. Side gates were not used in this experiment. (c) Drain current (I_{ds}) oscillations as a function of the back gate voltage (V_{gs}).

http://dx.doi.org/10.1371/journal.pone.0125142.g001

In the electric mode of operation we use the back-gate voltage to control the induced-charge in the island of the MSET. Fig 1(c) shows the drain current (I_{ds}) versus back-gate voltage (V_{gs}) at 300 mK. The plot shows a periodic oscillatory behaviour of the current, namely the Coulomb blockade oscillations. Several devices were studied with typical charging energies E_C = 100–300 μeV and gate capacitances C_G = 0.4–1.0 fF. We obtain a I_{ON}/I_{OFF} contrast ratio of up to 1.33 and typical tunnel junction resistances of R_T = 500 kΩ.

The magnetic mode of operation is based on the dependence of the GaMnAs band structure on the orientation of the magnetization due to spin orbit coupling. The magnetization reorientation yields a shift in the GaMnAs work function $\Delta\mu$ which in itself does not produce a response on the MSET. However, when the back gate is held at constant potential a change inμ causes an inward or outward flow of charge in the gate which in turn offsets the Coulomb oscillations. If the work function shift in the magnetic layer is adequate, the phase shift becomes half a period of the Coulomb blockade oscillations and complementary MSETs are obtained. The equation that defines the necessary work function shift for maximum on-off current ratio can be expressed as,

$$\Delta\mu(B)_{max} = \frac{e^2}{2C_G} \tag{1}$$

As well, the minimum potential shift for complementary operation at temperature T can be defined as a gate voltage shift equal to the FWHM of the Coulomb peak [23],

$$\Delta\mu(B)_{min} = \frac{4.35k_B TC_\Sigma}{C_G} \tag{2}$$

These two equations set the operational parameters in terms of temperature, magnetic field and physical dimensions of the structure.

The complementary operation of the MSET is experimentally demonstrated in Fig 2 for a device with E_C = 300μeV and C_G = 1.0 fF. To switch between complimentary modes we rotate continuously a saturating magnetic field from φ = 90°, where the magnetization resides Mparallel to the plane along the easy-axis [110] (Fig 2(a)) to φ = 0° where the magnetization is out-of-plane (Fig 2(b)). This can be seen in Fig 2(c) where we plot the source-drain current I_{ds}as a function of the magnetization angle φ and gate voltage V_{gs} for V_{ds} = 15μV and B = 0.7 T. When the magnetization is in-plane, we select V_{gs} = -61 μV as the logic 0 and V_{gs} = +86 μV as the logic 1 for the n-type SET (see Fig 2(d)). As the magnetic field is rotated to φ = 0°, the magnetization is re-oriented in the [001] direction and the GaMnAs work function increases producing a gate voltage shift of ΔV_{gs} = 150 μV [24]. The combination of work function potential shift in the $Ga_{0.94}$ $Mn_{0.06}$ As layer with the 300 μV gate voltage period of the device result in that the magnetically-induced gate voltage shift coincides with half a Coulomb oscillation (fulfilling Eq (1)). The MSET behaves effectively as a p-type SET, Fig 2(e). Therefore we can arbitrarily select the function of the SET from an n-SET for in-plane magnetization φ = 90° to a p-type for out-of-plane magnetization φ = 0°.

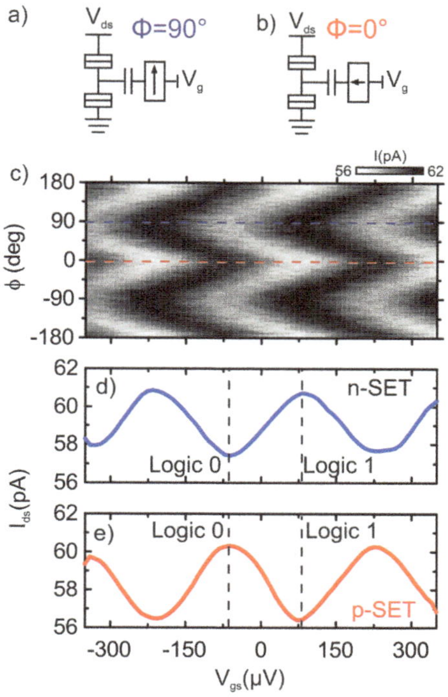

Figure 2: Complementary operation.

Schematic diagram of the MSET for different magnetization orientations. (a) φ = 0° the magnetization is in-plane and (b) φ = 90° the magnetization is out-of-plane. (c) Coulomb blockade oscillations as a function of the direction of the back-gate voltage V_{gs} and the applied magnetic field orientation φ for B = 0.7 T. The-dashed blue and red lines indicate the operating points. (d) MSET Ids-Vgs transfer function at φ = 0°. The logic 0 (1) has been selected at a low (high) current level, n-type SET. (e) MSET Ids-Vgs transfer function at φ = 90°. The logic outputs have been inverted, p-type SET.

http://dx.doi.org/10.1371/journal.pone.0125142.g002

We now focus on the logic design that could be implemented at the single device level. It has been shown that any Boolean function of two variables can be implemented on a SET [25]. In this letter, by making use of the magnetic mode of operation, we demonstrate two sets of reprogrammable logic gates. The inputs in this case are the drain voltage (input A) and the gate voltage (input B) and the output is the drain current. In Fig 3(a) we plot the $V_{ds} - V_{gs}$ diagram of the MSET showing the characteristic Coulomb diamonds at φ = 0°. Overimposed, we schematically show two logic gates framed in red AND and NAND. The output is drawn as en empty dot if the results is 0 (low current level) and a full dot if the results is 1 (high current level). Upon changing the magnetization angle to φ = 90°, the whole diagram is shifted horizontally by ΔV_{gs} = -150 μV and, therefore, at the same input voltages the gate output changes to the two blue-framed logic gates, from (N) AND to (N)OR as depicted by the arrow. The reconfigurable logic gates are demonstrated in the histograms in Fig 3(b)–3(e). In order to discriminate between logic outputs 0 and 1 we select the low(high) current threshold at I_{ds} = 80(90) pA. In Fig 3(b) we represent a histogram of the current output for an AND gate implemented on the edge of the Coulomb diamond. By rotating the magnetization to the in-plane direction (φ = 90°) the Coulomb diamonds shift and the logic gate switches function to OR (Fig 3(c)). Moreover, in Fig 3(d) and 3(e) we demonstrate the logical complement set of reprogrammable Boolean gates. The set switches between NAND (φ = 0°) and NOR (φ = 90°) logic gates.

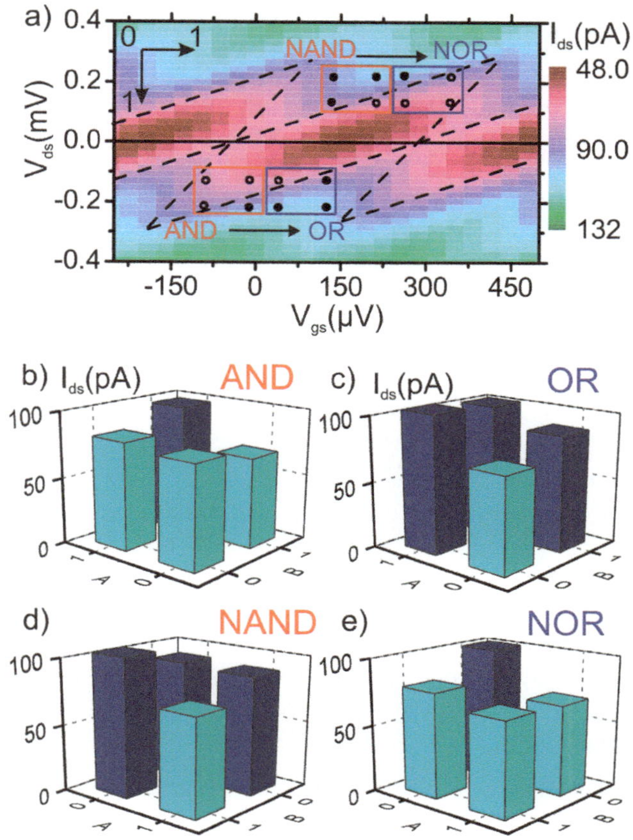

Figure 3: Single-device logic.

(a) $V_{ds} - V_{gs}$ map of the drain current for φ = 0° showing the characteristic Coulomb diamonds. Red and blue frames sketch the implemented logic gates for φ = 0° and 90° respectively. (b-c) AND-OR set of reprogrammable logic gates. AND gate implemented at φ = 0° (b) and OR gate at φ = 90° (c) with V_{ds} (input A) 0(1) defined as −132(−220) μV and V_{gs} (input B) 0(1) defined as −96(0) μV. (d-e) NAND-NOR set of reprogrammable logic gates. NAND gate implemented at φ = 0° (d) and NOR gate at φ = 90° (e) with V_{sd}(input A) 0(1) defined as 220(132) μV and V_{gs} (input B) 0(1) defined as 128(224) μV.

http://dx.doi.org/10.1371/journal.pone.0125142.g003

Finally, we briefly comment on the possibilities for multi-device reconfigurable logic design. As an example, we focus on pull-down networks formed by two elements with a common magnetic back-gate and independently addressable electrical gates. We concentrate on the case of MSET with identical I_{ds}-V_{gs} transfer functions and logic inputs defined by Fig 2(c) and2(d). The circuit in Fig 4(a) consists of a resistor load and two MSETs in series. In the state φ= 0° both MSET are in the p-type state and the network realizes the operation OR. However this gate can be reconfigured by rotating the magnetization, φ = 90°. In this situation both MSET are in the n-type state and the logic operation becomes NAND. Similarly, for a pull-down network consisting of two MSETs in parallel, Fig 4(b) the logic gate can be programmed to switch between AND (φ = 0°) and NOR (φ = 90°) operations.

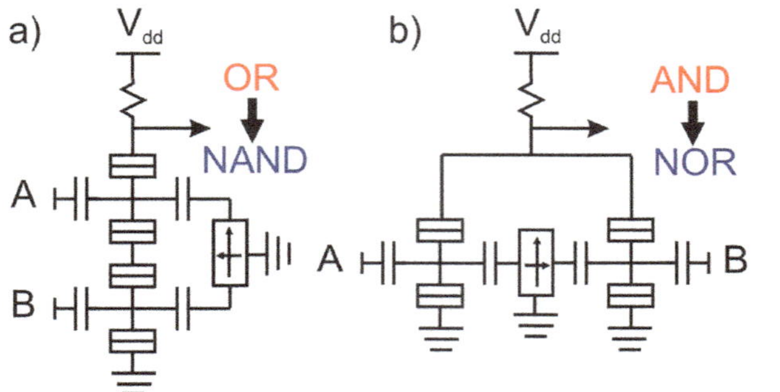

Inputs A	Inputs B	Output 0°	Output 90°
0	0	0	1
0	1	1	1
1	0	1	1
1	1	1	0

Inputs A	Inputs B	Output 0°	Output 90°
0	0	0	1
0	1	0	0
1	0	0	0
1	1	1	0

Figure 4: Logic at the multiple device level considering identical SETs and the logic inputs defined in Fig 2.

The inputs A and B are defined as taken as the SET gate values. (a) A series pull-down network performs the OR operation at $\varphi = 0°$ and NAND at $\varphi = 90°$. (b) Parallel pull-down network performs the AND operation at $\varphi = 0°$ and NOR at $\varphi = 90°$.

http://dx.doi.org/10.1371/journal.pone.0125142.g004

DISCUSSION

In conclusion, we have demonstrated the complementary logic operation of an aluminium MSET by making use of magnetization-dependent work function of the GaMnAs magnetic back-gate. By using the source-drain voltage and the gate voltage as binary inputs we showed a set of Boolean gates and its logical complement implemented at the single device level, AND-OR and NAND-NOR gates. Finally, we have described a route to integrate multiple MSET in to integrated reconfigurable logic circuits. Although here we present results on MSETs, magnetic gating could be exploited to switch function in more conventional circuit elements such as CMOS field-effect transistors or novel electronic devices such as graphene ambipolar transistors [26] or spin-hot carrier transistors [27]. For reliable room temperature switching, magnetic materials with work function shifts of the order of 60 mV will be required [28]. Materials with large work function anisotropy such as CoPt alloys could provide that functionality [21]. Moreover, non-volatile logic reconfiguration could be also achieved by making use of the stability of the magnetic moments along non collinear "magnetic easy axes" orientations without applying external magnetic fields or electrical currents.

ACKNOWLEDGMENTS

The authors thank M. Klein, D.A. Williams for fruitful discussion.

AUTHOR CONTRIBUTIONS

Conceived and designed the experiments: MFGZ CC TJ AF JW. Performed the experiments: CC. Analyzed the data: MFGZ. Contributed reagents/materials/analysis tools: LZ AI RC BG. Wrote the paper: MFGZ CC JW.

REFERENCES

1. Strukov DB, Snider GS, Stewart DR, Williams RS. The missing memristor found. Nature. 2008;453: 80–83. doi: 10. 1038/nature06932. pmid:18451858
2. Joo S, Kim T, Shin SH, Lim JY, Hong J, Song JD, et al. Magnetic-field-controlled reconfigurable semiconductor logic. Nature. 2013;494: 72–76. doi: 10. 1038/nature11817. pmid:23364687
3. Heinzig A, Slesazeck S, Kreupl F, Mikolajick T, Weber WM. Reconfigurable silicon nanowire transistors. Nano Letters. 2012;12: 119–124. doi: 10. 1021/nl203094h. pmid:22111808
4. Yu WJ, Kim UJ, Kang BR, Lee IH, Lee EH, Hee Y, et al. Adaptive logic circuits with doping-free ambipolar carbon nanotube transistors. Nano Letters. 2009;9: 1401–1405. doi: 10. 1021/nl803066v. pmid:19281215
5. Mol JA, Verduijn J, Levine RD, Remacle F, Rogge S. Integrated logic circuits using single-atom transistors PNAS 2011;108: 13969–13972. doi: 10. 1073/pnas. 1109935108
6. Kastner MA. The single-electron transistor. Rev Mod Phys. 1992;64: 849–858. doi: 10. 1103/RevModPhys. 64. 849.
7. Gonzalez-Zalba MF, Heiss D, Podd G, Ferguon AJ. App Phys Lett. 2012;101: 103504. doi: 10. 1063/1. 4750251.
8. Zhu C, Gu Z, Shang L, Dick RP, Knobel RG. Towards an ultra-low-power architecture using single-electron tunneling transistors. Proceedings of the 44th annual Design Automation Conference. 2007; 312.
9. Ono K, Shimada H, Ootuka Y. Enhanced magnetic valve effect and magneto-Coulomb oscillations in ferromagnetic single electron transistors. J Phys Soc Japan. 1997;66: 1261. doi: 10. 1143/JPSJ. 66. 1261.
10. Barnas J, Martinek J, Michalek G, Bulka B R, Fert A. Spin effects in ferromagnetic single-electron transistors. Phys Rev B. 2000;62: 12363. doi: 10. 1103/PhysRevB. 62. 12363.
11. Shirakashi J, Takemura Y. Ferromagnetic single-electron transistor with multiple tunnel junctions. J Mag Soc Japan. 2001;25: 783. doi: 10. 3379/jmsjmag. 25. 783.
12. Jalil MBA, Tan SG, Ma MJ. Enhanced magneto-Coulomb effect in asymmetric ferromagnetic single electron transistors. J App Phys. 2009;105: 07C905 doi: 10. 1063/1. 3055271.
13. Takiguchi M, Shimada H, Mizugaki Y. Correlation between polarity of magnetoresistance ratio and tunnel resistance in ferromagnetic single-electron transistor with superconductive island. J J App Phys. 2014;53: 043101 doi: 10. 7567/JJAP. 53. 043101.
14. Kirchner S, Si Q. Quantum criticality out of equilibrium: Steady state in a magnetic single-electron transistor. Phys Rev Lett. 2009;103: 206401 doi: 10. 1103/PhysRevLett. 103. 206401. pmid:20365995
15. Tucker JR. Complementary digital logic based on Coulomb blockade. J App Phys. 1992;72: 4399–4413. doi: 10. 1063/1. 352206.
16. Ishikuro H, Hiramoto T. Influence of quantum

confinement effects on single electron and single hole transistors. In: Electron Devices Meeting, 1998. IEDM '98. Technical Digest. , International. 1998; 119–122.

17. Uchida K, Koga J, Ohba R, Toriumi A. Programmable single-electron transistor logic for low-power intelligent Si LSI. In: Solid-State Circuits Conference, 2002. Digest of Technical Papers. ISSCC. 2002 IEEE International. 2002;1: 206–460.
18. Wunderlich J, Jungwirth T, Irvine AC, Kaestner B, Shick AB, Campion RP, et al. Coulomb blockade anisotropic magnetoresistance and voltage controlled magnetic switching in a ferromagnetic GaMnAs single electron transistor. J Magn Magn Mater. 2007;310: 1883–1888. doi: 10. 1016/j. jmmm. 2006. 10. 676.
19. Dery H, Dalal P, Cywinski L, Sham LJ. Spin-based logic in semiconductors for reconfigurable large-scale circuits. Nature. 2007;447: 573–576. doi: 10. 1038/nature05833. pmid:17538616
20. Hai PN, Sugahara S, Tanaka M. Reconfigurable logic gates using single-electron spin transistors. Japn J Appl Phys. 2007;46: 6579–6585. doi: 10. 1143/JJAP. 46. 6579.
21. Wunderlich J, Jungwirth T, Kaestner B, Irvine AC, Shick AB, Stone N, et al. Coulomb blockade anisotropic magnetoresistance effect in a (Ga,Mn)As single-electron transistor. Phys Rev Lett. 2006;97: 077201. doi: 10. 1103/PhysRevLett. 97. 077201. pmid:17026268
22. Ciccarelli C, Zarbo LP, Irvine AC, Campion RP, Gallagher BL, Wunderlich J, et al. Spin gating electrical current. App Phys Lett. 2012;101: 122411–122414. doi: 10. 1063/1. 4752013.
23. Beenakker CWJ. Theory of coulomb-blockade oscillations in the conductance of a quantum dot. Phys Rev B. 1991;44: 1646–1656. doi: 10. 1103/PhysRevB. 44. 1646.
24. In-plane rotations between [110] and [1–10] directions were explored leading to $\Delta\mu$ = 46 μeV insufficient for complementary operation.
25. Klein M, Lansbergen GP, Mol JA, Rogge S, Levine RD, Remacle F. Reconfigurable logic devices on a single dopant atom-operation up to a full adder by using electrical spectroscopy. ChemPhysChem 2009;10: 162–173. doi: 10. 1002/cphc. 200800568. pmid:19115322
26. Wang H, Hsu A, Wu J, Kong J, Palacios T. (2010.) Graphene-based ambipolar RF mixers. IEEE Elec Dev Lett. 2010;31: 906 doi: 10. 1109/LED. 2010. 2052017.
27. Mizuno Y, Ohya S, Hai PN, Tanaka M. Spin-dependent transport properties in GaMnAs-based spin hot-carrier transistors. Appl Phys Lett. 2007;90: 62505/1–3 doi: 10. 1063/1. 2724771.
28. Betz AC, Barraud S, Wilmart Q, Placais B, Jehl X, Sanquer M, et al. High frequency characterisation of thermionic charge transport in silicon-on-insulator nanowire transistors. App Phys Lett 2014;104: 043106. doi: 10. 1063/1. 4863538.

CHAPTER 7 LOGSUM USING GARBLED CIRCUITS

José Portêlo[1,2], Bhiksha Raj[3], Isabel Trancoso[1,2]

[1]INESC-ID, Lisbon, Portugal

[2]Instituto Superior Técnico, Universidade de Lisboa, Lisbon, Portugal

[3]Language Technologies Institute, Carnegie Mellon University, Pittsburgh, Pennsylvania, United States of America

ABSTRACT

Secure multiparty computation allows for a set of users to evaluate a particular function over their inputs without revealing the information they possess to each other. Theoretically, this can be achieved using fully homomorphic encryption systems, but so far they remain in the realm of computational impracticability. An alternative is to consider secure function evaluation using homomorphic public-key cryptosystems or Garbled Circuits, the latter being a popular trend in recent times due to important breakthroughs. We propose a technique for computing the logsum operation using Garbled Circuits. This technique relies on replacing the logsum operation with an equivalent piecewise linear approximation, taking advantage of recent advances in efficient methods for both designing and implementing Garbled Circuits. We elaborate on how all the required blocks should be assembled in order to obtain small errors regarding the original logsum operation and very fast execution times.

INTRODUCTION

An increasing number of server-based applications perform tasks such as classification, processing and analysis of user data. Often, these data are private, and should not be exposed to the server. Likewise, the server's inputs to these computations are also private and may not be exposed to the user. Hence, it becomes necessary to perform the computations in a privacy preserving manner, such that the user's data and the server's inputs are not revealed to one another, while yet ensuring that the appropriate party gets the correct result from the computations.

Stated in abstract terms, a user Alice possesses data $\mathbf{x}$ (e.g. email, a biometric measurement, or a voice sample). System Bob is capable of computing a function $f(\mathbf{x};\theta)$ with parameter θ(e.g. a spam filter or a biometric classifier). Alice desires to obtain $f(\mathbf{x};\theta)$ from Bob; however, she is unwilling to expose $\mathbf{x}$ to him. Bob too is unwilling to expose θ to Alice. The challenge of privacy-preserving computation is to enable Alice to obtain $f(\mathbf{x};\theta)$ such that Bob learns nothing of $\mathbf{x}$ and Alice learns nothing more of θ besides what she may glean from $f(\mathbf{x};\theta)$ itself.

In principle, such transactions could be achieved using fully homomorphic encryption [1], which permits computation to be performed on encrypted data. However, practical fully homomorphic encryption schemes remain computationally impractical in spite of recent advances [2, 3].

Instead, Alice and Bob must achieve their goals through secure function evaluation (SFE) [4,5]. SFE involves repeated

Citation: Portêlo J, Raj B, Trancoso I (2015) Logsum Using Garbled Circuits. PLoS ONE 10(3): e0122236. doi:10.1371/journal.pone.0122236.

exchange of partial information between Alice and Bob using protocols which mask the exchanged values using a combination of techniques such as public-key encryption [6, 7], oblivious transfer (OT) [8], etc., until the appropriate party gets the desired result.

Arguably the most popular approach to SFE is to cast the function to be evaluated as agarbled Boolean circuit (GC) [4, 9, 10]. Like logic circuits, GCs are composed of logic gates, with each gate represented by a truth table. However, while in regular logic circuits the inputs and outputs of each truth table are binary values (0/1), in GCs each possible entry of the truth table is an encrypted value which can only be "unlocked" by a input-specific key pair. The entries, in turn, too are keys which may be used to unlock other gates downstream the circuit.

Although in principle any function could be computed in this manner, practical realizations face computational challenges. Evaluation of a GC requires exchange of information through OT and repeated decryption of gate outputs, both of which are highly expensive operations. Ideally the circuit itself must be designed to minimize this overhead.

Traditionally, circuit optimization has largely been viewed as the problem of minimizing the number of gates in the circuit. The problem of finding the smallest circuit to perform a given Boolean function is known to be intractable in general [11], although a number of effective algorithms have nevertheless been proposed [12–15].

In the case of GCs, we must also consider other factors. Not all gates are equally expensive to compute: gates such as NOT and XOR do not require explicit encryption in a GC and are orders of magnitude cheaper to compute than other gates, and must preferentially be used. Elsewhere, where feasible, complex operations may need to be replaced with cheaper approximations or or encrypted table lookups. GCs do not support looping operations. In addition, in situations where the underlying computation is floating-point, the effect of the loss of resolution from fixed-point implementation must also be considered.

The design of Garbled Circuits is thus a complex problem. A number of toolkits have been developed to support the conversion of algorithmic circuits to their garbled forms [16–19]. Typically, they include several optimized garbled components for "primitives", such as multiplication, maximum-value computation, vector inner-products, etc., which may be employed in the design of larger circuits. All of these primitives are simple, in the sense that they embody linear arithmetic computation over one or more inputs.

In this paper we develop a GC for a very important complex non-linear primitive, the logsumoperation, a key building block in many statistical and signal processing computations. We build upon previous work [20, 21]. In [20] the authors use a GC to compute the logarithm as an integer operation, i.e., for input $a > 0$, the logarithm is computed as $b = \lfloor \log_2 a \rfloor + 1$. Because bis always an integer approximating the real value of the logarithm, this technique would lead to large errors if it were used for computing the logsum operation. In [21] the authors use a homomorphic encryption system to compute the logsum by successively performing secure protocols for exponentiation and logarithms. While this technique has reduced error with appropriate pre-scaling of the inputs, the encryption imposes large computational overhead and performing all the required protocols requires substantial amounts of time. In order to avoid the large computation errors or large execution times of the previous approaches, our design casts the logsum operation as a combination of simple operations such as table lookups, additions and multiplications, arranged to minimize the loss of resolution in the output while simultaneously retaining efficiency through the use of low-cost computational units, such that the actual garbled computation may be performed in reasonable time.

This paper starts with necessarily brief overviews of the different operations involved in logsum computations, and the basic idea underlying Garbled Circuit protocols. We propose a new method for computing the logsum operation using GCs, starting with an array of just two values, and later generalizing it to any number of input values. The performance of the proposed method is then analyzed for two different GC toolkits in terms of execution time for different numbers of input values, different lengths (in bits) of these values, and different operation parameters. The paper ends with some concluding remarks.

METHODS

Logsum Operation

In many signal processing and pattern classification problems it is necessary to compute the occurrence probability of an event X. The probability models employed are commonly mixture models of the form $P(X) = \sum_{i=1}^{N} P(X, Y_i)$. In order to avoid underflow/overflow problems, the actual computations are usually performed with log probabilities. Rather than computing P(X), one works with logP(X), which is in turn expressed as in terms of the logarithm of the component probabilities as

$$z(m) = \log\left(\sum_{i=1}^{N} \exp(m_i)\right), \quad (1)$$

where z(m) = log(P(X)), and $m_i = \log(P(X,Y_i))$. The equation above is a logsum operation. To further avoid overflow/underflow problems that arise when all the m_i terms become too large in magnitude, Equation 1 is usually further recast as

$$z(m) = m_X + \log\left(\sum_{i=1}^{N} \exp(m_i - m_X)\right), \quad (2)$$

with $m_X = \max_i m_i$.

Logsum operations are found in a variety of signal processing and statistical computation tasks, such as sparse signal recovery [22], matrix recovery [23], and in the evaluation of a large variety of statistical models such as latent Dirichlet allocation [24] and Gaussian mixture models [25]. Thus, the logsum operation may in fact be considered one of the key building blocks in statistical and signal processing computations.

Garbled Circuits

Garbled Circuits are best explained with an example. Bob has Boolean input a and Alice has inputs b and c. They wish to compute the function f(a,b,c) = (a⊕b)∧c. The logic circuit forf(a,b,c) and the truth tables for the gates in it are shown in Fig. 1. However, Bob does not want to reveal a to Alice, and Alice does not want to reveal b or c to Bob. In order to still be able to compute f(a,b,c), they must hence garble the circuit.

a	b	r
0	0	0
0	1	1
1	0	1
1	1	0

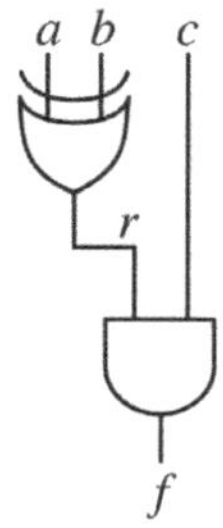

r	c	f
0	0	0
0	1	0
1	0	0
1	1	1

Figure 1: Example of a logic circuit and corresponding truth tables.

The circuit implements an XOR gate (r = a⊕b) followed by an AND gate (f = r∧c). The binary truth tables for the XOR and AND gates are presented on the left hand side and right hand side of the circuit, respectively.

http://dx.doi.org/10.1371/journal.pone.0122236.g001

Bob generates the logic circuit implementing f(·). Thereafter, for each input and intermediate value (a, b, c, r in the example), Bob generates two private keys, one for each bit value, totaling eight keys in our example: $K_a^{0/1}, K_b^{0/1}, K_c^{0/1}, K_r^{0/1}$. For gates that generate intermediate values (r = a⊕b in our example), he replaces the outputs of the truth table with the encryption of the key corresponding to the output, performed with the keys corresponding to the inputs. For example, for a = 0, b = 1 the output is r = 1; the corresponding input keys are K_a^0, K_b^1 and the encrypted output is $\mathcal{E}_{K_a^0}(\mathcal{E}_{K_b^1}(K_r^1))$. For the output gate (f = r∧c in our example) he encrypts the output value itself. The garbled values for our example are presented in Table 1.

Table 1: Example of garbled truth tables

http://dx.doi.org/10.1371/journal.pone.0122236.t001

a	*b*	*r*
K_a^0	K_b^0	$\mathcal{E}_{K_a^0}(\mathcal{E}_{K_b^0}(K_r^0))$
K_a^0	K_b^1	$\mathcal{E}_{K_a^0}(\mathcal{E}_{K_b^1}(K_r^1))$
K_a^1	K_b^0	$\mathcal{E}_{K_a^1}(\mathcal{E}_{K_b^0}(K_r^1))$
K_a^1	K_b^1	$\mathcal{E}_{K_a^1}(\mathcal{E}_{K_b^1}(K_r^0))$
r	***c***	***f***
K_r^0	K_c^0	$\mathcal{E}_{K_r^0}(\mathcal{E}_{K_c^0}(0))$
K_r^0	K_c^1	$\mathcal{E}_{K_r^0}(\mathcal{E}_{K_c^1}(0))$
K_r^1	K_c^0	$\mathcal{E}_{K_r^1}(\mathcal{E}_{K_c^0}(0))$
K_r^1	K_c^1	$\mathcal{E}_{K_r^1}(\mathcal{E}_{K_c^1}(1))$

[Left] Garbled table for the XOR gate ($r = a \oplus b$). The garbled values of the keys $K_r^{0/1}$ associated with output r, $\mathcal{E}_{K_a^{0/1}}(\mathcal{E}_{K_b^{0/1}}(K_r^{0/1}))$, are encrypted using the keys $K_a^{0/1}$ and $K_b^{0/1}$ associated with inputs *a* and *b*, respectively. [Right] Garbled table for the AND gate ($f = r \wedge c$). The garbled values of output f, $\mathcal{E}_{K_r^{0/1}}(\mathcal{E}_{K_c^{0/1}}(0/1))$, are encrypted using the keys $K_r^{0/1}$ and $K_c^{0/1}$ associated with inputs *r* and *c*, respectively.

doi:10.1371/journal.pone.0122236.t001

To compute the function Bob transmits the keys corresponding to his bit choice for his inputs $K_a^?$ to Alice. Alice recovers the keys corresponding to the bit choice for her inputs $K_b^?, K_c^?$, as well as the truth tables for the gates of the circuit from Bob using oblivious transfer [8]. Because of the properties of OT, Bob does not learn either the value of Alice's bits or the portion of the truth table she actually needs to perform her computations. To evaluate the circuit Alice successively evaluates each of the gates in the circuit. The nature of the computation is such that for each gate, Alice will possess two keys, one for each input. She decrypts all four encrypted output in the truth table for the gate, using her two keys. The keys will be inappropriate for three of the four outputs; hence Alice can only correctly decipher one of the four encrypted values, which will be a key for the next gate. After repeating this process for all gates, she finally obtains the desired value f. Alice and Bob never learn each others' inputs.

For a long time it was believed that GCs were of purely theoretical interest, but many advances have made GC much more efficient and practical to use. For instance, all the OTs for transferring the input ciphers may be pre-computed [26], meaning that all the computationally expensive operations regarding the OT are performed offline, and only an additional yet simple operation per OT is performed while evaluating the GC. Furthermore, all the OT may be computed efficiently by using shorter ciphers [27, 28], which can be implemented using elliptic curve cryptography [29]. In the gate decryption phase, the point and permute technique [16] may be used to reduce the total number of required decryptions from 4 to 1 by associating a permutation bit to the input ciphers chosen using OT. Finally, all the XOR gates of a GC may be evaluated for "free" [30], i.e., without significant computational cost, as the XOR gate does not need a garbled table and its evaluation consists of XOR-ing its garbled input values. NOT gates are similarly "free" as well. Custom designing circuit operations to preferentially utilize these gates will result in faster evaluation of the resulting Garbled Circuit [31].

On top of these implementation tricks, other useful properties of GC have been found. For example, it is possible to reuse the same GC several times under some conditions [32] by using functional encryption [33, 34], which means that if the same operation must be computed several times we only need a single circuit instead of one circuit for each time we want to compute that function. Moreover, several privacy and security guarantees regarding GC have been proven [9, 35–37].

In implementing the basic operations using GC, some issues different from the ones faced when using hardware circuits arise. Decision making operations such as comparisons (greater/lesser-than, maximum/minimum) and multiplexing can be implemented similarly to digital circuits. However, branching (if-then-else) and cycles/recursion (for loop, do-while) should be avoided. The first requires building and evaluating circuits for all possibilities in order to hide which one was actually chosen. The second requires explicitly expanding the cycles, one circuit per iteration, which may result in a huge intractable circuit. Another problem arises when computing non-linear functions, since the usual trick of computing the initial elements of the corresponding Taylor series is impractical to implement using GCs. An efficient work-around is to replace these operations with piecewise linear approximations (PLA), whose coefficients can be stored in look-up tables. An additional restriction exists on the number representation of the ciphered values. Operations on floating-point values usually require some normalization on them before they are performed, introducing an additional and unnecessary computational cost. To counter this, a fixed-point representation is preferred, despite a possible loss of precision in representing those same values. Finally, given the wide-spread popularity of GCs, a huge variety of software implementations of all the required protocols and encryption systems are available. One of the first practical implementations of GC was Fairplay [16], and since then a number of different implementations have been developed in order to further improve their efficiency and practicality [17–19].

Logsum Operation using Garbled Circuits

We now show how to perform the computation of the logsum of an array of values using GC. We start by describing how the logsum operation can be computed using a piecewise linear approximation, then we describe the full circuit for the simple situation where there are only two input values, and finally we generalize it to any number of input values.

Piecewise Linear Approximation.

Many algorithms rely on evaluating non-linear functions in order to compute an output. These functions are normally replaced by the corresponding Taylor series, whose terms are computed until the error between iterations falls below a pre-determined threshold. However, when limited computational resources are available, the non-linear function is often be replaced by a PLA. A method for computing piecewise linear approximations using GCs has already been proposed [38], but it only focuses on a 1st-order approach (ramp quantization (RQ)), discarding a 0th-order approach (step quantization (SQ)). It is nevertheless instructive to evaluate both options in the context of GCs, where computation is at a premium. Although it is expected that the error obtained with the SQ approach should be larger than the one obtained with the RQ approach, the computational benefits of the lower-order approximation can outweigh the increased error from loss of resolution.

A standard way to implement a PLA is to split the domain of the input function f(·) into kintervals, and then compute linear approximations f(·) for each interval, adjusting all the interval limits t_i and linearization parameters $n_{1,i}$ and $n_{2,i}$ in order to obtain the minimum error possible regarding f(·). These parameters are then placed in the form of a look-up table such as the one presented in Table 2.

Table 2: Look-up table with k entries for the piecewise linear approximation

http://dx.doi.org/10.1371/journal.pone.0122236.t002

$m{:}t$	$f(m){:}n_1 \cdot m+n_2$	
$m \leq t_1$	$n_{1,1}$	$n_{2,1}$
$t_1 < m \leq t_2$	$n_{1,2}$	$n_{2,2}$
…	…	…
$t_{k-2} < m \leq t_{k-1}$	$n_{1,k-1}$	$n_{2,k-1}$
$t_{k-1} < m$	$n_{1,k}$	$n_{2,k}$

Note that the domain of the ith interval is given by $]t_{i-1},t_i]$.

doi:10.1371/journal.pone.0122236.t002

For obtaining the correct parameter pair $(n_{1,i},n_{2,i})$ for each value of m, it is necessary to compare m with all the t in order to find an index j such that $t_j < m < t_{j+1}$, allowing for computing $f_{PLA}(m) = n_{1,j+1} \cdot m + n_{2,j+1} \approx f(m)$, an approximation of the desired value. In the situations where $m < t_1$ or $m > t_{k-1}$, the approximation of the desired value is obtained by computing $f_{PLA}(m) = n_{1,1} \cdot m + n_{2,1}$ or $f_{PLA}(m) = n_{1,k} \cdot m + n_{2,k}$, respectively. This last scenario is not unlikely, since many functions have an unbounded domain, meaning that it is necessary to force artificial bounds when optimizing the locations of t. Therefore, special care is required while computing $n_{*,1}$ and $n_{*,k}$, so that they do not provide increasingly erroneous outputs as the difference between m and t_1 or t_{k-1} increases.

Higher Order Approximations

We did not consider any higher order polynomial approximations for the logsum operation for a variety of reasons. Let d be the degree of the approximation polynomial for the logsum operation. Preliminary experiments showed that although there was some small improvement in terms of the obtained error as d increased, the improvement rate decreased sharply. In particular, for $d \geq 3$ (cubic and higher order approximations) the error obtained is almost the same as for the situation where $d = 2$ (quadratic approximation). Also, increasing d leads to a quadratic increase in the number of required multiplications for evaluating the look-up table. Even if $d = 2$ is considered, this would require computing three multiplications instead of just one. Since the multiplication is by far the most computationally demanding operation in our GC approach, increasing d would lead to a significant and undesirable increase in the overall execution time. Finally, and most importantly, increasing d leads to having to perform several multiplications in cascade, which, given a GC approach, causes an exponential increase of the number of bits required to represent the output.

Logsum of $N = 2$ values.

Recalling Equation 2, the logsum of an array of numbers can be written as

$$z(m) = m_X + \log\left(\sum_{i=1}^{N} \exp\left(m_i - m_X\right)\right), \tag{3}$$

where $m_X = \max_i m_i$. For the simplest case where $N = 2$, this equation can be simplified to

$$z(m) = m_X + \log(1 + \exp(m_N - m_X)), \tag{4}$$

with $m_X = \max(m_1,m_2)$, $m_N = \min(m_1,m_2)$. After performing PLA, the approximation for the logsum operation is given by

$$z(m) \approx m_X + (n_1 \cdot (m_N - m_X) + n_2). \tag{5}$$

This is the equation implemented in our approach. Fig. 2 contains a circuit diagram showing how the logsum is computed using GC, with $N = 2$ inputs and $k = 4$ look-up table entries.

There are three major blocks in this diagram, separated from each other by horizontal dotted lines. In each block, the individual boxes correspond to a simple operation. The first block contains the computation of all the necessary elements for obtaining the logsum z(m), namely m_X, m_N and $m_{diff} = m_N - m_X$, by means of computing a minimum (MIN), a maximum (MAX) and a subtraction (SUB), respectively. The second block contains the circuit for obtaining the parameters for the PLA of the logsum operation. It starts by performing comparisons (CMP) between m_{diff} and the look-up table entries t, then the corresponding decisions i are used as control inputs to the hierarchical multiplexer (4-MUX), and finally the unknown piecewise linear approximation parameters$n_{1,?}$ and $n_{2,?}$ are outputted. The third block performs the linear approximation of the logsum operation, outputting the final result. It performs a multiplication (MUL) and an addition (ADD) between m_{diff} and $n_{1,?}$ and $n_{2,?}$, respectively. The final operation is an addition between m_X and $f_{PLA}(m)$, thus completing the computation of the logsum.

http://dx.doi.org/10.1371/journal.pone.0122236.g002

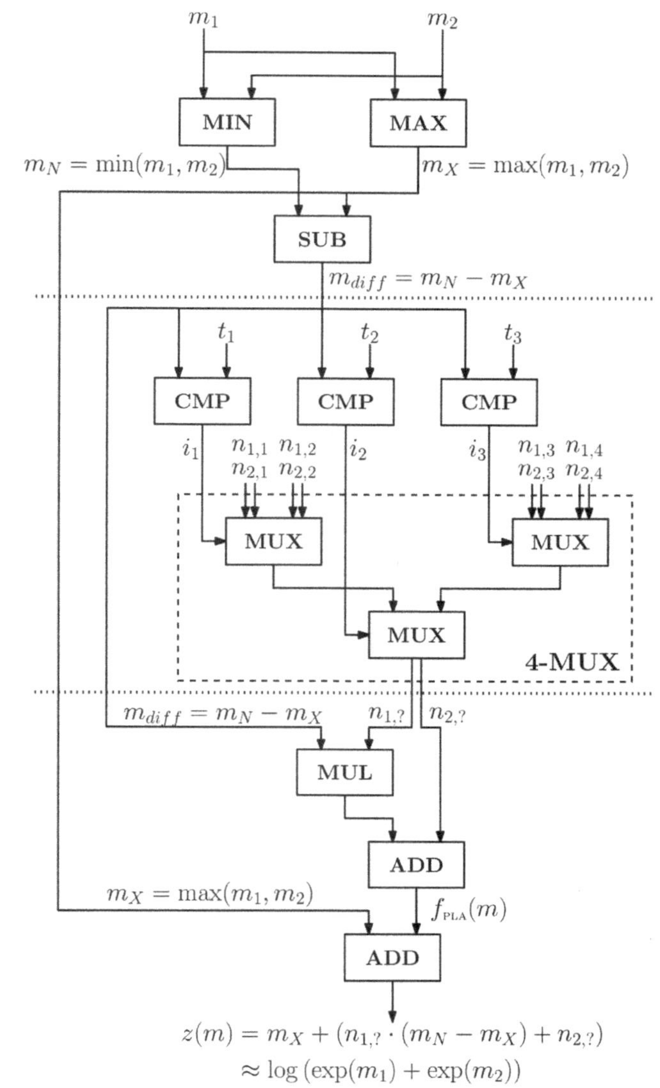

Figure 2: Circuit diagram for the logsum operation with N = 2 inputs and k = 4 look-up table entries.

This diagram raises several implementation concerns. One of the most relevant is the number of bits (ℓ)—how it increases with each computation and what can be done to keep it as small as possible, since more bits mean more computational costs and consequently longer execution times. The operations where the number of bits of the output is larger than the number of bits of the inputs are ADD (1 extra bit), SUB (1 extra bit) and MUL (ℓ extra bits), so the other operations will be ignored in this analysis. Regarding the SUB operation, since the output will always be a negative number, the sign bit is uninformative, and therefore can be discarded. As for the MUL and the first ADD operations, the ℓ least significant bits of $f_{PLA}(m)$ need to be truncated, as otherwise one may end up with a scaling factor different from the one of m_X (recall that all numbers are being represented with fixed-point instead of floating-point). We also remove the most significant bit, as preliminary experiments suggest that it never contains any useful information, i.e., it is always 0. On the last ADD we do not remove any bits, thus increasing the total number of bits by 1 every time the logsum operation is computed.

A second concern is the choice between the SQ and RQ approaches to the PLA. Both of them were implemented following the diagram in Fig. 2, but whereas the RQ approach requires all the blocks, for the SQ approach the MUL and the first ADD block can be ignored, since in this situation $n_{1,?} = 0$. This also means that for the SQ approach there is no need to reduce the number of bits of the output.

Logsum of any N Values

The Garbled Circuit described previously implements the logsum operation for the simplest case of N = 2 input values. Here we present the generalization of this methods to any given N. Let us start by recalling the original formulation of the logsum in Equation 3, which for N = 4 can be written as

$$\begin{aligned} z(m) &= \log\left(\exp(m_1) + \exp(m_2) + \exp(m_3) + \exp(m_4)\right) \\ &= \log\left(e^{\log(\exp(m_1)+\exp(m_2))} + e^{\log(\exp(m_3)+\exp(m_4))}\right) \\ &= \log\left(e^{z(m')} + e^{z(m'')}\right), \end{aligned} \qquad (6)$$

with $m' = (m_1, m_2)$ and $m'' = (m_3, m_4)$. This means that the logsum for N = 4 input values can be obtained by computing two logsum operations, each of two input values. This process can be repeated as many times as needed, leading to a hierarchical tree structure for computing the logsum operation, as illustrated in Fig. 3. With this structure, the difference in the number of bits between the output and the inputs is $p = \lceil \log_2 N \rceil$.

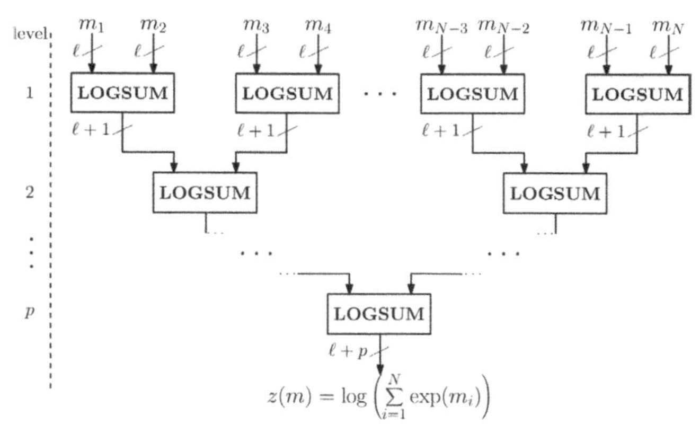

Figure 3: Circuit diagram for the logsum operation for any N inputs.

The logsum of N elements is computed using a hierarchical structure, computing the logsum of two elements at a time. After each level the number of bits required to represent the output increases by 1. If the inputs are represented by ℓ bits, the output will be represented by ℓ+p bits, with $p = \lceil \log_2 N \rceil$.

http://dx.doi.org/10.1371/journal.pone.0122236.g003

RESULTS AND DISCUSSION

In order to evaluate the performance of our implementation of the logsum operation, two sets of experiments were performed. The first set was designed to analyze the effect in the absolute error of changing the number of bits of the input (ℓ) and the number of entries on the look-up table (k) when computing the logsum of N = 2 input values. The second set was designed to analyze the effect in terms of absolute error and execution time of computing the logsum of input arrays of different sizes N, for selected values of ℓ and k. We considered two different GC toolkits: the one from the VIPP group [18] and the JustGarble toolkit [19]. Both toolkits implement many techniques mentioned earlier for improving the efficiency of GC, namely pre-computing OT, using the permute technique for gate decryption, free XOR gate evaluation, etc. However, the VIPP toolkit was designed for performing tasks where only a very small number of gates (i.e. a few thousand) are required and not-so-small execution times (i.e. a few seconds) are acceptable. Examples of such tasks include the approaches for iris matching and ECG signal quality evaluation, described in [39] and [20], respectively. Therefore, the VIPP toolkit was not optimized in terms of circuit file size and memory usage. As for the JustGarble toolkit, it was designed in order to both maximize the number of gates that can be used in a single circuit and minimize execution time and memory usage. This means that special care was taken in terms of memory management during the software implementation phase, which has a profound impact in terms of the overall execution time. The advantage of the VIPP toolkit is that, unlike the JustGarble toolkit, it comes with efficient basic block designs in terms of the amount of non-XOR gates for most of the operations we require for computing the logsum. The errors obtained in all experiments are always the same regardless of the GC toolkit considered; the difference occurs only when the execution times are analyzed.

The results obtained in terms of error in the first experiment for both the step quantization (SQ) and the ramp quantization (RQ) approaches are presented in Tables 3 and 4, respectively. In every cell of both tables, the first and second rows contain the mean and maximum absolute error with respect to the true value of the logsum, respectively. Each experiment was performed by generating one million uniformly distributed pairs of numbers and computing the logsum over each pair, both with our approach and the exact equation. Analyzing the results for the SQ approach, we notice that the error obtained depends much less on ℓ than on k, as increasing the number of bits beyond $\ell = 12$ does not visibly reduce the error, but increasing kleads to a significant reduction of the error, specially for larger values of ℓ. This is expected, since the maximum error in this approach is directly governed by the size of the steps, which is much greater than the precision of the values. In the RQ approach, both increasing ℓ and kleads to large reductions in the error. In some cases, even small changes in the parameters result in differences of up to one order of magnitude. Although the error obtained using the SQ approach may be expected to always be larger than the one obtained using the RQ approach, for $\ell = 8$ this is not the case. The reason may be that the effect of removing the least significant bits after the MUL block is more significant if there is a small number of bits to start with.

Table 3: Absolute error for different values of ℓ and k; step quantization (SQ) approach

http://dx.doi.org/10.1371/journal.pone.0122236.t003

	k		
ℓ	**32**	**64**	**128**
8	1.17×10^{-2}	1.01×10^{-2}	1.03×10^{-2}
	(5.49×10^{-2})	(3.74×10^{-2})	(4.05×10^{-2})
12	0.53×10^{-2}	0.28×10^{-2}	0.15×10^{-2}
	(2.74×10^{-2})	(1.52×10^{-2})	(0.89×10^{-2})
16	0.52×10^{-2}	0.27×10^{-2}	0.13×10^{-2}
	(2.68×10^{-2})	(1.46×10^{-2})	(0.83×10^{-2})
24	0.52×10^{-2}	0.27×10^{-2}	0.13×10^{-2}
	(2.68×10^{-2})	(1.45×10^{-2})	(0.82×10^{-2})
32	0.52×10^{-2}	0.27×10^{-2}	0.13×10^{-2}
	(2.68×10^{-2})	(1.45×10^{-2})	(0.82×10^{-2})

ℓ corresponds to the number of bits representing the inputs and *k* is the number of entries of the look-up table for the piecewise linear approximation. In each cell, the first and second rows contain the mean and maximum absolute error with respect to the true value of the logsum, respectively.

doi:10.1371/journal.pone.0122236.t003

Table 4: Absolute error for different values of ℓ and k; ramp quantization (RQ) approach

http://dx.doi.org/10.1371/journal.pone.0122236.t004

	k		
ℓ	**32**	**64**	**128**
8	1.69×10^{-2}	1.66×10^{-2}	1.57×10^{-2}
	(5.02×10^{-2})	(5.42×10^{-2})	(5.01×10^{-2})
12	1.10×10^{-3}	1.01×10^{-3}	1.02×10^{-3}
	(3.58×10^{-3})	(3.33×10^{-3})	(3.50×10^{-3})
16	1.45×10^{-4}	0.77×10^{-4}	0.68×10^{-4}
	(4.40×10^{-4})	(2.66×10^{-4})	(2.30×10^{-4})
24	1.20×10^{-4}	0.30×10^{-4}	0.08×10^{-4}
	(2.54×10^{-4})	(0.66×10^{-4})	(0.20×10^{-4})
32	1.20×10^{-4}	0.30×10^{-4}	0.07×10^{-4}
	(2.54×10^{-4})	(0.66×10^{-4})	(0.19×10^{-4})

ℓ corresponds to the number of bits representing the inputs and *k* is the number of entries of the look-up table for the piecewise linear approximation. In each cell, the first and second rows contain the mean and maximum absolute error with respect to the true value of the logsum, respectively.

doi:10.1371/journal.pone.0122236.t004

The results obtained in terms of error in the second experiment for the SQ and RQ approaches are presented in Tables 5 and 6, respectively. Similarly to the previous experiment, in every cell of all the tables the first and second rows contain the mean and maximum absolute error regarding the logsum, respectively, and one million uniformly distributed arrays of size N were considered. Regarding the SQ approach, we observe that increasing N leads to only slightly increasing the error accumulated by successively computing the logsum, meaning that this technique scales well with N for all values of ℓ and k. Observing the results for the RQ approach, we notice that it only scales efficiently with N for larger values of ℓ and k. However, since this is the most interesting situation because it is the one where smaller errors are obtained, scenarios with small ℓ and k may be ignored.

Table 5: Absolute error for selected values of N, ℓ and k; step quantization (SQ) approach

http://dx.doi.org/10.1371/journal.pone.0122236.t005

N	8	32	128	512
ℓ = 8, *k* = 32	0.17×10^{-1}	0.24×10^{-1}	0.41×10^{-1}	0.66×10^{-1}
	(1.00×10^{-1})	(1.14×10^{-1})	(1.22×10^{-1})	(1.41×10^{-1})
ℓ = 12, *k* = 128	0.27×10^{-2}	0.35×10^{-2}	0.40×10^{-2}	0.46×10^{-2}
	(1.69×10^{-2})	(1.80×10^{-2})	(2.13×10^{-2})	(2.32×10^{-2})
ℓ = 24, *k* = 128	0.26×10^{-2}	0.32×10^{-2}	0.35×10^{-2}	0.37×10^{-2}
	(1.39×10^{-2})	(1.79×10^{-2})	(1.84×10^{-2})	(2.06×10^{-2})

N corresponds to the number of inputs for the logsum, ℓ corresponds to the number of bits representing the inputs and *k* is the number of entries of the look-up table for the piecewise linear approximation. In each cell, the first and second rows contain the mean and maximum absolute error with respect to the true value of the logsum, respectively.

doi:10.1371/journal.pone.0122236.t005

Table 6: Absolute error for selected values of N, ℓ and k; ramp quantization (RQ) approach

http://dx.doi.org/10.1371/journal.pone.0122236.t006

N	8	32	128	512
ℓ = 8, *k* = 32	0.48×10^{-1}	0.79×10^{-1}	1.11×10^{-1}	1.43×10^{-1}
	(1.12×10^{-1})	(1.37×10^{-1})	(1.66×10^{-1})	(1.90×10^{-1})
ℓ = 12, *k* = 128	0.28×10^{-2}	0.48×10^{-2}	0.68×10^{-2}	0.89×10^{-2}
	(0.69×10^{-2})	(0.83×10^{-2})	(1.01×10^{-2})	(1.20×10^{-2})
ℓ = 24, *k* = 128	1.10×10^{-5}	1.40×10^{-5}	1.80×10^{-5}	2.40×10^{-5}
	(4.50×10^{-5})	(5.40×10^{-5})	(6.50×10^{-5})	(7.00×10^{-5})

N corresponds to the number of inputs for the logsum, ℓ corresponds to the number of bits representing the inputs and *k* is the number of entries of the look-up table for the piecewise linear approximation. In each cell, the first and second rows contain the mean and maximum absolute error with respect to the true value of the logsum, respectively.

doi:10.1371/journal.pone.0122236.t006

We now analyze the execution time of our approach using GC. Currently, it takes a few nanoseconds for a microprocessor to compute a logsum over N = 2 elements. Because of the additional operations required by GC (encryption/decryption, shuffling, etc.), our approach requires substantially larger amounts of time to perform the logsum compared to the non-private counterpart. Nevertheless, for many real-time applications, this approach is still fast enough. Taking advantage of all techniques for efficient implementation of GC, all steps but the actual circuit evaluation can be performed offline; therefore, we only present the execution times for that particular step. We confirmed experimentally that if these techniques were not considered, the overall execution times would increase at least by one order of magnitude.

We considered two different processor and memory settings for our experiments: 1) an Intel Core i7-3630QM CPU @ 2.40GHz with a 6MB L3 cache memory and an 8GB DDR3 RAM memory and 2) an Intel Xeon E5530 CPU @ 2.40GHz with a 8MB L3 cache memory and a 48GB DDR3 RAM memory. The results obtained for the SQ and RQ approaches for the first setting are presented in Figs. 4 and 5, respectively. The execution times obtained for both GC toolkits when the second setting was considered are around 10% to 15% slower than the ones from the first setting, but have otherwise a very similar behavior to them. Therefore, and in order to avoid overloading this section, we decided not to reproduce them. Each of the results was obtained by averaging the execution time of one thousand consecutive runs of the logsum algorithm. As expected, for both GC toolkits and using the same values of ℓ and k, the RQ approach takes longer than the SQ approach to compute a logsum. We observe that for the VIPP toolkit there is a near quadratic increase of the execution time with increasing values ofN, probably a consequence of the successively larger circuits that need to be loaded into memory. We also notice that for the JustGarble toolkit there is a constant linear relationship between how long it takes to compute the logsum and the number of inputs for all values of ℓ and k, meaning that our approach also scales nicely with N regarding the execution time. Finally and more importantly, the JustGarble toolkit is consistently two or three orders of magnitude faster than the VIPP toolkit, definitely showing that when larger garbled circuits are considered, it is extremely important to use a toolkit specially designed for optimal memory management.

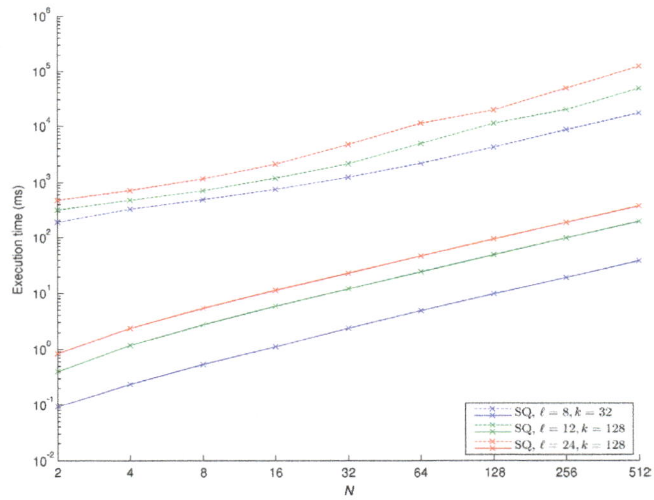

Figure 4: Execution time for selected values of N, ℓ and k, step quantization (SQ) approach, when an Intel Core i7-3630QM CPU @ 2.40GHz with a 6MB L3 cache memory and an 8GB DDR3 RAM memory is considered.

N corresponds to the number of inputs for the logsum, ℓ corresponds to the number of bits representing the inputs and k is the number of entries of the look-up table for the piecewise linear approximation. The dashed lines correspond to the experiments performed with the VIPP toolkit; the solid lines correspond to the experiments performed with the JustGarble toolkit.

http://dx.doi.org/10.1371/journal.pone.0122236.g004

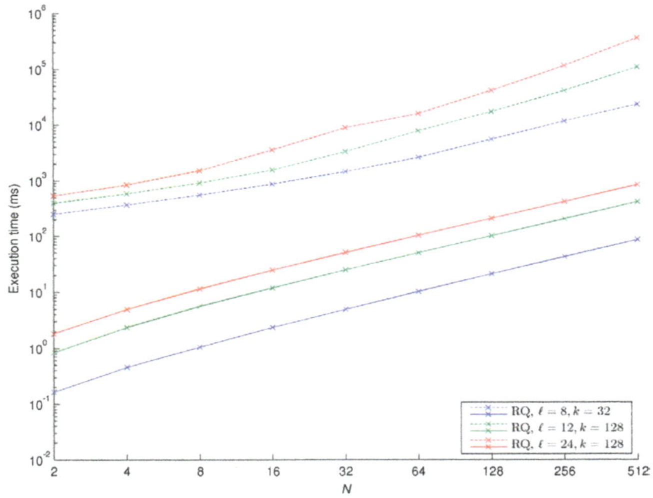

Figure 5: Execution time for selected values of N, ℓ and k, ramp quantization (RQ) approach, when an Intel Core i7-3630QM CPU @ 2.40GHz with a 6MB L3 cache memory and an 8GB DDR3 RAM memory is considered.

N corresponds to the number of inputs for the logsum, ℓ corresponds to the number of bits representing the inputs and k is the number of entries of the look-up table for the piecewise linear approximation. The dashed lines correspond to the experiments performed with the VIPP toolkit; the solid lines correspond to the experiments performed with the JustGarble toolkit.

http://dx.doi.org/10.1371/journal.pone.0122236.g005

CONCLUSIONS

This paper presents a technique for computing the logsum operation using Garbled Circuits. We describe a simple technique that approximates the logsum with a piecewise linear approximation, which is then expressed in the form of a garbled circuit that can be computed efficiently, while maintaining the privacy of its inputs. Furthermore, even within this format, unlike previous approaches, the solution is scalable, in the sense that the accuracy of the approximation can be traded off for computation time by increasing the parameters ℓ and k. The SQ and RQ options offer the following choice—the SQ approach can be used for fast computation where the underlying problem can tolerate lower precision, and RQ can be used where greater precision is required. Even with the increased computation, this method is still much faster than previously proposed homomorphic encryption approaches. We evaluated two different GC toolkits designed for distinctive ends, and concluded that memory management should be a prime concern when implementing a general purpose GC toolkit. The logsum operation is an important cornerstone for many signal processing applications. The technique proposed here presents an ideal primitive to be incorporated in the implementation of such applications in a privacy-preserving scenarios.

AUTHOR CONTRIBUTIONS

Conceived and designed the experiments: JP BR IT. Performed the experiments: JP. Analyzed the data: JP BR IT. Contributed reagents/materials/analysis tools: JP BR IT. Wrote the paper: JP BR IT.

REFERENCES

1. Gentry C. A Fully Homomorphic Encryption Scheme. Ph. D. Thesis, Stanford University. 2009. Available: http:// crypto. stanford. edu/craig/craig-thesis. pdf.
2. Coron J-S, Mandal A, Naccache D, Tibouchi M. Fully Homomorphic Encryption over the Integers with Shorter Public Keys. Advances in Cryptology—CRYPTO 2011. Springer LNCS. 2011; 6841: 487–504.
3. van Dijk M, Gentry C, Halevi S, Vaikuntanathan V. Fully Homomorphic Encryption over the Integers. Advances in Cryptology—EUROCRYPT 2010. Springer LNCS. 2010; 6110: 24–43.
4. Yao A. Protocols for Secure Computation (Extended Abstract). FOCS 1982: Proceedings of the IEEE Symposium on Foundations of Computer Science; 1982 Nov 3–5; Chicago, IL, USA. 1982. p. 160–164. doi: 10. 1109/SFCS. 1982. 38
5. Lindell Y, Pinkas B. Secure Multiparty Computation for Privacy-Preserving Data Mining. Journal of Privacy and Confidentiality. 2009; 1(1): 59–98. doi: 10. 4018/978-1-59140-557-3. ch189
6. ElGamal T. A Public Key Cryptosystem and a Signature Scheme based on Discrete Logarithms. IEEE Trans Inf Theory. 1985; 31(4): 469–472. doi: 10. 1109/TIT. 1985. 1057074.
7. Paillier P. Public-key Cryptosystems based on Composite Degree Residuosity Classes. Advances in Cryptology—EUROCRYPT'99. Springer LNCS. 1999; 1592: 223–238.
8. Naor M, Pinkas B. Oblivious Transfer and Polynomial Evaluation. STOC 1999: Proceedings of the ACM Symposium on Theory of Computing. 1999 May 1–4; Atlanta, GA, USA. 1999. p. 245–254. doi: 10. 1145/301250. 301312
9. Bellare M, Hoang V, Rogaway P. Foundations of Garbled Circuits. CCS 2012: Proceedings of the ACM Conference on Computer and Communications Security. 2012 Oct 16–18; Raleigh, NC, USA. 2012. p. 784–796. doi: 10. 1145/2382196. 2382279
10. Lazzeretti R, Barni M. Private Computing with Garbled Circuits [Applications Corner]. IEEE Signal Process Mag. 2013; 30(2): 123–127. doi: 10. 1109/MSP. 2012. 2230540.
11. Kabanets V, Cai JY. Circuit Minimization Problem. STOC 2000: Proceedings of the ACM Symposium on Theory of Computing. 2000 May 21–23; Portland, OR, USA. 2000. p. 73–79. doi: 10. 1145/335305. 335314
12. Karnaugh M. The Map Method for Synthesis of Combinational Logic Circuits. Transactions of the American Institute of Electrical Engineers, Part I: Communication and Electronics. 1953; 72(5): 593–599. doi: 10. 1109/tce. 1953. 6371932
13. McCluskey E. Detection of Group Invariance or Total Symmetry of a Boolean Function. Bell System Technical Journal. 1956; 35(6): 1445–1453. doi: 10. 1002/j. 1538-7305. 1956. tb03836. x.
14. Rudell R. Multiple-Valued Logic Minimization for PLA Synthesis. Memorandum No. UCB/ERL M86/65.

University of California. 1986. Available:http://www. eecs. berkeley. edu/Pubs/TechRpts/1986/ERL-86-65. pdf.

15. de Micheli G. Synthesis and Optimization of Digital Circuits. McGraw-Hill Higher Education. 1994.
16. Malkhi, D, Nisan N, Pinkas, B, Sella, Y. Fairplay—A Secure Two-Party Computation System. USENIX 2004: Proceedings of the USENIX Security Symposium. 2004 9–13 Aug; San Diego, CA, USA. 2004. p. 287–302.
17. Huang, Y, Evans, D, Katz, J, Malka, L. Faster Secure Two-Party Computation using Garbled Circuits. USENIX 2011: Proceedings of the USENIX Security Symposium. 2011 8–12 Aug; San Francisco, CA, USA. 2011. p. 539–554. Software available:http://mightbeevil. com/framework.
18. Pignata T. Garbled Circuit Designer and Executer from the Visual Information Processing and Protection (VIPP). Software available:http://clem. dii. unisi. it/~vipp/index. php/software/135-garbledcircuit.
19. Bellare, M, Hoang, V, Keelveedhi, S, Rogaway, P. Efficient Garbling from a Fixed-Key Blockcipher. SP 2013: Proc IEEE Symp Secur Priv. 2013 May 19–22; Berkeley, CA, USA. 2013. p. 478–492. doi: 10. 1109/SP. 2013. 39. Software available:http://cseweb. ucsd. edu/ groups/justgarble/.
20. Barni, M, Guajardo, J, Lazzeretti, R. Privacy Preserving Evaluation of Signal Quality with Application to ECG Analysis. WIFS 2010: Proceedings of the IEEE International Workshop on Information Forensics and Security. 2010 Dec 12–15; Seattle, WA, USA. 2010. p. 1–6. doi: 10. 1109/WIFS. 2010. 5711460
21. Pathak M, Rane S, Sun W, Raj B. Privacy Preserving Probabilistic Inference with Hidden Markov Models. ICASSP 2011: Proc IEEE Int Conf Acoust Speech and Signal Process. 2011 May 22–27; Prague, Czech Republic. 2011 p. 5868–5871.
22. Shen Y, Fang J, Li H. Exact Reconstruction Analysis of Log-Sum Minimization for Compressed Sensing. IEEE Signal Process Lett. 2013; 20(12): 1223–1226. doi: 10. 1109/LSP. 2013. 2285579.
23. Dend Y, Dai Q, Liu R, Zhang S, Hu S. Low-Rank Structure Learning via Nonconvex Heuristic Recovery. IEEE Trans Neural Netw Learn Syst. 2013; 24(3): 383–396. doi: 10. 1109/TNNLS. 2012. 2235082.
24. Blei DM, Ng AY, Jordan MI, Lafferty J. Latent Dirichlet Allocation. J Mach Learn Res. 2003; 3: 993–1022.
25. Reynolds DA, Rose RC. Robust Text-Independent Speaker Identification using Gaussian Mixture Speaker Models. IEEE Transactions on Speech and Audio Processing. 1995; 3(1): 72–83. doi: 10. 1109/89. 365379.
26. Beaver D. Precomputing Oblivious Transfer. Advances in Cryptology–CRYPTO'95. Springer LNCS. 1995; 963: 97–109.
27. Aiello B, Ishai Y, Reingold O. Priced Oblivious Transfer: How to Sell Digital Goods. Advances in Cryptology—EUROCRYPT 2001. Springer LNCS. 2001; 2045: 119–135.
28. Naor, M, Pinkas, B. Efficient Oblivious Transfer Protocols. SODA 2001: Proceedings of the Annual Symposium on Discrete Algorithms. 2001 Jan 7–9; Washington, DC, USA. 2001. p. 448–457.
29. Koblitz N. Elliptic Curve Cryptosystems. Math Comput. 1987; 48(177): 203–209. doi: 10. 1090/S0025-5718-1987-0866109-5.
30. Kolesnikov, V, Schneider, T. Improved Garbled Circuit: Free XOR Gates and Applications. ICALP 2008: Proceedings of the International Colloquium on Automata, Languages and Programming. 2008 Jul 6–13; Reykjavik, Iceland, 2008. p. 486–498. doi: 10. 1007/978-3-540-70583-3_40
31. Kolesnikov, V, Sadeghi A-R, Schneider, T. Improved Garbled Circuit Building Blocks and Applications to Auctions and Computing Minima. CANS 2009: Proceddings of the International Conference on Cryptology and Network Security. 2009 Dec 12–14; Kanazawa, Japan. 2009. p. 1–20. doi: 10. 1007/978-3-642-10433-6_1
32. Goldwasser, S, Kalai, Y, Popa, R, Vaikuntanathan, V, Zeldovich, N. Reusable Garbled Circuits and Succinct Functional Encryption. STOC 2013: Proceedings of the ACM Symposium on Theory of Computing. 2013 Jun 1–4; Palo Alto, CA, USA. 2013. p. 555–564. doi: 10. 1145/2488608. 2488678
33. Sahai A, Waters B. Fuzzy Identity-based Encryption. Advances in Cryptology—EUROCRYPT 2005. Springer LNCS. 2005; 3494: 457–473.
34. Boneh, D, Sahai, A, Waters, B. Functional Encryption: Definitions and Challenges. TCC 2011: Proceedings of the Theory of Cryptography Conference. 2011 Mar 28–30; Providence, RI, USA. 2011. p. 253–273. doi: 10. 1007/978-3-642-19571-6_16
35. Boyd S, Kim S-J, Patil D, Horowitz M. Digital Circuit Optimization via Geometric Programming. Oper Res. 2005; 53(6): 899–932. doi: 10. 1287/opre. 1050. 0254.
36. Järvinen, K, Kolesnikov, V, Sadeghi A-R, Schneider, T. Garbled Circuits for Leakage-Resilience: Hardware Implementation and Evaluation of One-Time Programs. CHES 2010: Proceedings of the International Workshop on Cryptographic Hardware and Embedded Systems. 2010 Aug 17–20; Santa Barbara, CA, USA. 2010. p. 383–397. doi: 10. 1007/978-3-642-15031-9_26
37. Mohassel P, Riva B. Garbled Circuits Checking Garbled Circuits: More Efficient and Secure Two-Party Computation. Advances in Cryptology—CRYPTO 2013. Springer LNCS. 2013; 8043: 36–53.
38. Pignata, T, Lazzeretti, R, Barni, M. General Function Evaluation in a STPC Setting Via Piecewise Linear Approximation. WIFS 2012: Proceedings of the IEEE International Workshop on Information Forensics and Security. 2012 Dec 2–5; Costa Adeje, Tenerife, Spain. 2012. p. 55–60. doi: 10. 1109/WIFS. 2012. 6412625
39. Luo Y, Cheung S, Pignata T, Lazzeretti R, Barni M. An Efficient Protocol for Private Iris-code Matching by Means of Garbled Circuits. ICIP 2012: Proc Int Conf Image Proc. 2012 Sep 30–Oct 3; Lake Buena Vista, Orlando, FL, USA. 2012 p. 2653–2656.

CHAPTER 8

HOW TO MAKE A SYNTHETIC MULTICELLULAR COMPUTER

Javier Macia[1, 2] and Ricard Sole[1, 2, 3]

[1]ICREA-Complex Systems Lab, Universitat Pompeu Fabra, Barcelona, Spain
[2]Institut de Biologia Evolutiva, UPF-CSIC, Barcelona, Spain
[3]Santa Fe Institute, Santa Fe, New Mexico, United States of America

ABSTRACT

Biological systems perform computations at multiple scales and they do so in a robust way. Engineering metaphors have often been used in order to provide a rationale for modeling cellular and molecular computing networks and as the basis for their synthetic design. However, a major constraint in this mapping between electronic and wet computational circuits is the wiring problem. Although wires are identical within electronic devices, they must be different when using synthetic biology designs. Moreover, in most cases the designed molecular systems cannot be reused for other functions. A new approximation allows us to simplify the problem by using synthetic cellular consortia where the output of the computation is distributed over multiple engineered cells. By evolving circuits in silico, we can obtain the minimal sets of Boolean units required to solve the given problem at the lowest cost using cellular consortia. Our analysis reveals that the basic set of logic units is typically non-standard. Among the most common units, the so called inverted IMPLIES (N-Implies) appears to be one of the most important elements along with the NOT and AND functions. Although NOR and NAND gates are widely used in electronics, evolved circuits based on combinations of these gates are rare, thus suggesting that the strategy of combining the same basic logic gates might be inappropriate in order to easily implement synthetic computational constructs. The implications for future synthetic designs, the general view of synthetic biology as a standard engineering domain, as well as potencial drawbacks are outlined.

INTRODUCTION

A fundamental trait of biological systems is their capacity to perform computations [1]. Although cells are composed of molecules and their viability relies on extracting and using energy to maintain them, they are not "just" matter and energy. Information, and how it is processed and used, is an essential ingredient of biology. Adaptation to environmental signals requires the processing and proper output to incoming information. This is of no surprise when we consider that life is strongly tied to genetic information [2]. Similarly, a computational picture of biological systems is at the core of important, unanswered questions on how organisms behave [3].

How do biological systems compute? Computation is present at multiple scales, from molecules to collective decisions [4]–[10]. Developmental processes [11], [12], collective intelligence [13], [14] and complex decision-making in cells [15]–[18] can be mapped into some class of formal computational framework. Early works in theoretical

Citation: Macia J, Sole R (2014) How to Make a Synthetic Multicellular Computer. PLoS ONE 9(2): e81248. doi:10.1371/journal.pone.0081248

biology, cybernetics, and Boolean dynamical systems widely emphasized the view of molecular phenomena within cells as the likely result of computational processes [19]–[22]. But beyond the classical theoretical approach to computation, one especially important avenue involves the engineering of cellular circuits in order to construct given computational functions or devices performing computations [23]–[31]. An example is given in figure 1, where we show the potential implementation of a simple logic gate using engineered regulatory networks. Here a NAND gate is built by combining a few basic components. Two input molecules (α and β) can be sensed by appropriate receptors or simply diffuse into the cell where they interact with operator sites. Only in the absence of both signals the output is produced.

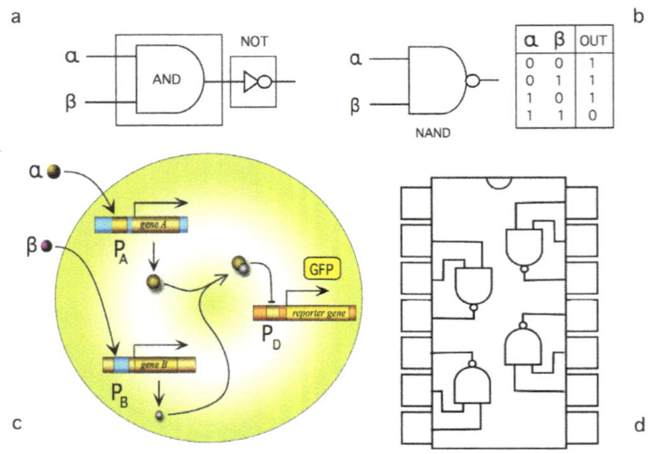

Figure 1: Simple logic gates can be implemented out from minimal sets of logic units.

The NAND gate (a) is obtained as a sequential combination of AND and NOT gates. The compressed symbol is shown in (b) along with the truth table. An example of a synthetic implementation of the NAND logic can be made (c) using genetic regulatory elements A and B forming a regulatory heterodimer complex that prevents the expression of the reporter gene. In conventional electronics, combinations of such gates allow to construct more complex circuits and chips (d), which are then used as basic modules for further circuit designs.

http://dx.doi.org/10.1371/journal.pone.0081248.g001

In this context, it has been suggested that complex computational tasks might be obtained by engineering biological structures (molecules and cells) in such a way that they can respond to given sets of inputs and generate a pre-defined output response. Using synthetic biology techniques, a great deal of examples involving logic gates and simple combinations of them have been obtained and some specific computational problems addressed (see [24], [32], [33]and references cited). Much is expected from these developments towards new approaches to complex diseases, for example. But the promise of a reliable, scalable, reusable, robust and predictable life-based technology that could allow constructing complex living machines has been shown to be much more limited than expected [34]. After a successful first wave of important results, the promise of arbitrarily complex constructs obtained in a LEGO-like fashion is far from achieved. A flexible toolbox of reusable elements is yet to be developed and all synthetic designs so far devised are lmited to specific tasks and can not be applied to other problems. In particular, the combinatorial potential implicit in standard circuit engineering has not yet been explored.

Although it is known from the basic theory of combinatorial circuits that some particular logic gates (such as the NOR or NAND) can be used to build any conceivable circuit, this extrapolation has failed to succeed when applied to synthetic biological designs. In that sense, although several authors have been able to build such special gates and claimed that they could *in principle* implement any potential, complex cellular circuit, the extra engineering required makes that claim far fetched. More specifically, the idea of constructing complex devices by just combining logic gates in a standard manner fails due to the so-called *wiring problem*. In contrast to electronics, where all wires are identical but physically isolated, in a cellular context each connection must be implemented by a different biochemical element, e.g. proteins. Even relatively simple devices, such is a MUX circuit (figure 2a), are difficult to obtain [35]. This circuit involves three inputs, one of which (α) is the so called *selector signal*. As can be seen, the state of this selector element determines which one of the two inputs (β and γ) is "chosen" as the final output. Its use is widespread in electronics and it is part of many different applications. In figure 2b we show a standard implementation of this circuit obtained by connecting several NAND gates (here shown as AND+NOT elements). Despite its simplicity, this circuit requires a considerable engineering effort in order to follow standard circuit design principles within a single cell. For illustrative purposes we show a possible implementation of this circuit in figure 2c. We can easily appreciate that the internal logic of our proposed circuit requires several promoters to be connected through different molecular "wires". Such limitations pose immediate constraints to the possibility of creating robust, scalable, and flexible devices with higher computational capacities. Hence, the development of decision-making circuits performing complex functions and, in general, the path towards living computers seems compromised by the failure of standard design principles.

Combinatorial circuits are constructed in conventional electronic design by using predefined gates and wiring them in order to execute a given input-output table. This is illustrated by the so called multiplexer (MUX) circuit, whose representation and logic table is shown in (a). Using AND and NOT gates, a standard implementation is displayed in (b). In (c) we show an example of a synthetic gene network implementing a single-cell multiplexer. The output signal is a GFP reporter. A very different design of the MUX system is shown in (d). Here the circuit can be easily designed by splitting the computation into two separated *and disconnected* engineered cells, both able to display the output signal. A simplified diagram that summarized the logic of (d) is shown in (e).

http://dx.doi.org/10.1371/journal.pone.0081248.g002

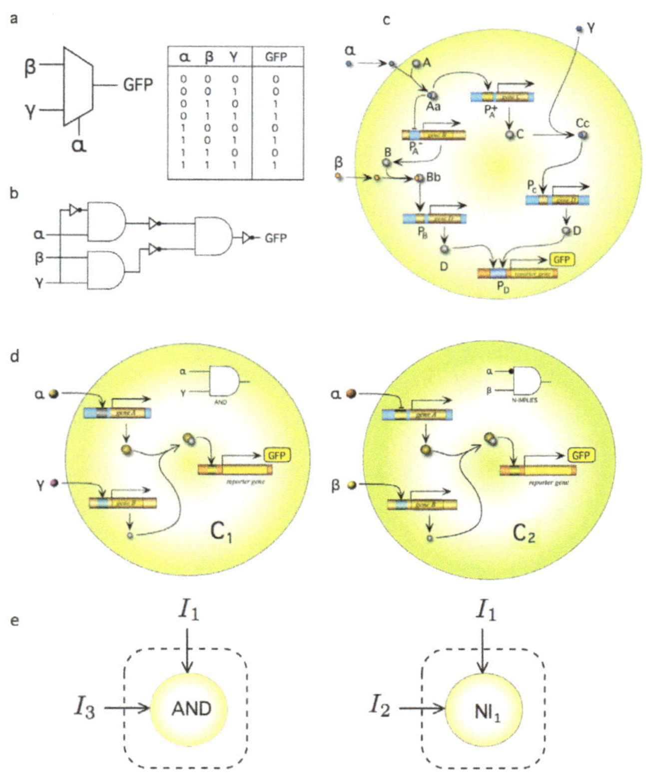

Figure 2: Standard and non-standard circuit design.

One way of approaching the problem of complex wiring is to use a cellular consortium, where different types of engineered cells are at work. Such cellular consortia are common in nature[37]–[39] and provide a more flexible scenario for building complex synthetic circuits [23],[40]–[43]. Once a library of engineered cells has been constructed, it is possible to combine them in different ways to obtain different circuits. This is illustrated by the work of Tamsir et al.[44] where the authors used a set of NOR-like gates constructed on *Escherichia coli* by using two specific promoters, where the inputs and outputs are controlled by expression of different quorum-sensing molecules without cross-talk among the different cells. Here, as in conventional circuits, the colonies were spatially distributed on agar plates and connected through quorum-sensing wires. A reporter colony is used to indicate the final output response. By arranging the colonies in different spatial configurations, all of the elementary two-input logic gates can be implemented. This example provides an instance of the application of standard engineering principles to synthetic multicellular systems. However, because of its construction, it is once again limited in complexity, scalability, and flexibility.

In [45], [46] we proposed a very different approach, which we named *distributed multicellular computation* (DMC). Under this approach, circuits are also divided into different cell types, but the similarities with standard electronics ends here. Roughly speaking, we allow circuits to be broken into pieces with the component indicating output scattered over the different pieces. Several types of reporter cells can be present and do not need to be connected. Upon this assumptions, it was shown [46] that complex circuits can be built from very simple cellular consortia. Each cell requires only a small amount of engineering and additional cell-cell communication molecules can be used (but are not always needed) to exchange information among cells. Multicellular implementation is conceptually appealing. It conceals the implementation details of each encapsulated logic gate, which can be individually designed and optimized. As such, it can facilitate circuit implementation and reduce interference with the host cell's physiology by minimizing the number of components introduced into each cell strain. Therefore, to assemble a complex multicellular circuit, the experimenter needs to be concerned with only two factors: the input-output function of each cellular gate and the output-input matching between layers. Moreover, another benefit of multicellular computing is that it allows the suppression of noise in each layer. Because the wiring-molecule output from each gate is mixed, and represents the sum over a population, spurious or 'noisy' responses within a small proportion of cells can be filtered out in subsequent layers [47].

Instead of using NAND gates, the reduction of wiring requirements was achieved by combining the standard AND and NOT gates with the non-standard logic gates called inverted implies (N-implies, NI). There are two possible NI logic gates (hereafter indicated as NI_1 and NI_2) defined in tables 1 and 2 which describe a Boolean function where the underlying circuit decides which of two bits is (strictly) larger than the other. In that case the first table would correspond to $I_1 < I_2$ whereas the second defines $I_1 > I_2$. Despite these gates not being commonly used in standard designs, they have a clear biological meaning found in many regulatory genetic networks, i.e. one of the inputs triggers the expression of an output gene whereas the other blocks this expression.

Table 1: Truth table for a NI_1 **Boolean** function

http://dx.doi.org/10.1371/journal.pone.0081248.t001

I_1	I_2	*GFP*
0	0	0
0	1	1
1	0	0
1	1	0

doi:10.1371/journal.pone.0081248.t001

Table 2: Truth table for a NI_2 Boolean function

http://dx.doi.org/10.1371/journal.pone.0081248.t002

I_1	I_2	*GFP*
0	0	0
0	1	0
1	0	1
1	1	0

doi:10.1371/journal.pone.0081248.t002

Additionally, under this new approach, the gates (NOT, AND and NI) are connected in a new way. Whereas in electronics design rules try to minimize the number of gates but do not address to the number of wires or the complexity of the resulting network of connections, here the number of wires is minimized and the network of connections is reduced to a simple, fixed topology. This was experimentally implemented using engineered yeast cells, which allowed building a library of cell types that could be combined in multiple ways in order to create different types of combinatorial circuits [45], [46].

As a example, in figure 2d we present an alternative implementation for the MUX circuit based in DMC with *distributed output production*. As we can see, the two-cell implementation is made by using two *disconnected* elements, *both* being able to express the reporter gene. In electronics, it would mean breaking the circuit into two pieces; each one having a light bulb to indicate the output. Even though it makes little sense for electronic engineering, it solves a problem when dealing with a living computational device.

Considering these previous results, the goal of this paper is two-fold. On the one had we want to present a general picture of optimized circuits implementing arbitrary Boolean functions based on the combination of DMC and *distributed output production*. Such a picture would be helpful in guiding the choice and development of components and wires in synthetic constructs. On the other hand, we also want to see how far the analogies made between standard electronic circuits and their cellular counterparts can be stretched.

The previous choices make the potential set of designed circuits simple by construction. As shown below, searching for circuit designs compatible with our proposal leads to simple solutions. Such solutions largely combine a subset of logic gates that departs from the standard designs in several ways. Previous work has obtained optimal solutions in different systems [48] by using evolutionary algorithms, and the networks which evolved were simpler in some ways to hand-designed synthetic biology networks. Similarly, here we perform our analysis by means of an evolutionary algorithm exploring the space of possible designs for logic circuits. It is important to mention that our approach implicitly assumes that the underlying engineering associated with these multicellular consortia is not affected by a number of relevant problems, including cross-talk, noise, and population dynamics. Some of these problems were addressed in [45], [46] where it was shown that the DMC approach is able to overcome some of these problems. The analog, noisy, and population-dependent extensions of the work presented here will be explored elsewhere.

MATERIALS AND METHODS

Boolean Models of Cellular Computation

The simplest theoretical framework to define biological computation is a Boolean approximation. In such framework, the set of possible states to be observed is limited to two, i. e. $\Sigma=\{0,1\}$. A given input string I made of zeros and ones can be written as an element of

$$\Sigma^N = \underbrace{\{0,1\} \times \ldots \times \{0,1\}}_{n} \tag{1}$$

Such an input string would correspond to a set of present or absent input signals, which can be molecules but also physical variables. Let us focus on a given Boolean function Φ involving N inputs and one output. This is formally indicated as a mapping

$$\Phi : \Sigma^N \to \Sigma^1 \tag{2}$$

This function defines a input-output mapping between any given binary string of N bits $\mathbf{I}\in\{0,1\}^N$ and the two possible output values $\{0,1\}$. Complex circuits can be obtained out from a combination of several smaller sub-circuits called *logic gates*, and this can be done in multiple ways. These logic gates are two particularly relevant subsets of Boolean functions, namely i) the 16 two-input one-output logic gates defining the set $\mathcal{G}^{(2,1)}=\{g_{ij}^k\}$ where g_{ij}^k represents the *k-st* gate responding to inputs i and j, and ii) the one-input one-output gates, i. e. the set $\mathcal{G}^{(1,1)}=\{g_{i0}^1,g_{i0}^2\}$ where g_{i0}^1 is the negation (NOT) and g_{i0}^2 is the identity function (Id) in response to i *-st* input. It is well known that multiple subsets of logic gates can be used to implement *any* possible combinatorial circuit. These are known as *functionally complete sets* (FCSs). Typical examples of these sets are the pairs {AND,NOT} and {OR,NOT} but since the NAND and NOR gates are obtained from the combination of these previous pairs, it actually occurs that the single-function sets {NAND} and {NOR} are themselves FCSs. This statement can be proven [49] using the rules of Boolean algebra.

As mentioned above, the single-cell implementation, although possible in principle, has two drawbacks. The first involves the unavoidable design problem associated with the use of several molecular wires associated with each gene-gene connection. The second is the limited flexibility of a single-purpose design. Most typical designs cannot be recycled in any way, but a flexible and scalable system should allow for combination among components such that multiple functions could be implemented. We must keep in mind, as we have pointed out before, that the standard methods for circuit design focus on the minimization of the number of logic gates but do not pay attention to the number of wires or to the complexity of the pattern of connections. This becomes a limitation for the application of standard rules for cellular circuit design, and hence a novel methodology for cellular circuit designs seems necessary.

To reach this goal, we propose a different approach according to the following criteria.

- the circuit can be distributed in a network of different engineered cell types (*distributed computation*),
- the output production can take place in differently cell types simultaneously (*distributed output production*),
- the set of *wires* connecting the different cell types must be minimal, and
- the pattern of connections between cells must be as simple as possible.

Distributed computation allows for a minimization of circuit complexity, since each cell carries a small amount of engineering, limited to implementing one given logic gate. Logic gates (cells) generate an output responding to an external input or at the most to an external input and to the output produced by other different cell, i.e. a wire, according to the logic of one of the possible functions from the set $\mathcal{G}^{(1,1)}\cup\mathcal{G}^{(2,1)}$. This output can be the final output of the circuit or a new wire. Of note, this assumption implies a significant difference with respect to standard methodologies because there are no hidden gates, i.e. gates responding only to internal signals (wires), typically present in standard circuits. Every feed-forward circuit implemented by our system is based on three basic motifs, shown in Figure 3. These motifs involve engineered cells (colour balls) and links α_i connecting them. By splitting different parts of the circuitry over different cells, we can take advantage of the intrinsic modularity of cells as units. Furthermore, distributed output production allows for a strong relaxation of wiring requirements. The connections between different cell types can be implemented by producing small diffusible molecules that can be secreted by a given cell type and sensed by another cell type. Our analysis is centred on the minimal scenario for combinational digital circuits, where feedback connections are not allowed.

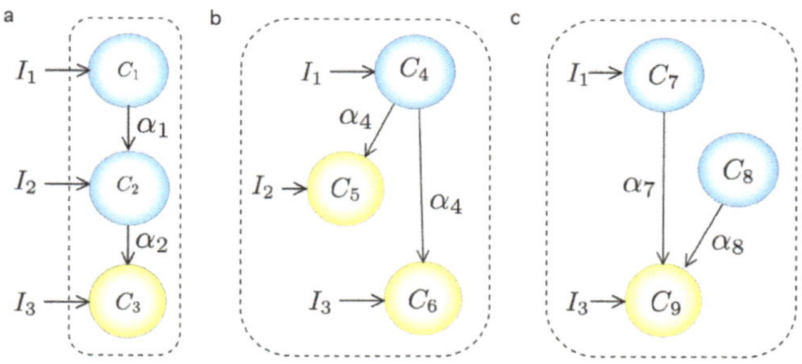

Figure 3: A general circuit design can be obtained by starting from a multicellular system where each virtual cell is a given logic gate.

Here each engineered cell is indicated as C_q and wires generated by C_q are indicated as α_q. Cell C_q will produce an output according with the logic defined by g^k_{ij}. Here g^k_{ij} represents the *k*-th logic gate responding to two inputs, the external one I_i and the internal wire α_j secreted by cell C_j. The upper layer involves single-input gates, i.e. g^k_{i0} (thus only the identity or NOT are possible). Different motifs of connections can emerge according with the criteria introduced, such as independent strings of connected cells, where each cell responds to different wire (a), the same wire can be sensed by more than one cell (b), or a given cell responds to wires produced in more than one cell (c). In this last case, due to cells only can sense two inputs (one external and one internal) all wires produced in different parts of the circuit but sensed by the same cell must be implemented using the same diffusible molecule, i.e. wire α_7 and α_8 are implemented by the same molecule, which can be produced in cells C_7 and C_8 independently. This situation corresponds to an implicit implementation in C_9 of the OR logic with respect to wires α_7 and α_8. Yellow cells (C_3, C_5, C_6 and C_9) can produce independently the final output signal e.g. a GFP.

http://dx.doi.org/10.1371/journal.pone.0081248.g003

Evolving Distributed Circuits

The efficient design of synthetic biocomputers faces a complex optimization problem. As the number of potential elements grows with circuit complexity, so does the potential number of solutions. Such combinatorial explosion can be managed by using automated methods of design [50]. This has been done in some special cases, most of which consider the analog nature of genetic systems. They include small memory devices (flip-flops, [51]), pulses and bandwidth detectors [52], and are generalized to different scenarios through simulated annealing [53], non-linear programming methods [54] standard growth and selection procedures[55], evolutionary optimization of a set of independent circuits [56], and in silico automated design inspired by standard minimization techniques borrowed from electronic design [57]. All of these methods exhibit advantages and limitations, and the predicted circuits, especially when dealing with a large number of biological parts, involve complex wiring diagrams and are not expected to be reused for multiple functions.

In this paper we show the potential of DMC for evolving complex decision-making circuits that would be very difficult to implement using inspiration from standard electronics. Along with the previously mentioned MUX circuit, we show the result of evolving five different standard circuits of increasing complexity. These are a binary comparator, a three-bit adder, the 3-bit parity circuit (see description below), and two 4-input circuits (see below). In order to compare the expected networks resulting from electronic design principles with our (much less complex) proposed DMC constructs, in figures 4a, b, c we show their traditional implementation using logic gates. The left and right columns correspond to the implemented wiring diagrams using different one- and two-input logic gates and NAND gates with arbitrary inputs, respectively. We can clearly appreciate how rapidly the wiring complexity increases once we consider three input functions. The dramatic increase in circuit complexity is especially well illustrated by the expected designs based on NAND gates (right column). Here we allow these gates to receive multiple inputs. If we force them to include only two inputs, the number of gates and links rapidly explodes.

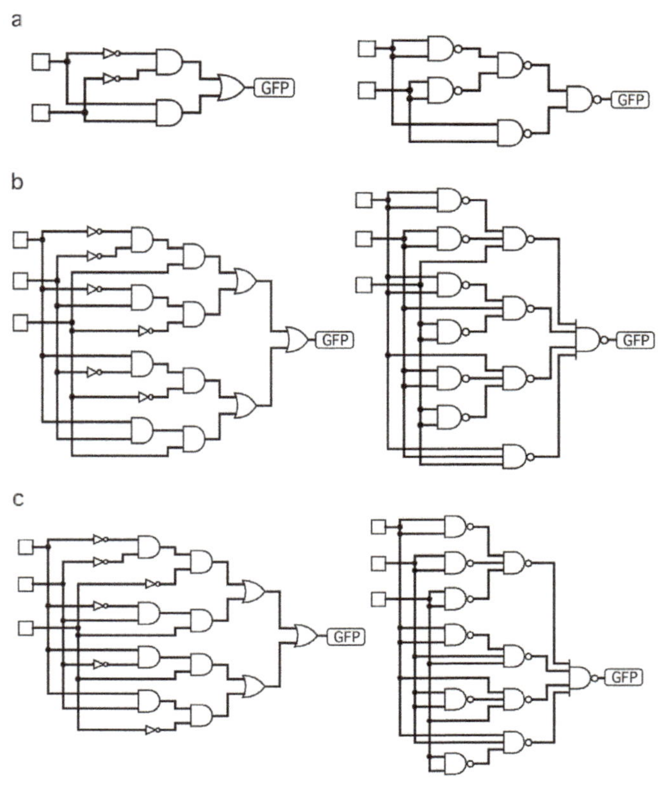

Figure 4: Examples of standard engineering designs of three cases studies.

(a) Two-bit comparator (b) the three-bit adder and (c) a 3-bit parity circuit. Here the set of inputs appears indicated as open squares and the single-output element is marked as GFP.

The left and right columns are different implementations of the same circuits design (Boolean table) but they have been constructed used diverse logic gates (left) or only using NAND gates with variable numbers of inputs. These circuits have been generated using the *Logisim* software package. Once the truth table is provided, it builds the logic circuits, either choosing the appropriate set of two-input logic gates or using just NAND gates. If the NAND gates were chosen such that they only include two inputs, the circuits would be much more complex.

http://dx.doi.org/10.1371/journal.pone.0081248.g004

In order to design a given circuit implementing an arbitrary Boolean function Φ, we start with a simple feed-forward architecture. We define a set of input signals $\{I_1,...,I_N\}$ with $I_i \in \{0,1\}$and a library $\Omega = \{C_q(g_{ij}^k)\}$ of different cells. Each cell implements the k -*st* Boolean functiong_{ij}^k, responding to both an external input I_i and to an internal wire α_j, or only responding to an external input, i.e. g_{i0}^k.

To explore the potential sets of gates to be used in designing arbitrary circuits, we have used an evolutionary algorithm as our search engine for optimal synthetic circuits. We start with a set $\Gamma = \{\Gamma_1,...,\Gamma_s\}$ of randomly wired circuits. Here s different potential solutions are considered and each circuit will have N inputs and one output. The total number of possible functions implementable by this circuits is thus 2^{2^N} . Each circuit will be composed by a random number of logic gates, which are also randomly chosen from the set $\mathcal{G}^{(1,1)} \cup \mathcal{G}^{(2,1)}$ which includes $2^1 + 2^4 = 18$ different potential choices (for general definitions, further information and classification, see [36]). Once a given input-output function is defined, many possible combinations can implement it.

Fitness Function

Each of these circuits is characterized by a fitness function $F(\Gamma_m)$. This fitness function is defined as the combination of two different terms, i.e. $F(\Gamma_m) = f(F_\Phi(\Gamma_m), F_c(\Gamma_m))$. The first term, $F_\Phi(\Gamma_m)$, measures how good the computation of Φ performed by the circuit Γ_m is. However, good computations are not enough but it is necessary minimize the complexity of the circuit's design. This can be achieved in several ways. The second term of the fitness function, $F_c(\Gamma_m)$, accounts for additional evolutionary scenarios satisfying the criteria presented in section 2 in order to safely translate them into an experimentally feasible construct.

Specifically, $F_\Phi(\Gamma_m)$ is defined as the normalized distance

$$F_\Phi(\Gamma_m) = 1 - \frac{1}{2^N} \sum_{j=1}^{N} |O_j^m - \phi_j| \tag{3}$$

where O_j^m and ϕ_j are the observed and expected outputs associated to the j-th input string (i. e. $\phi_j = \Phi(I_j)$), respectively. The highest value, $F_\Phi(\Gamma_m) = 1$, will be obtained when perfect matching is achieved.

We have explored different scenarios imposing a selection force described by the second term of the fitness function, $F_c(\Gamma_m)$, namely:

Evolutionary Algorithm

The evolutionary process starts from a set Γ of s circuits. Each one of these circuits is formed by a random number τ of logic gates, with $\tau \in \{1, N \cdot 2^N\}$, where N is the number of external inputs. These gates are randomly wired and the process follows several different steps.

Step 1: For each circuit the computational term of the fitness ($F_\Phi(\Gamma_m)$) is evaluated according to expression (3) upon the different input strings I_j.

Step 2: For all circuits with highest $F_\Phi(\Gamma_m)$ values, the second term $F_c(\Gamma_m)$ is evaluated. This criterion prevents possible biases associated with the constrained component $F_c(\Gamma_m)$, favouring good computation.

Step 3: The total fitness is calculated according to

$$F(\Gamma_m) = \begin{cases} F_c(\Gamma_m) & \text{if } F_\Phi(\Gamma_m) = \max\{F_\Phi(\Gamma_i)\} \\ 0 & \text{otherwise} \end{cases} \tag{4}$$

where $F_\Phi(\Gamma_i) = \{F_\Phi(\Gamma_1),...,F_\Phi(\Gamma_s)\}$. Each on of these circuits has a probability to pass to the next round proportional to the total fitness, i.e.

$$\wp(\Gamma_m) = \frac{F(\Gamma_m)}{\sum_{i=1}^{s} F(\Gamma_i)} \tag{5}$$

Step 4: Random mutations are introduced in the circuits. Random mutations can occur at different levels, namely i) addition of a new wire with probability μ_a, ii) deletion of a wire with probability μ_d, iii) addition of a new logic gate with probability θ_a, iv) deletion of an existing gate with probability θ_d, and v) modification of a logic gate with probability γ. More specifically, modifications of a logic gate are implemented by inverting a randomly chosen output bit in the truth table that defines the logic function.

Step 5: In order to maintain a constant population s of circuits in each round, the set of circuits coming from the previous round is completed by new randomly generated circuits.

Step 6: Go to to Step 1.

In order to illustrate the potential for constructing simple, nonstandard circuits from the DMC metaphor, let us consider several representative examples. These examples belong to the most standard set of components used within electronic devices, but they are also relevant to potential applications of synthetic consortia. The circuits shown below are, among others, the minimal designs obtained simultaneously under the different scenarios described by the function $F_c(\Gamma_m)$ in section 3.1. Simulations were run 500 times for each $F_c(\Gamma_m)$ conditions with the set of parameters $\{\mu_a, \mu_d, \theta_a, \theta_d, \gamma\}$ randomly chosen in the interval $(0.001, 0.1)$ in a population of $s = 1000$ circuits. The logic circuits obtained have been tested using the *Logisim* software package (see http://ozark.hendrix.edu/~burch/logisim/ for detailed information).

1. Minimal gate number: Given a library of cells to be used, our goal is to make the total number of circuit elements as small as possible. Of note, thinking in terms of wet lab implementation, minimization of gate diversity can be useful in order to reduce the li-

brary of engineered cells. Here, $F_c(\Gamma_m)$ is defined as

$$F_c(\Gamma_m) = \frac{1}{\xi_m} \quad (6)$$

where ξ_m is the total number of gates forming in the circuit Γ_m.

2. Minimal wire number: The implementation of wires, i.e. connections between elements (e.g. cells), is one of the strongest limitations for synthetic designs [34], [46]. Here we minimize the total number of wires used in the circuit by defining $F_c(\Gamma_m)$ as

$$F_c(\Gamma_m) = \frac{1}{1+\omega_m} \quad (7)$$

where ω_m is the total number of different wires present in Γ_m.

3. Minimal complexity: This optimization procedure looks for a *structurally* minimal wiring pattern [46], [58] as the fitness function, in terms of information processed by the circuit. Roughly speaking, this complexity measure C_m weights the contribution of modularity versus integration [58]. We define $F_c(\Gamma_m)$ as

$$F_c(\Gamma_m) = \frac{1}{1+C_m} \quad (8)$$

where C_m is defined as [41]

$$C_m = \frac{1}{2}\sum_{i=1}^{Z} \langle I(X_i^q, X - X_i^q) \rangle \quad (9)$$

Here, Z is the number of gates in the circuit and $I(X_i^q, X - X_i^q)$ is the mutual information between the q $-st$ subset of the circuit formed by i gates, i.e. X_i^q, and the rest of the circuit (i. e. $X - X_i^q$). The symbols $\langle\rangle$ indicate the average among all possible subsets formed by $i$gates. Finally, the index i covers all possible sizes, form subsets formed by one gate ($i = 1$) to the whole circuit ($i = Z$).

Interestingly, as shown below, all these different constraints converge into essentially the same basic space of possible solutions.

RESULTS

Multiplexer

Our first example was the MUX circuit. The MUX circuit obtained following our method is extremely simple. The resulting construct requires just two different engineered cell types. Instead of a complex set of engineered regulatory interactions within a single cell, we can create a consortium involving just two cells. The most important feature to be noticed (aside from the simple design) is that the cells *are not connected* as a consequence of the distributed output. The whole system performs the computation and thus both pieces are required, but there is no need to couple them.

Comparator

In many relevant applications, a given device might need to evaluate when two or more signals are equal or not. A comparator circuit performs such an evaluation. Table 3 describes the Boolean function for a problem involving two bits (corresponding to the XNOR logic).

Table 3: Truth table for a Two-bits magnitude comparator

http://dx.doi.org/10.1371/journal.pone.0081248.t003

I_1	I_2	*GFP*
0	0	1
0	1	0
1	0	0
1	1	1

doi:10.1371/journal.pone.0081248.t003

The minimal implementation in terms of number of logic gates is a single XNOR gate. However, in a biological context, implementation of complex gates such as XOR and XNOR require the layering of multiple genetic circuits, thus necessitating substantial efforts in circuit construction and tuning [59]. As a standard alternative, this circuit can be built by combining several NOT, AND and OR gates. Comparators are a widespread component in most electronic devices and are commonly used in converters, detectors, and oscillators. Their potential for synthetic biology applications seems clear if we consider that most potential decision-making circuits performing complex tasks are likely to require this type of operation.

Here, we have evolved this circuit imposing a minimization of circuit complexity. The minimal configuration obtained involves three different cell types implementing the AND, NOT and NI gates, and one wire (see figure 5a). As it occurs with the MUX system, two subsets of cells that do not exchange signals can be used to implement the Boolean function, whereas five NAND gates would be needed in the NAND-based logic. Of note, implementation of logic gates such as NOT, AND and NI are easier than implementation of XNOR gates [59], and the circuit implementation combining these gates (despite one wire being required) can be easier from an experimental point of view. The coloured balls represent the basic set of engineered cells indicating their internal Boolean functions. The dashed boxes indicate subsets of cells linked through the same communication signal (wire) represented by coloured dashed arrows. In (a) the binary comparator circuit is shown, after all simplifications have been performed. Figure (b) shows a DMC circuit implementing a binary three-bits adder using distributed logic. Finally, in (c) we display the minimal three-bit parity circuit is shown. Although the standard circuit is quite complex, a cell consortia involving six different cell types is enough to implement this complex function. Of note, these examples involve two different wires at the most in the most complex circuit and hence a real wet lab implementation is feasible.

http://dx.doi.org/10.1371/journal.pone.0081248.g005

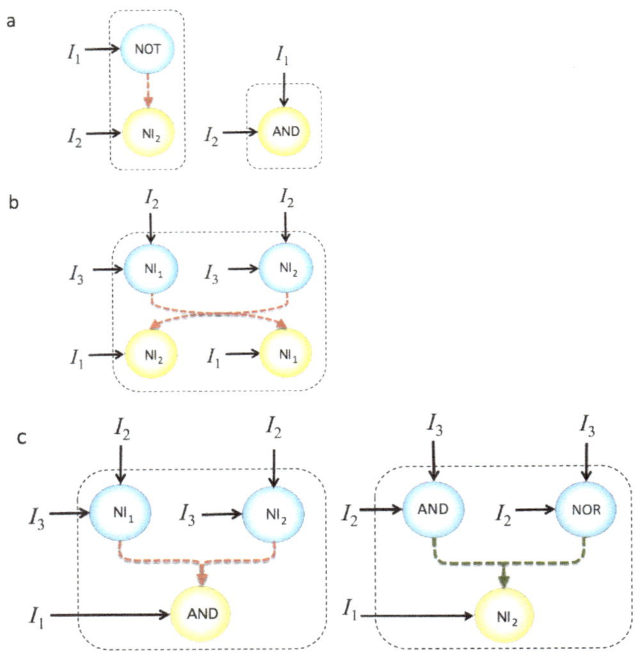

Figure 5: Examples DMC designs obtained by using an evolutionary algorithm to find solutions to given computational functions, as defined by Boolean tables of the examples shown in figure 4.

Three-bit Adder

An additional operation that is essential in most devices is the binary addition of several input numbers described by the truth table. Let us consider the three-bit adder, as defined in table 4.

Table 4: Truth table for a Three-bit adder

http://dx.doi.org/10.1371/journal.pone.0081248.t004

I_1	I_2	I_3	$I_1 + I_2 + I_3$
0	0	0	0
0	0	1	1
0	1	0	1
0	1	1	0
1	0	0	1
1	0	1	0
1	1	0	0
1	1	1	1

doi:10.1371/journal.pone.0081248.t004

This circuit requires a large number of gates (in both scenarios) and wires that make it almost prohibitive for real scenarios. Because of this, it provides a perfect illustration of the potential for strongly reducing circuit complexity by combining non-standard gates with distributed output production. Here the standard implementation would be highly difficult to implement, whereas the consortium solution is completely feasible (see figure 5b).

Parity Bit Circuit

Our fourth example is another three-input, one-output circuit that implements the parity bit circuit, described in table 5. An even parity bit circuit generates an output of 0 if the number of 1 *s* in the input sequence is even and 1 if the number of 1 *s* in the input sequence is odd. Because of its simplicity, parity is used in many applications where an operation can be repeated in case of difficulty, or where simply detecting an error is helpful. Parity can be used as a control to check whether the input string is the expected one or some error (wrong bit) is present. Figure 5c shows the minimal implementation found by the evolutionary algorithm. In this case two different wires are required.

Table 5: Truth table for a 3-Parity bit circuit

http://dx.doi.org/10.1371/journal.pone.0081248.t005

I_1	I_2	I_3	Paritybit
0	0	0	1
0	0	1	0
0	1	0	0
0	1	1	1
1	0	0	0
1	0	1	1
1	1	0	1
1	1	1	0

doi:10.1371/journal.pone.0081248.t005

Inputs: Two Illustrative Examples

In order to emphasize the advantages of DMC combined with distributed output production, we have analyzed one of the most complex functions involving four inputs. This function, described by the truth table shown in figure 6a, is complex due to the fact that the canonical form does not allow for Boolean simplifications, as analysed in [57]. The function involves a rather complex computation, which cannot be reduced nor decomposed in simpler components and, as a result, has a complex associated standard circuit. Here 17 two-input gates (and four NOT gates) are needed to construct the circuit based on electronic design rules. As we can see, the number of wires rapidly explodes. Such circuit complexity makes a mapping between these designs and a synthetic construct highly unlikely to even be possible.

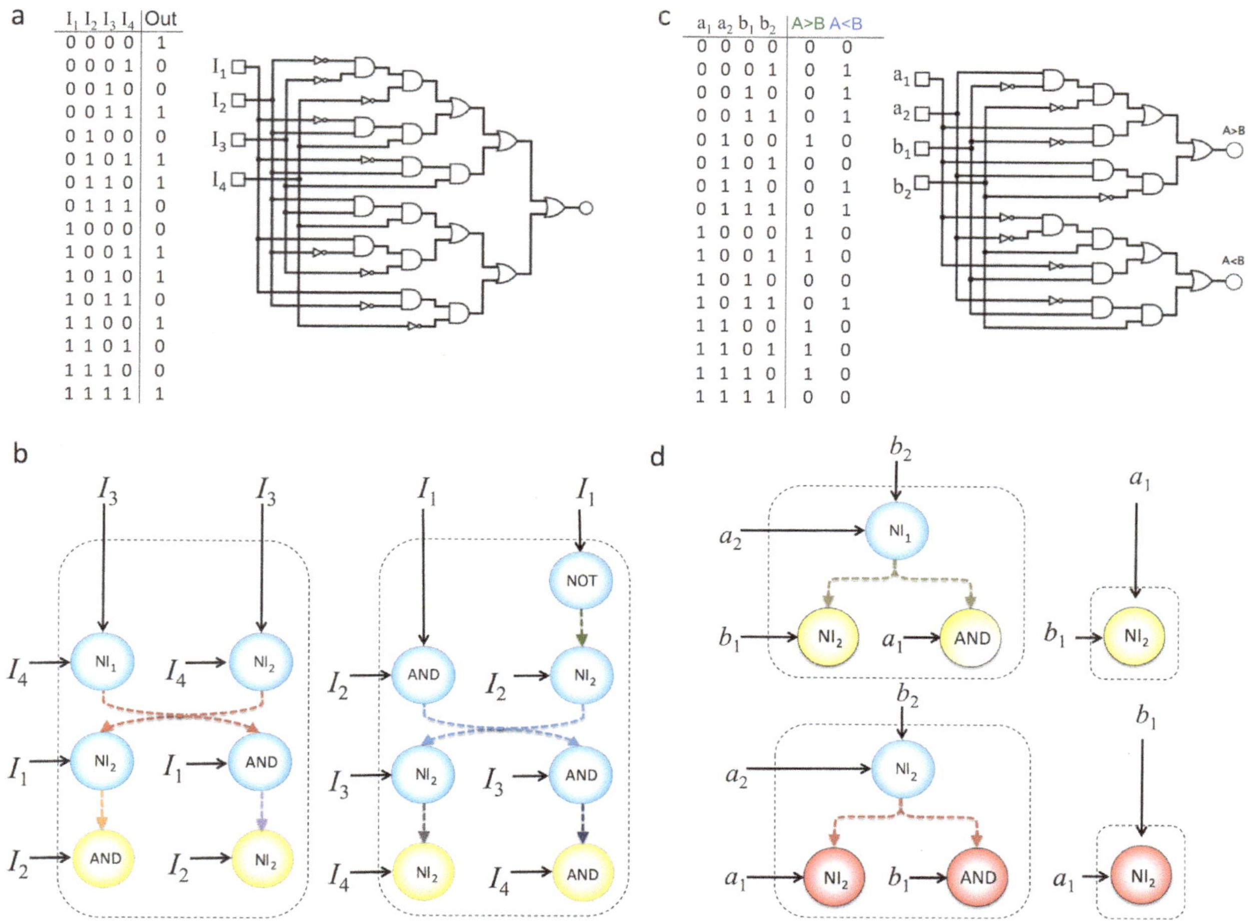

(a)

I_1	I_2	I_3	I_4	Out
0	0	0	0	1
0	0	0	1	0
0	0	1	0	0
0	0	1	1	1
0	1	0	0	0
0	1	0	1	1
0	1	1	0	1
0	1	1	1	0
1	0	0	0	0
1	0	0	1	1
1	0	1	0	1
1	0	1	1	0
1	1	0	0	1
1	1	0	1	0
1	1	1	0	0
1	1	1	1	1

(c)

a_1	a_2	b_1	b_2	A>B	A<B
0	0	0	0	0	0
0	0	0	1	0	1
0	0	1	0	0	1
0	0	1	1	0	1
0	1	0	0	1	0
0	1	0	1	0	0
0	1	1	0	0	1
0	1	1	1	0	1
1	0	0	0	1	0
1	0	0	1	1	0
1	0	1	0	0	0
1	0	1	1	0	1
1	1	0	0	1	0
1	1	0	1	1	0
1	1	1	0	1	0
1	1	1	1	0	0

Figure 6: Scaling up DMC.

In (a) we consider a complex, nontrivial 4-input 1-output Boolean function analysed in Marchisio and Stelling 2011 [57]. Figure shows the truth table and the standard design using two-input gates along with NOT gates. Below (b) we display a minimal circuit implementing this function. In (c) a two-bits comparator circuit is implemented according with the standard methodology, whereas in (d) an alternative design obtained by evolution is shown involving less gates an wires Only two communication signals (wires) are required for this implementation. Two colours (red and yellow) are used to indicate the two reporter molecules associated to each possible output (either $A > B$ or $A < B$).

http://dx.doi.org/10.1371/journal.pone.0081248.g006

Again the solutions found by using our evolutionary algorithm minimizing the number of wires involve a reduced number of wires and cells. Figure 6b shows the minimal circuit obtained, involving seven wires and thirteen cells, all of them with the minimal engineering required for our method. This is the worst case scenario that can be found. Another circuit of great importance in electronic engineering, the so called two-bit comparator, which processes 4-bit input strings and has a two-bit output, is shown in figure 6c. The standard circuit is again very large, with 14 two-input gates and four inverters. In figure 6d the corresponding DMC design is shown, requiring only eight cells. Of note, it is interesting to see that the minimal circuits are formed by a non-standard combination of AND and N-Implies gates. The large number of the later type is a characteristic trait of DMC evolved circuits, as shown below.

Bias towards Non-standard Gates

A more general analysis has been performed by considering the whole set of possible random 3-input, 1-output functions. Our evolutionary algorithm was used to search for evolved DMC designs minimizing circuit size and circuit complexity. The resulting optimized circuits involve different abundances of different gates and have different topological arrangements. Our results are summarized by means of a graph that captures both the frequency of gates used in the final design and how often they are found together within a given solution.

Again, let us start with the MUX circuit, which is a good representation of a complex design. Figure 7a shows the weighted graph associated to the frequency of gates used in the generation of evolved MUX circuits imposing wires minimization. These frequencies are the average of 500 independent runs of the evolutionary algorithm evolving MUX circuits.

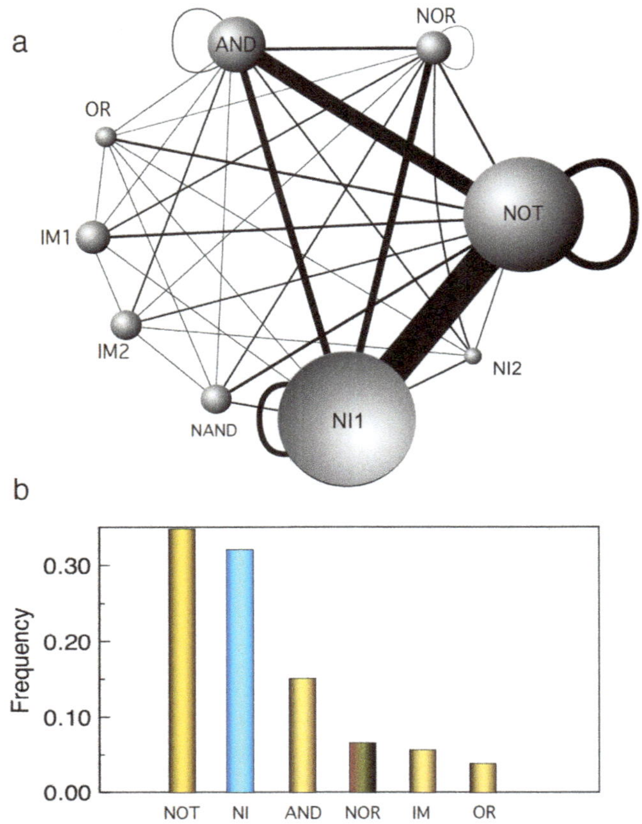

Figure 7: Graphs showing the frequency of gates in a MUX circuit.

(a) Weighted graph showing the frequency of gates used in the generation of evolved MUX circuits under DMC. The diameter of the nodes is proportional to the frequency of the gate. Links indicate that two gates have been wired together within a circuit. The weights provide a measure of how frequently a given pair has been used. Note the disproportionate frequency of N-implies (NI_1) which are typically connected to NOT and AND gates, much less with NOR gates. Note also the vanishing frequency of NOR-NOR links (see text). In (b) we summarize the relative frequencies of different gates. The two N-Implies gates (here NI=N1+IN2) have been added together. These and the NOR abundances are highlighted with distinctive colours. Circuits were evolved imposing minimal number of wires.

http://dx.doi.org/10.1371/journal.pone.0081248.g007

The diameter of the nodes is proportional to the frequency of its abundance and the thickness of the links represents the probability that two gates are present in the same circuit. The graph allows us to draw to important conclusions.

We can observe an overabundance of a few gates. The histogram shown in figure 7b allows comparing these frequencies easily. The most common 2-input 1-output gate is clearly non-standard from the point of view of engineering principles. The second is that neither combinations of the standard gates NOR or NAND seem to be very relevant here. Their frequency is much smaller than other gates, particularly the non-conventional N-IMPLIES (NI_1), which is the most common choice, followed by AND and NOT gates. The fact that the most common gates define a non-standard set of logic functions (which has been shown to be complete, see [45]) suggests that we must consider a whole alternative landscape of combinatorial logic designs under our distributed computation paradigm.

This analysis can be extended to the whole set of 3-input 1-output functions, i.e. for each possible function the evolutionary algorithm has found a possible solution under different constraints. This experiment has been repeated 500 times and the frequencies shown in figure 8 are the average of these independent runs of the evolutionary algorithm. Figure 8a shows the different frequency of appearance of logic gates in circuits evolved without additional constrains (red bars), i.e. the unique imposition is a proper computation, and the corresponding results obtained by evolution imposing minimal size (green bars) and minimal complexity (blue bars). As the figure shows, the patterns are significantly different. In the absence of additional constraints, the most frequent gates are NOT, AND, and not surprisingly NOR and NAND gates. However, in the presence of additional constrains, patterns change and the abundance of NOR and NAND gates is reduced, whereas the NI_1 and NI_2 gates emerge as an alternative for improved circuits.

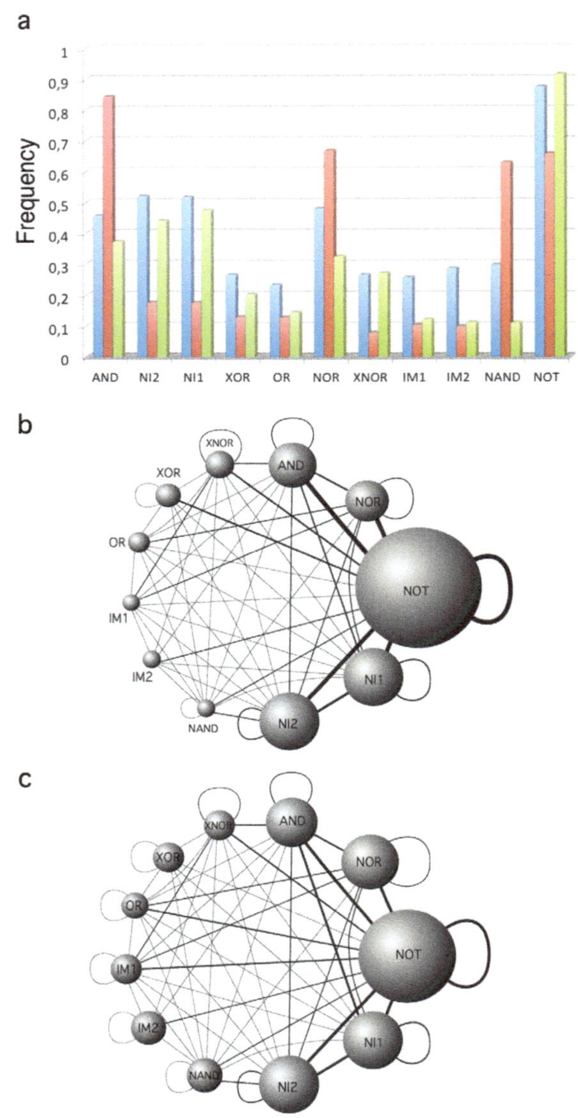

Figure 8: Frequency of appearance of different logic gates for circuits evolved without evolutionary pressure.

In (a) the graph displays the frequency of appearance of different logic gates for circuits evolved without any specific evolutionary pressure (red), imposing minimum size (green), and under minimal circuits complexity (blue). These frequencies have been calculated as the average results for 500 runs. In (b) the graph displays the average results for 500 runs of an evolutionary pressure towards minimum size. The frequency of a given gate is proportional to node diameter, whereas the probability of finding two gates in the same circuit is given by the link weights. In (c) we show the corresponding results for evolved networks under minimal circuits complexity. In both cases, the most represented gates are NOT, AND, NI_1, NI_2 and NOR.

http://dx.doi.org/10.1371/journal.pone.0081248.g008

Despite the NOR gate is a functional complete set by itself, circuits involving only, or dominantly, NOR gates are not the optimal solutions found as the thin auto-link in the NOR node of the graphs indicate (figures 8b–c). Similar arguments can be applied to the NAND gate. Of note, the similarity between graphs obtained imposing minimal size and minimal informational complexity is remarkable, suggesting that the optimization of one aspect could be related with an implicit optimization of the other. Future work should be devoted to analyse these implications carefully.

DISCUSSION

The continuous advance towards the design and synthesis of biocomputers shows that there is an enormous potential for innovation [60], [61]. One of the goals of ongoing synthetic designs is the construction of complex living circuits able to perform complex decision-making tasks. Here we present a promising approach to this problem, which requires reliable and flexible approximations and the potential for extensive reuse and combination of basic units.

We have explored the landscape of circuit designs associated to a number of complex decision-making constructs based on distributed multicellular computation [45], [46]. Such circuits include a large number of potentially important scenarios, including among them many related to biomedical applications, bioremediation scenarios, and several kinds of bioengineering problems. Engineering biofilms, tissue architecture, and growing biomaterials all deal with multiple interacting cell types and it is precisely its multicellular character that makes our method scalable and easy to implement. Our results suggest that the potential associated with multicellular consortia, which have been explored in different scenarios [62]–[65], can be used for designing complex computational devices.

The results described above suggest that, according to the design rules proposed, other functional complete sets, such as AND and NI, can be more optimal for cellular implementation than the standard ones, indicating that the optimal functional complete set dependents on constraints imposed on the circuit topology. Our approximation strongly departs from standard electronic design and in doing so we are able to greatly reduce the connectivity requirements as well as provide a source of flexible combinatorial power.

Our work provides a robust approach to building more complex computations, is predictable and scalable, and opens the door to the future design of multicellular chips. It also allows for re-thinking the way cells and tissues process information beyond man-made metaphors. Its general nature makes it easily extendable to other forms of molecular interactions and model organisms or even protocellular systems [25], [66] and can take advantage of additional techniques, such as microencapsulation [67], [68]. Our results provide a proof of concept that the expectations of synthetic biology in terms of creating complex computational functions from simple ones can be reached, and allows for the exploration novel forms of optimization beyond standard engineering [69].

Despite the promising previous results obtained so far under the multicellular consortia approximation [44], [45], future work should be devoted to analyse possible drawbacks that can limit the potentiality of multicellular circuits, such as stochastic effects on low cell number populations, the impact of different growth rates in a multicellular consortia, circuit stability in long term experiments, and experimental implementation of different wire molecules. These limitations will be strongly dependent upon the specific embodiment of the future synthetic devices. For instance, the same logic circuit can be implemented by using transcriptional or post-transcriptional regulatory genetic circuits hosted in prokaryotes or eukaryotes organisms, and depending on these aspects the same circuit can behave in a different way.

Moreover, other layers of the hierarchy of computational systems, including memory and more complex integrated circuits, should also be explored. Although still far from a complex architecture that we could identify with a "computer", our method shows enormous potential to achieve such goal.

ACKNOWLEDGMENTS

We thank Alfred Borden and the members of the Complex Systems Lab for useful discussions.

AUTHOR CONTRIBUTIONS

Conceived and designed the experiments: JM RS. Performed the experiments: JM RS. Analyzed the data: JM RS. Contributed reagents/materials/analysis tools: JM RS. Wrote the paper: JM RS.

REFERENCES

1. Hopfield M (1994) Physics, Computation, and Why Biology Looks so Different. J. Theor. Biol. 171: 53–60. doi: 10. 1006/jtbi. 1994. 1211
2. Nurse P (2008) Life, logic and information. Nature 454: 424–426. doi: 10. 1038/454424a
3. Brenner S (2012) Turing centenary: Life's code script. Nature 482: 461. doi: 10. 1038/482461a
4. Bray D (1995) Protein molecules as computational elements in living cells. Nature 376: 307–312. doi: 10. 1038/376307a0
5. Amos M (2004) *Cellular Computing*. Oxford University Press, New York.
6. Amos M (2005) Theoretical and Experimental DNA Computation. Springer, New York.
7. Reed MA, Tour MJ (2000) Computing with molecules. Sci. Am. 282: 86–93. doi: 10. 1038/

scientificamerican0600-86

8. Simpson ML, Sayler GS, Fleming JT, Applegate B (2001) Whole-cell biocomputing. Trends Biotechnol. 19: 317–323. doi: 10. 1016/s0167-7799(01)01691-2
9. Sauro HH, Khodolenko BN (2004) Quantitative analysis of signaling networks. Prog. Biophys. Mol. Biol. 86: 5. 43.
10. Istrail S, Ben-Tabou S, Davidson EH (2007) The regulatory genome and the computer. Dev. Biol. 310: 187–195. doi: 10. 1016/j. ydbio. 2007. 08. 009
11. Hogeweg P (2002) Computing an organism: on the interface between informatic and dynamic processes. Biosystems 64: 97–109. doi: 10. 1016/s0303-2647(01)00178-2
12. Deneubourg JL (1989) Collective patterns and decision-making. Ethol. Ecol. and Evol. 1: 295–311. doi: 10. 1080/08927014. 1989. 9525500
13. Solé RV, Delgado J (1996) Universal Computation in Fluid Neural Networks. Complexity 2: 49–56. doi: 10. 1002/(sici)1099-0526(199611/12)2:2<49::aid-cplx13>3. 0. co;2-t
14. Fernando CT, Liekens AM, Bingle LE, Beck C, Lenser T, et al. (2009) Molecular circuits for associative learning in single-celled organisms. J. R. Soc. Interface 6: 463–469. doi: 10. 1098/rsif. 2008. 0344
15. Perkins TJ, Swain PS (2009) Strategies for cellular decision-making. Mol. Syst. Biol. 5: 326. doi: 10. 1038/msb. 2009. 83
16. Hjelmfelt A, Weinberger ED, Ross J (1991) Chemical implementation of neural networks and Turing machines. Proc Natl Acad Sci USA. 88: 10983–10987. doi: 10. 1073/pnas. 88. 24. 10983
17. Wang B, Kitney RI, Joly N, Buck M (2011) Engineering modular and orthogonal genetic logic gates for robust digital-like synthetic biology. Nature Communications 2: 508 doi:10. 1038/ncomms1516.
18. Wang B, Barahona M, Buck M (2013) A modular cell-based biosensor using engineered genetic logic circuits to detect and integrate multiple environmental signals. Biosensors and Bioelectronics 40: 368–376. doi: 10. 1016/j. bios. 2012. 08. 011
19. Kauffman SA (1993) The origins of order. Oxford U. Press, New York.
20. Kauffman SA, Peterson C, Samuelsson S, Troein C (2003) Random Boolean network models and the yeast transcriptional network. Proc. Natl. Acad. Sci. USA 100: 14796–14799. doi: 10. 1073/pnas. 2036429100
21. Klemm K, Bornholdt S (2005) Topology of biological networks and reliability of information processing, Proc. Natl. Acad. Sci. USA 102: 18414–18419. doi: 10. 1073/pnas. 0509132102
22. Andrianantoandro E, Basu S, Karig D, Weiss R (2006) Synthetic biology: new engineering rules for an emerging discipline. Mol Syst Biol. 2: E1–E14. doi: 10. 1038/msb4100073
23. Goni-Moreno A, Amos M, de la Cruz F (2013) Multicellular Computing Using Conjugation for Wiring. PLoS ONE 8(6): e65986. doi: 10. 1371/journal. pone. 0065986
24. Purnick PEM, Weiss R (2009) The second wave of synthetic biology: from modules to systems. Nature Reviews Molecular Cell Biology 10: 410–422. doi: 10. 1038/nrm2698
25. Solé RV, Munteanu A, Rodriguez-Caso C, Macia J (2007) Synthetic protocell biology. From reproduction to computation. Phil. Trans. R Soc. B Biol Sci. 362: 1727–1739. doi: 10. 1098/rstb. 2007. 2065
26. Kobayashi H, Kaern M, Araki M, Chung K, Gardner TS, et al. (2004) Programmable cells: Interfacing natural and engineered gene networks. Proc. Natl. Acad. Sci. USA 101: 8414–8419. doi: 10. 1073/pnas. 0402940101
27. Kramer BP, Fischer C, Fussenegger M (2004) BioLogic gates enable logical transcription control in mammalian cells. Biotechnol. Bioeng. 87: 478–484. doi: 10. 1002/bit. 20142
28. Benenson Y, Gil B, Ben-Dor U, Adar R, Shapiro E (2004) An autonomous molecular computer for logical control of gene expression. Nature 429: 423–429. doi: 10. 1038/nature02551
29. Tan CM, Song H, Niemi J, You LC (2007) A synthetic biology challenge: making cells compute. Mol. Biosyst. 3: 343–353. doi: 10. 1039/b618473c
30. Benenson Y (2009) Biocomputers: from test tubes to live cells- Mol. BioSyst. 5: 675–685. doi: 10. 1039/b902484k
31. Friedland AE, Lu TK, Wang X, Shi D, Church G, et al. (2009) Synthetic gene networks that count. Science 324: 1199–1202. doi: 10. 1126/science. 1172005
32. Ruder WC, Lu TK, Collins JJ (2011) Synthetic Biology Moving into the Clinic. Science 333: 1248–1252. doi: 10. 1126/science. 1206843
33. Weber W, Fussenegger M (2012) Emerging biomedical applications of synthetic biology. Nature Rev. Gen. 13: 21–35. doi: 10. 1038/nrg3094
34. Kwok R (2010) Five hard truths for synthetic biology. Nature 463: 288–290. doi: 10. 1038/463288a
35. Moon TS, Clarke EJ, Groban ES, Tamsir A, Clark RM, et al. (2011) Construction of a genetic multiplexer to toggle between chemosensory pathways in Escherichia coli. J Mol Biol. 406: 215–27. doi: 10. 1016/j. jmb. 2010. 12. 019
36. Tinder RF (2000) Engineering digital design: Revised Second Edition 317319. ISBN 0-12-691295-5.
37. VerBerkmoes NC, Denef VJ, Hettich RL, Banfield JF (2009) Systems Biology: Functional analysis of natural microbial consortia using community proteomics Nature Reviews Microbiology. 7: 196–205. doi: 10. 1038/nrmicro2080
38. Bernstein HC, Carlson RP (2012) Microbial Consortia Engineering for Cellular Factories: in vitro to in silico systems Computational and structural biotechnology journal. 3: e201210017 Available: http://dx. doi. org/10. 5936/csbj. 201210017.
39. O'Toole G, Kaplan HB, Kolter R (2000) Biofilm formation as microbial development. Annu Rev Microbiol. 54: 49–79. doi: 10. 1146/annurev. micro. 54. 1. 49

40. You L, Cox RS, Weiss R, Arnold FH (2004) Programmed population control by cell-cell communication and regulated killing. Nature 428: 868–871. doi: 10. 1038/ nature02491

41. Basu S, Gerchman Y, Collins CH, Arnold FH, Weiss R (2005) A synthetic multicellular system for programmed pattern formation. Nature 434: 1130–1134. doi: 10. 1038/nature03461

42. Shou W, Ram S, Vilar JMG (2006) Synthetic cooperation in engineered yeast populations. Proc. Natl. Acad. Sci. USA 104: 1877–1882. doi: 10. 1073/pnas. 0610575104

43. Brenner K, Karig DK, Weiss R, Arnold FH (2007) Engineered bidirectional communication mediates a consensus in a microbial biofilm consortium. Proc. Natl. Acad. Sci. USA 104: 17300–17304. doi: 10. 1073/pnas. 0704256104

44. Tamsir A, Tabor JJ, Voigt CA (2010) Robust multicellular computing using genetically encoded NOR gates and chemical wires. Nature 469: 212–215. doi: 10. 1038/ nature09565

45. Regot S, Macia J, Conde N, Furukawa K, Kjelln J, et al. (2011) Distributed biological computation with multicellular engineered networks. Nature 469: 207–211. doi: 10. 1038/nature09679

46. Macia J, Posas F, Solé RV (2012) Distributed Computation: The New Wave of Synthetic Biology Devices. Trends Biotech. 30: 342–349. doi: 10. 1016/j. tibtech. 2012. 03. 006

47. Li B, You L (2011) Division of logic labour. Nature 496: 171–172. doi: 10. 1038/469171a

48. McGregor S, Vasas V, Husbands P, Fernando C (2012) Evolution of Associative Learning in Chemical Networks. PLoS Comput Biol 8(11): e1002739 doi:10. 1371/journal. pcbi. 1002739.

49. Enderton H (2001) A Mathematical Introduction to Logic, second edition. Harcourt Academic Press.

50. Marchisio MA, Stelling J (2009) Computational design tools for synthetic biology. Curr. Opin. Biotech. 20: 479–485. doi: 10. 1016/j. copbio. 2009. 08. 007

51. Rodrigo G, Jaramillo A (2007) Computational design of digital and memory biological devices. Syst. Synth. Biol. 1: 183–195. doi: 10. 1007/s11693-008-9017-0

52. Cao H, Romero-Campero FJ, Heeb S, Camara S, Krasnogor N (2010) Evolving cell models for systems and synthetic biology. Syst. Synth. Biol. 4: 55–84. doi: 10. 1007/s11693-009-9050-7

53. Rodrigo G, Carrera J, Jaramillo A (2007) Genetdes: automatic design of transcriptional networks. Bioinformatics 23: 1857–1858. doi: 10. 1093/bioinformatics/btm237

54. Huynh L, Kececioglu J, Koppe M, Tagkopoulos I (2012) Automatic Design of Synthetic Gene Circuits through Mixed Integer Non-linear Programming. PLoS ONE 7(4): e35529. doi: 10. 1371/journal. pone. 0035529

55. François P, Hakim V (2004) Design of genetic networks with specified functions by evolution in silico. Proc. Natl. Acad. Sci. USA 101: 580–585. doi: 10. 1073/pnas. 0304532101

56. Dasika MS, Maranas CD (2008) Optcircuit: an optimization based method for computational design of genetic circuits. BMC Syst Biol 2: 24. doi: 10. 1186/1752-0509-2-24

57. Marchisio MA, Stelling J (2011) Automatic Design of Digital Synthetic Gene Circuits. PLoS Comput. Biol. 7: e1001083. doi: 10. 1371/journal. pcbi. 1001083

58. Tononi G, Sporns O, Edelman GM (1999) Measures of degeneracy and redundancy in biological networks. Proc. Natl. Acad. Sci. USA 96: 3257–3262. doi: 10. 1073/pnas. 96. 6. 3257

59. Siuti P, Yazbek J, Lu TK (2013) Synthetic circuits integrating logic and memory in living cells. Nature Biotechnology. doi:10. 1038/nbt. 2510.

60. Benenson Y (2012) Biomolecular computing systems: principles, progress and potential. Nature Rev. Genet. 13: 455–468. doi: 10. 1038/nrg3197

61. Ausländer S, Ausländer D, Muller M, Wieland M, Fussenegger M (2012) Programmable single-cell mammalian biocomputers. Nature 487: 123–127. doi: 10. 1038/nature11149

62. Weber W, Daoud-El Baba M, Fussenegger M (2007) Synthetic ecosystems based on airborne inter- and intrakingdom communication. Proc. Natl. Acad. Sci. USA 104: 10435–10440. doi: 10. 1073/pnas. 0701382104

63. Brenner K, Arnold FH (2008) Engineering microbial consortia: a new frontier in synthetic biology. Trends Biotech. 28: 483–489. doi: 10. 1016/j. tibtech. 2008. 05. 004

64. Wintermute EH, Silver PA (2010) Dynamics in the mixed microbial concourse. Genes Dev. 24: 2603–2614. doi: 10. 1101/gad. 1985210

65. Song H, Payne S, Gray M, You L (2009) Spatiotemporal modulation of biodiversity in a synthetic-mediated ecosystem. Nat. Chem. Biol. 5: 929–935. doi: 10. 1038/ nchembio. 244

66. Smaldon J, Romero-Campero FJ, Fernandez Trillo F, Gheorghe M, Alexander C, et al. (2010) A computational study of liposome logic: towards cellular computing from the bottom up. Syst. Synth. Biol. 4: 157–179. doi: 10. 1007/s11693-010-9060-5

67. Chang TMS (2005) Therapeutic applications of polymeric artificial cells. Nature Rev. Drug Discov. 4: 221–235. doi: 10. 1038/nrd1659

68. Ausländer S, Wieland M, Fussenegger M (2012) Smart medication through combination of synthetic biology and cell microencapsulation. Metab. Eng. 14: 252–260. doi: 10. 1016/j. ymben. 2011. 06. 003

69. Banga JR (2008) Optimization in computational systems biology. BMC Syst. Biol. 2: 47. doi: 10. 1186/1752-0509-2-47

CHAPTER 9

NOISE-AIDED LOGIC IN AN ELECTRONIC ANALOG OF SYNTHETIC GENETIC NETWORKS

Edward H. Hellen[1], Syamal K. Dana[2], Ju¨rgen Kurths[3], Elizabeth Kehler[1], Sudeshna Sinha[4]

[1]Department of Physics and Astronomy, University of North Carolina Greensboro, Greensboro, North Carolina, United States of America

[2]Council of Scientific and Industrial Research-Indian Institute of Chemical Biology, Kolkata, India

[3]Potsdam Institute for Climate Impact Research, Potsdam, Germany

[4]Indian Institute of Science Education and Research Mohali, SAS Nagar, Mohali, Punjab, India

ABSTRACT

We report the experimental verification of noise-enhanced logic behaviour in an electronic analog of a synthetic genetic network, composed of two repressors and two constitutive promoters. We observe good agreement between circuit measurements and numerical prediction, with the circuit allowing for robust logic operations in an optimal window of noise. Namely, the input-output characteristics of a logic gate is reproduced faithfully under moderate noise, which is a manifestation of the phenomenon known as Logical Stochastic Resonance. The two dynamical variables in the system yield complementary logic behaviour simultaneously. The system is easily morphed from AND/NAND to OR/NOR logic.

INTRODUCTION

Realization of logic functions in different physical systems is one of the key questions that commands widespread research interest in science and engineering. Universal general-purpose computing devices can be constructed entirely from NOR/NAND logic gates [1], [2]. It is particularly interesting to investigate if systems of biological relevance can also yield logic outputs consistent with the truth tables of different logic functions (see Table 1). Biological systems are capable of stochastic resonance[3]–[6], a process in which a small signal is amplified due to the presence of an appropriate level of noise, leading to the possibility of a biological system performing robust noise-aided logic operations in response to weak input signals.

Citation: Hellen EH, Dana SK, Kurths J, Kehler E, Sinha S (2013) Noise-Aided Logic in an Electronic Analog of Synthetic Genetic Networks. PLoS ONE 8(10): e76032. doi:10.1371/journal.pone.0076032.

Table 1: Logic Table

http://dx.doi.org/10.1371/journal.pone.0076032.t001

Input Set (I_1,I_2)	OR	AND	NOR	NAND
(0,0)	0	0	1	1
(0,1)/(1,0)	1	0	0	1
(1,1)	1	1	0	0

Relationship between the two inputs and the output of the fundamental OR, AND, NOR and NAND logic operations. Note that the four distinct possible input sets (0,0), (0,1), (1,0) and (1,1) reduce to three conditions as (0,1) and (1,0) are symmetric. Note that *any* logical circuit can be constructed by combining the NOR (or the NAND) gates [1,2].
doi:10.1371/journal.pone.0076032.t001

A new idea in this direction uses the interplay between noise and nonlinearity constructively to enhance the robustness of logic operations. Namely, in an optimal window of noise, the input-output characteristics of a logic gate is reproduced faithfully. This phenomenon is termed *Logical Stochastic Resonance* (LSR) [7]–[12]. Specifically, in LSR we consider the state of a nonlinear system when driven by input signals, consisting of two randomly streaming square waves. It was observed that the response of such a system shows a remarkable feature: in an optimal band of noise, the output of the system, determined by its state, is a logical combination of the two input signals in accordance with the truth tables of fundamental logic operations.

An important motivation for further studying LSR stems from an issue that is receiving widespread attention currently. The number of transistors in an integrated circuit has approximately doubled every year in accordance with Moore's law. The rapid shrinking of computing platforms with smaller power supplies has brought with it problems of smaller noise margins and higher error rates. Namely, as computational devices and platforms continue to shrink in size, we encounter fundamental noise that cannot be suppressed or eliminated. Hence an understanding of the cooperative behavior between a device noise-floor and its nonlinearity plays an increasingly crucial role in paving the way for smart computing devices. In this direction, LSR indicates a way to turn potentially performance degrading noise to assist the desired operation. Further, it is of far reaching interest to obtain analogous behaviour, not merely in human engineered physical systems, but also in systems of chemical and biological relevance, in order to explore the information processing capacity of naturally occurring systems where noise is ubiquitous.

Since the idea of LSR was first introduced [7], several systems implementing and displaying LSR have been found. To begin with, the basic electronic realizations of simple bistable potentials were reported [7], [8]. Subsequently, noise-aided reprogrammable logic gates have been implemented with noisy nanomechanical oscillators [12], chemical systems [12] and optical systems [13], [14].

Most recently, in the context of biological systems, theoretical ideas have been proposed[15]–[18] on the implementation of LSR in a synthetic genetic network [19]. Now, in this work, we will provide *experimental realizations* of these ideas in an electronic analog of a noisy synthetic gene network. Specifically then, we will investigate the possibility of obtaining reliable logic outputs, and *explicitly demonstrate the pivotal role of noise in the optimization of the logic performance in this circuit.* Further, we will show that the system is easily changed from AND/NAND logic to OR/NOR demonstrating potential for re-programmability [15], [16]. Our results will thus provide verification and further understanding of noise aided logic in systems that are of considerable importance in biology.

Since understanding the intracellular processes in a network of interacting biomolecules is difficult, an alternative approach has been started recently [20], to design artificial genetic networks to derive desired functional behaviors. One important early design is a clock using three genes inhibiting each other in a cyclic order [21]. Taking into account the standard chemical kinetics for expression, degradation and inhibition, a dynamical system model was proposed where the repressor-protein concentrations and mRNA concentrations were expressed as dynamical variables. Another design is a synthetic genetic toggle-switch network[22] whose potential for noise-aided logic operation is investigated here. The significance of using both numerical simulation and electronic circuits to model a potential synthetic genetic network is two-fold. Firstly, the numerical and circuit methods provide *different* imperfect models of a potential biological system. Agreement between these two models indicates robustness in the system and therefore greater likelihood that the same behavior could be realized in the proposed biological system. The biological system is generally much more difficult to construct, and therefore investigating proposed networks in simpler systems is prudent. Secondly, modeling with stochastic differential equations is nontrivial compared to ordinary differential equations, so that the addition of experimental measurements from a physical system such as an analog circuit provides valuable verification. Thus the circuit is an additional tool for investigating potentially interesting biological networks in the presence of noise.

Here we use two repressors and constitutive promoters as our model system for implementing logic functions. We begin with a brief description of LSR, then we describe the synthetic gene network model, and define what constitutes logic inputs and logic outputs in this system. We then go on to present the electronic analog of the system followed by a comparison of numerical simulation and experimental measurement.

METHODS

We begin with a short description of the general principle of LSR. Consider a general nonlinear dynamic system, given by

$$\frac{dx}{dt} = F(x) + I + D\eta(t) \tag{1}$$

where $F(x)$ is a generic nonlinear function which has or nearly has two distinct stable energy wells. I is a low amplitude input signal and $\eta(t)$ is an additive zero-mean Gaussian noise with unit variance with D being the noise strength.

We achieve a logical input-output correspondence with such a system by encoding N inputs in N square waves. Specifically, for two logic inputs, we drive the system with a low amplitude signal I, taken to be the sum of two pulse trains: $I_1 + I_2$, where I_1 and I_2 encode the two logic inputs. Now the logic inputs can be either 0 or 1, giving rise to 4 distinct logic input sets (I_1, I_2): (0,0), (0,1), (1,0) and (1,1). Since the input sets (0,1) and (1,0) give rise to the same I, the input signal generated by adding two independent input signals is a 3-level aperiodic waveform.

The output of the system is determined by its state. For instance, for a bistable system with wells at $x = x_1$ and x_2, the output can be considered a logical 1 if it is in the well at x_1, and logical 0 if it is in x_2. If we consider the opposite assignment, namely logical 1 if the state is in well x_2 and logical 0 if the state is in well x_1, we obtain a complementary logic operation. Specifically we can have an output determination threshold x^*, located near the barrier between the wells, and the logical outputs are then simply given by the state being greater than or less than x^*. It is possible that the input I induces the appearance of the second energy well if it was not already there.

The central result of LSR is as follows: for a given set of inputs (I_1, I_2), a logical output, in accordance with the truth tables of the basic logic operations, is consistently obtained only in an optimal window of noise. Namely, under very small or very large noise the system does not yield reliable logic outputs, while in a band of moderate noise it produces the desired output.

Synthetic Genetic Network Model and Logic Operation

We consider the previously used variation [15] of the genetic toggle switch model comprised of two genes inhibiting each other [22]. The concentrations of the two expressed proteins are x and y, and their rates of change are:

$$\frac{dx}{dt'} = \frac{\alpha_1}{1+y^n} - \beta_1 x + g_1 + D\eta(t') \quad (2)$$

$$\frac{dy}{dt'} = \frac{\alpha_2}{1+x^n} - \beta_2 y + g_2 + I_1 + I_2 + D\xi(t') \quad (3)$$

where β_1, β_2 are the rates of decay of each expressed protein and n is the Hill coefficient. The α_1, α_2 describe the maximum expression rates in absence of inhibitor and they are used here as tunable parameters. In the original model g_1 and g_2 represent the basal synthesis rates of the promoters [23], however we use them as constant bias. The additive noise has strength D and η and ξ are chosen from unit variance zero mean Gaussian distributions. Such an additive noise source alters the background repressor production and represents the inherent stochasticity of biochemical processes such as transcription and translation, and the fluctuations in the concentration of a regulatory protein. I_1 and I_2 are two *low amplitude* inputs providing independent parallel production pathways of repressor y. The t' indicates dimensionless time.

The system above may have two stable configurations in the xy-plane: one state has a high value of x (x_u) and a low value of y (y_l); the other state has a low value of x (x_l) and a high value of y (y_u). That is, the two dimensional potential underlying this system has two wells, (x_u, y_l) and (x_l, y_u), in the xy-plane. Varying the parameters changes the depth and position of these wells, and also determines whether there are one or two wells. For example, Fig. 1shows that for the case $(g_i, I_i, D)=0$ the system in Eqs. (2)–(3) is bistable and therefore has two stable wells only when β_1 is close to 1.

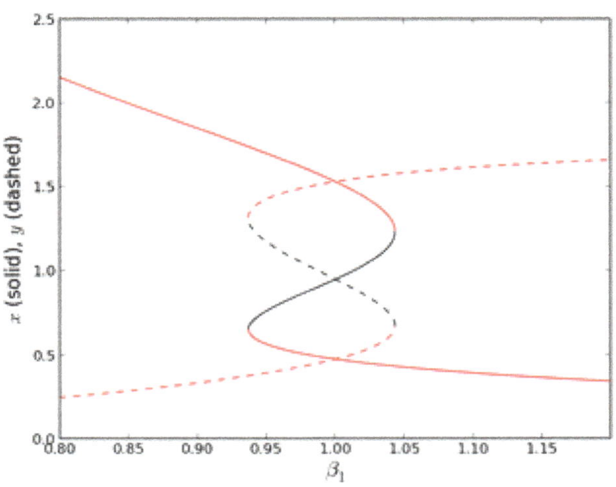

Figure 1: Bifurcation diagram for state-variables x **(solid) and** y **(dashed).**

x and y are complementary outputs, when one is high the other is low. Red (black) indicate stable (unstable) fixed points. System is bistable for $0.937<\beta_1<1.043$. For $\beta_1=0.9$, x is high and y is low. Calculated for Eqs. (2)–(3) with $n=2.4$, $\alpha_i=1.78$, $\beta_2=1$, $I_i=0$, $g_i=0$, and $D=0$.

http://dx.doi.org/10.1371/journal.pone.0076032.g001

Encoding Inputs

Here the low amplitude input signal is $I=I_1+I_2$, with I_1/I_2 equal to I_{ON} ($I_{ON}>0$) if the logic input is 1, and I_1/I_2 being 0 if the logic input is 0. So we have:

(i) $I_1+I_2=0$ corresponds to logic input set (0, 0).

(ii) $I_1+I_2=I_{ON}$ corresponds to logic input sets (0,1)/(1,0).

(iii) $I_1+I_2=2I_{ON}$ corresponds to logic input set (1,1).

Output

The outputs of the system are determined by the level of the dynamical variables $x(t)$ and $y(t)$. For instance the output can be considered a logical 1 if the state is at the high level, and logical 0 if it is at the lower level. That is:

(i) If $x < x^*$, then Logic Output is 0.

(ii) If $x > x^*$, then Logic Output is 1.

Here x^* is the *output determination threshold* that lies between the two states, e.g., at the position of the barrier between the wells. The results presented here are not sensitive to the specific value of x^*.

Specifically, in this work, we consider the logic output to be 1 when the state is close to the upper well, and 0 when the state is close to the lower well, for both x and y variables. So when the system switches wells, the output is "flipped" or "toggled".

The model in Eqs. (2)–(3) is based on the synthetic genetic toggle switch previously expressed in *E. coli* [22]. Parameter values used in [22] correspond here to $\alpha_i=(15.6,156)$, $n=(1,2.5)$, and $\beta_i=1$ in Eqs. (2)–(3). By comparison, here we use $\alpha_i=1.78$, $n=2.4$, $\beta_1=0.9$, and $\beta_2=1$. The bifurcation diagram in Fig. 1 indicates that these parameter values, along with $(g_i, I_i, D)=0$, result in a system with a single stable well at $x\approx1.8, y\approx0.35$. A non-zero input I can then "shift" the bifurcation diagram of Fig. 1 so that there is a stable state with low-x, high-y for $\beta_1=0.9$.

Circuit Realization

The circuit of a single inhibitory gene [24], [25] is shown in Fig. 2. The transistor current represents the rate of gene expression and the voltage V_{out} represents the concentration of expressed protein. V_{in} represents the concentration of repressor, and the V_{cth} adjusts the affinity of the repressor binding to the gene's DNA. The Hill function inhibition in Eqs. (2)–(3) is accounted for by the dependence of the transistor current on repressor concentration voltage V_{in}. The synthetic genetic network shown in Fig. 3 is comprised of two individual gene circuits connected in a loop, each inhibiting the other. For the model in Eqs. (2)–(3), the encoding inputs I_1 and I_2 add to production of y which is accounted for in Fig. 3 by the two logic-driven transistors sourcing current to V_y. Initially parameters g_1 and g_2 are taken to be zero.

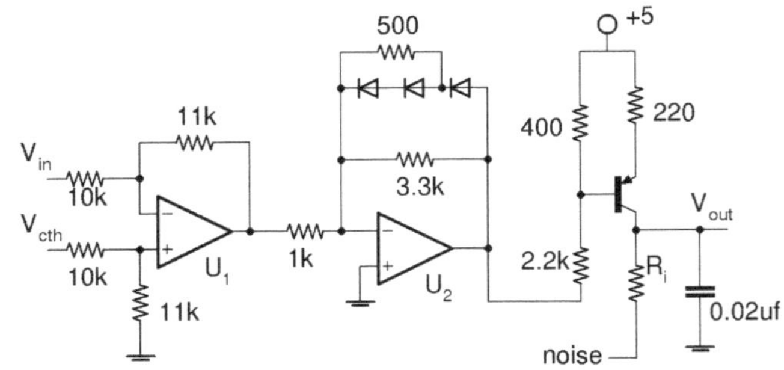

Figure 2: Circuit for single gene.

Inhibitory input at V_{in}. Expressed protein concentration is represented by V_{out}. R_i=470 Ω for gene-*y*, 520 Ω for gene-*x*. Dual op-amp is LF412 supplied by +/−5 V. The *pnp*transistor is 2N3906. The input noise has a mean of 0 V (gnd) and controllable standard deviation.

http://dx.doi.org/10.1371/journal.pone.0076032.g002

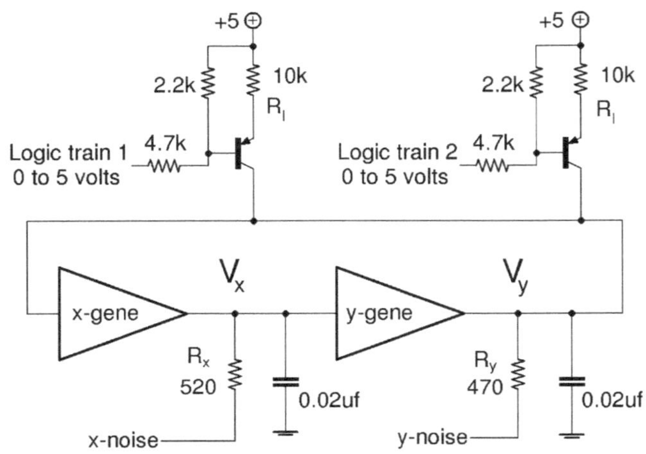

Figure 3: Circuit for synthetic genetic network.

Encoding inputs are 0 to 5 V pulse trains creating transistor currents of 0.1 and 0 mA, respectively, for $R_l = 10k\Omega$. The *x* and *y* gene circuits are shown in Fig. 2. Each noise input is connected to its own noise circuit shown in Fig. 4.

http://dx.doi.org/10.1371/journal.pone.0076032.g003

The circuit equations are obtained by applying Kirchoff's laws to V_x and V_y, the voltages across the capacitors in Fig. 3 [24], [25]. Multiplying both equations by R_y results in equations for V_x and V_y;

$$R_y C\frac{dV_x}{dt} = -\frac{R_y}{R_x}V_x + \frac{R_y}{R_x}V_{x\_noise} + R_y i_t \tag{4}$$

$$R_y C\frac{dV_y}{dt} = -V_y + V_{y\_noise} + R_y i_t + R_y i_1 + R_y i_2 \tag{5}$$

where i_t are the Fig. 2 transistor currents for each gene, and i_1 and i_2 are currents from the logic train transistors in Fig. 3. A noise generation circuit shown in Fig. 4 based on breakdown of a reverse biased base-emitter junction produces noise V_{noise} with zero mean and variable amplitude. We use a well regulated supply for the noise circuit to avoid adding AC signals from the building's electrical system into the noise. Two of these noise circuits are used to supply noisy voltages to each individual gene at locations indicated in Fig. 3.

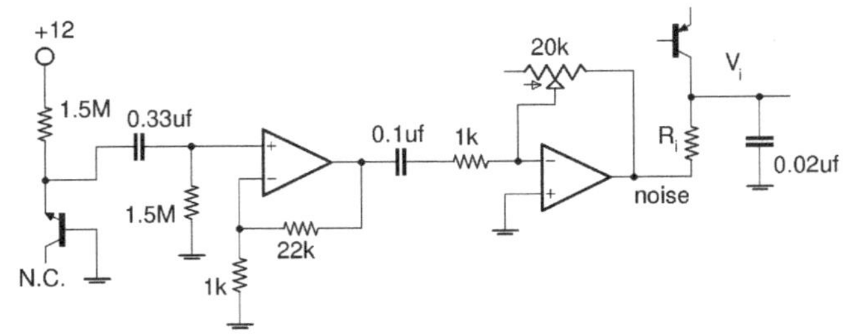

Figure 4: Noise circuit and its connection to resistor of gene circuit.

Source of noise is the reverse-biased base-emitter junction of the 2N3904 *npn* transistor on left. OPA2228 dual op-amps supplied from +/−12 V regulators. OPA2228 has gain-bandwidth product of 33 MHz.

http://dx.doi.org/10.1371/journal.pone.0076032.g004

The connections between model parameters ($\alpha_i, \beta_i, n, I_i, g_i, D$) and circuit parameters are presented in this section using relevant numerical values, with derivations of these connections given in the next section. Readers may go directly to *Results and Discussion* without loss of continuity. The connections are found by relating circuit Eqs. (4)–(5) to the model Eqs. (2)–(3) and by adjusting the dependence of the transistor current i_t on V_{in} in Fig. 2to match the Hill function inhibition. The dimensionless state-variables (x,y) in Eqs. (2)–(3) are related to voltages V_x and V_y by: $x = \frac{V_x}{V_{th}}, y = \frac{V_y}{V_{th}}$where V_{th} corresponds to the repressor's half-maximal inhibition binding constant K_i. The maximal expression rate is

$$\alpha_1 = \alpha_2 = \frac{i_{max}R}{V_{th}} = \frac{1.35}{0.76} = 1.78 \tag{6}$$

where the voltage $i_{max}R$ is $(3mA)(0.45k\Omega) = 1.35V$ and $V_{th} = 0.76V$. Protein decay rates are $\beta_1 = \frac{R_y}{R_x} = \frac{470}{520} = 0.90$, and $\beta_2 = 1$. The Hill coefficient n comes from

$$n\alpha = 1.17 G_1 G_{-2} \tag{7}$$

where $G_1 = -1.1$ and $G_{-2} = -3.3$ are closed loop gains of U1 and U2 in Fig. 2 resulting in $n = (1.17)(1.1)(3.3)/1.78 = 2.4$. The characteristic time is $R_y C = (470\ W)(2 \times 10^{-8} f) = 9.4\mu s$, so the dimensionless time is $t = \frac{t}{R_y C}$.

The high value I_{ON} of the encoding signals I_1, I_2 is given by

$$I_{ON} = \frac{R_y \times (1\ volt)}{R_l V_{th}} = \frac{470}{R_l(0.76)} = \frac{618}{R_l} \tag{8}$$

where I_{ON} is changed by varying R_l in Fig. 3, i.e. $R_l = 10k\Omega$ gives $I_{ON} = 0.062$. A non-zero value of g_2 in Eq. 3 is achieved in the circuit by including a third current-sourcing transistor in Fig. 3 in the same way as the two encoding signal transistors, but with the emitter resistor labelled R_g and the input grounded so that the transistor provides a constant current $i_g = (1volt)/R_g$. g_2 is changed by varying R_g in the same way R_l controls I_{ON}. The non-zero g_2 adds a term $R_y i_g$ to Eq. 5, where

$$g_2 = \frac{R_y i_g}{V_{th}} = \frac{470}{R_g(0.76)} = \frac{618}{R_g}. \tag{9}$$

$R_g = 10\text{k}\Omega$ gives $g_2 = 0.062$.

The V_{noise} terms in Eqs. (4)–(5) approximate white noise voltages. Each V_{noise} is characterized by its measured rms value V_{Nrms} and bandwidth. V_{Nrms} is controlled by changing the gain via the potentiometer in Fig. 4. Noise strength D in Eqs. (2)–(3) is given by

$$D = \frac{V_{Nrms}}{V_{th}\sqrt{\gamma RCf_{c1}}} = \frac{V_{Nrms}}{0.76\sqrt{\gamma(9.4\mu s)(1.5\text{MHz})}} \tag{10}$$

where $f_{c1} = 1.5$ MHz is the cut-off frequency for the amplifier in Fig. 4 and γ decreases from $\pi/2$ at low gain to $\pi/4$ at high gain (when the potentiometer is set to $20\text{k}\Omega$ in Fig. 4).

Circuit Analysis and Simulation

Here we describe the circuit analysis and derive the connections between the model parameters used in Eqs. (2)–(3) and the circuit parameters. Further details are given in Refs.[24], [25].

The single gene circuit in Fig. 2 is designed to reproduce the Hill function inhibition in Eqs. (2)–(3). The op-amp U1 is configured as a subtraction amplifier with gain $G_1 = -\frac{11}{10} = -1.1$. Replication of the Hill function behavior is achieved by allowing saturation of the output of the op-amp U2 and by having different unsaturated gains G_{+2} and G_{-2} for $V_{in} > V_{cth}$ and $V_{in} < V_{cth}$, respectively, due to the diodes in the feedback for U2. G_{-2} is the gain of U2 when its output goes negative, in which case the diodes are not conducting, and therefore $G_{-2} = -\frac{3.3}{1.0} = -3.3$. G_{+2} is a diminishing gain when the output of U2 becomes increasingly positive causing the diodes to go into conduction. An increasing repressor concentration corresponds to V_{in} surpassing V_{cth} which causes the unsaturated output at U2 to change from a negative voltage of $G_1 G_{-2}(V_{in} - V_{cth})$ to a positive voltage $G_1 G_{+2}(V_{in} - V_{cth})$. The increasing voltage at the output of U2 turns the transistor off ($i_t \to 0$) which corresponds to complete inhibition of protein expression. Maximal protein expression $\alpha_{1,2}$ in Eqs. (2)–(3) corresponds to the maximum value of i_t, designated i_{max}. $i_t = i_{max}$ occurs when $V_{in} = 0$ (no repressor) because the output of U2 saturates at $V_{-sat} = -3.5$ V (for the LF412 op-amp supplied by ± 5 V), resulting in a 0.65 V drop across the 220W and therefore $i_{max} = 3$ mA.

Circuit parameters for α and β are found by using V_{th} to convert Eqs. (4)–(5) to a dimensionless form for comparison to Eqs. (2)–(3). The $0.45\text{k}\Omega$ used for R in Eq. (6) comes from the $R_y = 470\text{W}$ being nearly in parallel with the resistance $(10\text{k}\Omega)$ at the input to the subtraction amplifier U1 for gene-x. The relation for Hill coefficient n is found by adjusting the dependence of i_t on V_{in} to match the slope of the Hill function $1/(1+x^n)$ at $x = 1$ resulting in the relation [25]:

$$\frac{-fV_{th}G_1G_{-2}}{f(5 - V_{-sat}) - 0.6} = \frac{-n}{4}. \tag{11}$$

In Fig. 2 the voltage divider fraction $f = 0.4/2.6 = 0.154$ and $V_{-sat} = -3.5$ V. Using Eq. (6) in Eq. (11) yields Eq. (7).

Parameter V_{th}'s correspondence to binding affinity of repressor to DNA is seen by noting thatEqs. (2)–(3) are dimensionless, meaning that in the process of going from chemical kinetic equations to Eqs. (2)–(3) the maximal expression rate α has been scaled by the repressor's inhibition binding constant K_i, and by a mRNA degradation rate [21], [25]. From Eq. (6) it follows that V_{th} is proportional to K_i since α is inversely proportional to K_i due to the scaling. To find the relation between V_{cth} in Fig. 2 and V_{th} we note that the Hill function equals 0.5 when $x = 1$. Therefore i_t must be half its maximum value when $V_{in} = V_{th}$ which gives [25]

$$\frac{i_t}{i_{max}} = \frac{f(5 - G_1 G_{-2}(V_{th} - V_{cth})) - 0.55}{f(5 - V_{-sat}) - 0.6} = 0.5.$$

Solving gives $V_{cth} \approx V_{th} + \frac{1}{G_1 G_{-2}}$. Using $n = 2.4$ and $\alpha = 1.78$ in Eqs. (6)–(7) gives: $G_1 G_{-2} = 3.65$, satisfied by $G_1 = -1.1$ and $G_{-2} = -3.3$; $V_{th} = 1.35/1.78 = 0.76$ V; and $V_{cth} = 0.76 + 1/3.65 = 1.03$ V.

Comparing Eqs. (3) and (5) shows that the encoding signals I_1, I_2 are related to the transistor currents i_1, i_2 by:

$$I_{1,2} = \frac{R_y i_{1,2}}{V_{th}}. \tag{12}$$

The encoding currents $i_{1,2}$ take on two possible values depending on whether their logic train input in Fig. 3 is high or low. When the input is high (>4V) the transistor is off so the current is zero. When the input is zero, the voltage divider consisting of the $4.7\text{k}\Omega$ and $2.2\text{k}\Omega$produces one volt across the R_1 connected to the emitter of the *pnp* transistor creating current$i_{1,2} = (1\text{volt})/R_1$. Equation (12) then gives Eq. (8) for I_{ON}. Results of an analysis for a non-zero value of parameter g_2 are the same as for encoding signals I_1, I_2 and currents i_1, i_2 because g_2's sourcing transistor is set up in the same way as the transistors in Fig. 3 for the encoding signals. Thus the non-zero value of g_2 is Eq. (9).

Here we show how to use simulations to predict the circuit results. In the process we find Eq. (10), the connection between circuit parameters and the noise strength D in Eqs. (2)–(3). A standard Euler-Maruyama simulation of Eq. (2) is

$$x_{i+1} - x_i = \left(\frac{\alpha_1}{1 + y_i^n} - \beta_1 x_i + g_1\right)\Delta t' + D\sqrt{\Delta t'}\,\eta_i(0,1) \tag{13}$$

where $\eta_i(0,1)$ is a unit variance zero mean normal random distribution. The noise circuit inFig. 4 produces a measurable *rms* voltage V_{Nrms} consisting of contributions from all the frequency components present in the noise. The variance of the noise is the integral of its spectral density function $SD(f)$ over frequency, and is obtained from a measurement of V_{Nrms};

$$V_{Nrms}^2 = \int SD(f)\,df. \tag{14}$$

Idealized white noise assumed in Eqs. (2)–(3) has a $SD(f)$ which is uniform over an infinite bandwidth. However for the real noise from the 2-stage noise amplifier circuit in Fig. 4 each op-amp's gain-bandwidth product produces a high frequency cut-off, f_{c1} and f_{c2}. The resulting $SD(f)$ has the form

$$SD(f) = \frac{SD_0}{\left(1+(f/f_{c1})^2\right)\left(1+(f/f_{c2})^2\right)} \tag{15}$$

where SD_0 is a constant related to the strength of the noise. The OPA228 op-amp has a gain-bandwidth product 33 MHz, therefore the first stage in Fig. 4 with fixed gain 22 × has cut-off, $f_{c1} = 33/22 = 1.5$ MHz. The second stage's cut-off f_{c2} varies from 33 to 1.5 MHz depending on the gain setting determined by the potentiometer in the feedback of the second stage op-amp.

The integration in Eqs. (14)–(15) gives

$$V_{\mathrm{Nrms}}^2 = \int_0^\infty \frac{SD_0}{\left(1+(f/f_{c1})^2\right)\left(1+(f/f_{c2})^2\right)} df$$
$$= \frac{f_{c1}f_{c2}}{f_{c1}+f_{c2}} \frac{\pi}{2} SD_0. \tag{16}$$

There are two limiting cases for the integral: for small gain $f_{c2} \approx 33$ MHz $>> f_{c1}$ giving $(\pi/2) f_{c1} SD_0$; and for large gain (potentiometer $\to 20\mathrm{k}\Omega$ in Fig. 4) $f_{c2} \approx f_{c1}$ giving $(\pi/4) f_{c1} SD_0$. Thus the integral in Eq. (16) is $\gamma f_{c1} SD_0$ where $f_{c1} = 1.5$ MHz and γ varies from $\pi/2$ for small noise to $\pi/4$ for large noise. For frequencies within the noise bandwidth (< 1.5 MHz) the amplitude spectral density ASD (units Volt Hz
-1/2) has a constant value given by

$$ASD_0 = \sqrt{SD_0} = \frac{V_{\mathrm{Nrms}}}{\sqrt{\gamma f_{c1}}} \tag{17}$$

ASD_0 is a good approximation of the white noise strength provided that the Fig. 3 genetic circuit's characteristic response rate $1/(RC)$ is much less than the noise bandwidth, meaning that $1/(RC) << 2\pi f_{c1}$. This condition is ensured since the RC used here is $(470\mathrm{W})(0.02\mu\mathrm{f}) = 9.4\mu s$, and the bandwidth of the noise is $f_{c1} \approx 1.5$ MHz.

Figure 5 shows the measured frequency content from the noise circuit when the potentiometer at the second stage is set for gain 11 × producing $V_{\mathrm{Nrms}} = 0.90$ V and $f_{c2} = 33/11 = 3$ MHz. In this case $f_{c2} = 2f_{c1}$ and $\gamma = \pi/3$. Figure 5 shows that the frequency content is relatively flat out to the cut-off near 1.5 MHz and therefore is a good approximation to white noise for the genetic network circuit. $V_{\mathrm{Nrms}} = 0.90$ V is on the high end of the noise amplitudes used here.

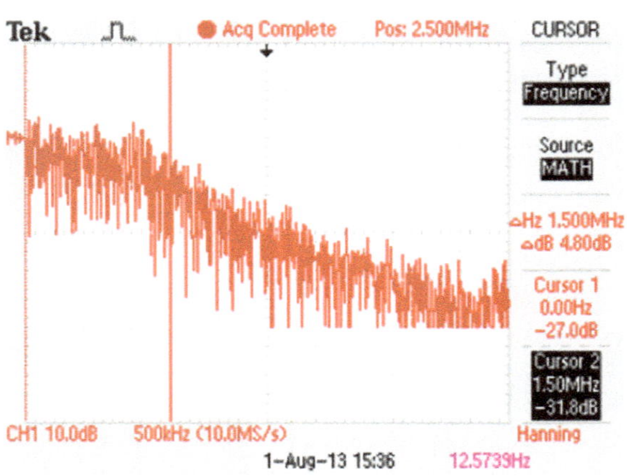

Figure 5: Measured frequency spectrum of noise.

For 2-stage noise amplifier shown in Fig. 4 with second stage gain 11 ×. Horizontal scale is 500 kHz/Div, so cursor indicates 1.5 MHz as location of cut-off frequency. From FFT function on Tektronix TDS 2024B oscilloscope.

http://dx.doi.org/10.1371/journal.pone.0076032.g005

The circuit Eqs. (4)–(5) which need to be simulated are of the form

$$\frac{dV}{dt} = \frac{f(V)}{RC} + \frac{V_{\mathrm{noise}}}{RC} \tag{18}$$

where V_{noise} approximates a white noise voltage. V_{noise} is characterized by its measured *rms*value V_{Nrms} and bandwidth, and Eq. (17) gives the noise's amplitude spectral density. The Euler-Maruyama simulation for Eq. (18) is

$$V_{i+1} - V_i = \frac{f(V_i)}{RC}\Delta t + \frac{ASD_0}{RC}\sqrt{\Delta t}\,\eta_i(0,1). \tag{19}$$

Using dimensionless time $\Delta t' = \Delta t/(RC)$ and the measured noise amplitude V_{Nrms} gives

$$V_{i+1} - V_i = f(V_i)\Delta t' + \frac{V_{\mathrm{Nrms}}}{\sqrt{\gamma RC f_{c1}}}\sqrt{\Delta t'}\,\eta_i(0,1) \tag{20}$$

Normalizing by the voltage scale V_{th} puts Eq. (20) in the form of Eq. (13) and gives the connection between circuit parameters and dimensionless noise amplitude D shown in Eq. (10). For example, using $V_{th} = 0.76$V, $RC = (470\mathrm{W})(0.02\mu\mathrm{f})$, $\gamma = \pi/2$, and $f_{c1} = 1.5$MHz gives $D = 0.28 V_{\mathrm{Nrms}}$.

RESULTS AND DISCUSSION

Figure 6 shows simulations and circuit measurements for three values of noise using parameters $n = 2.4$, $\alpha = 1.78$, $\beta_1 = 0.90$, $\beta_2 = 1$, $I_{ON} = 0.067$, and $g_1 = g_2 = 0$. It is apparent that for an optimal noise level (Fig. 6b) the circuit indeed performs the logic AND/NAND function, and that for the smaller (Fig. 6a) and larger (Fig. 6c) noise values faithful logic response is lost. At the low noise (Fig. 6a) the outputs sometimes fail to respond to the 0 to 1 transition from the AND of I_1, I_2, and for the 1 to 0 transition the outputs often wait until*both* inputs go low before responding thereby causing a delayed response. In Fig. 6b the responses are quick for both the up and down transition. Figure 6c shows that at the high noise level the responses are again quick as in 6b, but the outputs also make erroneous transitions. The circuit behavior is seen to be in agreement with the simulations of Eqs. (2)–(3).

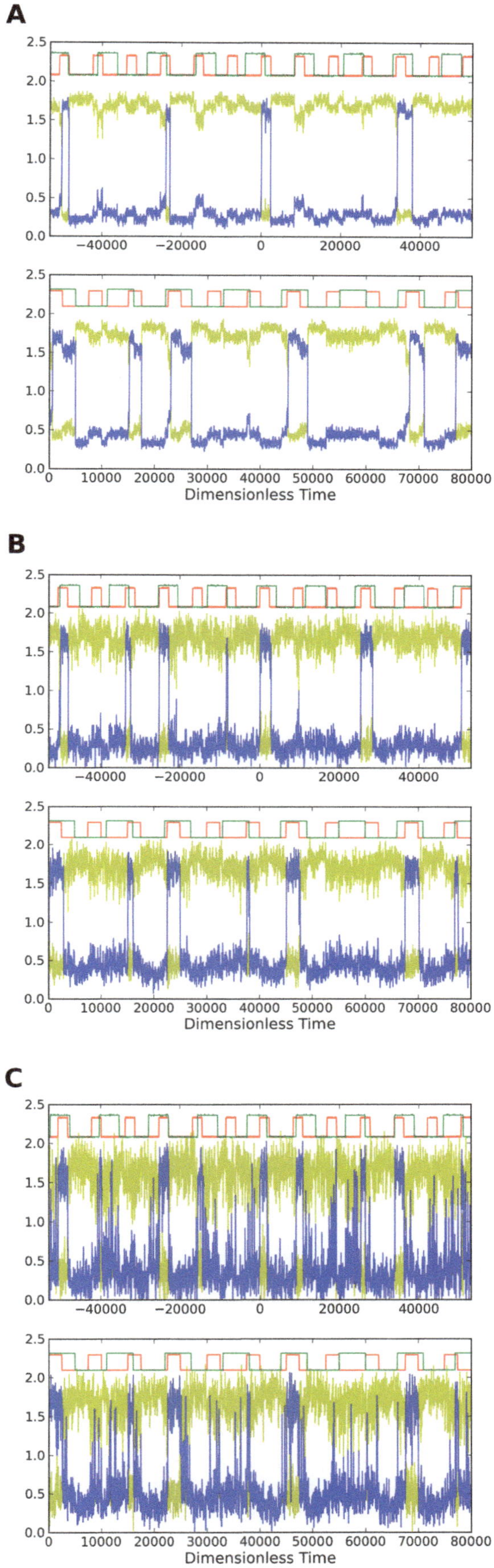

Figure 6: Time series for circuit measurements (upper graph) and simulations (lower) for different noise strengths showing AND/NAND LSR.

Circuit shown in Fig. 3. Simulations are of Eqs. (2)–(3). Upper red and green traces indicate logic states of the encoding inputs $I_{1,2}$ and lower traces show the complementary outputs x(yellow) and y(blue). Panel (b) has noise level within the optimal range for displaying AND/NAND characteristics. Noise strengths in simulation and in circuit are: (a) $D=0.043$ and $V_{\text{Nrms}}=200\text{mV}$, (b) $D=0.107$ and $V_{\text{Nrms}}=500\text{mV}$, (c) $D=0.15$ and $V_{\text{Nrms}}=700\text{mV}$. Voltages and times for the circuit measurements have been converted to dimensionless quantities as described in the text.

http://dx.doi.org/10.1371/journal.pone.0076032.g006

In order to investigate the range of noise strengths which produce accurate logic response, and to find optimal values for the amplitude I_{ON} of the small signal inputs $I_{1,2}$ we define an accuracy measure a,

$$a = a_{\text{L}} \times a_{\text{H}} \tag{21}$$

where a_{L} is the percent of time the x,y outputs are correct when the AND operation of I_1 and I_2 is low, and a_{H} is the percentage of time correct when the AND operation is high. This definition has the property that when the x,y outputs do not respond at all, then $a=0$ since $a_{\text{H}}=0$ even though $a_{\text{L}}=1$. The expectation then is that for no noise there should be no stochastic resonance response so that $a=0$, and that for extreme noise each accuracy would approach 50% so that $a \to 0.25$. If x,y respond immediately with no mistakes then $a=1$.Figure 7 shows accuracy a for simulations and circuit measurements as a function of noise strength for different values of encoding amplitude I_{ON}.

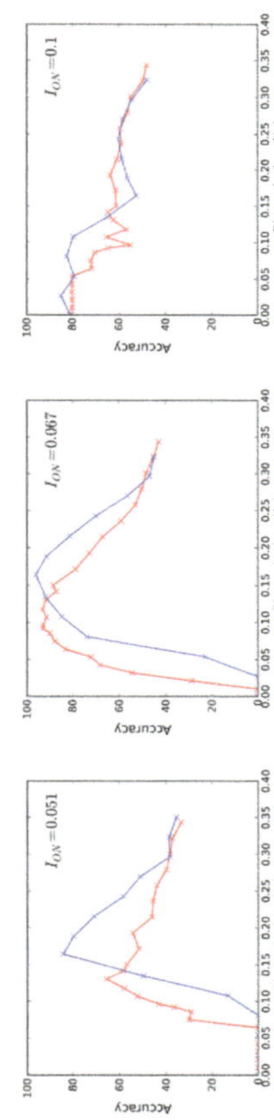

Figure 7: Accuracy a **of the AND/NAND logic response for simulations (red) and circuit measurements (blue) as a function of noise strength** D **and encoding amplitude** I_{ON}.

Noise strength V_{Nrms} for the circuit measurements has been converted to dimensionless strength D using Eq. (10) with $\gamma=\pi/2$.

http://dx.doi.org/10.1371/journal.pone.0076032.g007

Figure 7a shows that for a small value of encoding amplitude, $I_{ON}=0.051$ the network is not able to give a faithful logic response at any noise level. The response at low noise and at high noise are as predicted, $a=0$ and $a\rightarrow 0.25$, respectively, but the peak of the window of stochastic resonance response is well below 1. Figure 7b shows a window of noise providing faithful response for $I_{ON}=0.067$. The reason that the accuracy a is slightly below 1 in the window is that the x,y outputs do not respond immediately to the AND/NAND transitions. This time lag causes the percent of time with incorrect response to be non-zero and therefore $a_H$and a_L are less than 1. In principle an allowance for a time lag could be included in the calculation of a if it were deemed necessary. However, such an added complication would not make the noise window any more apparent than it already is in Fig. 7. Figure 7c shows that at a high value, $I_{ON}=0.10$, the outputs respond even with no noise, and the addition of noise only creates more errors. The relative shift between the simulation and circuit accuracies is due to assumptions made about the noise spectral density function and the integration in Eq. (16) leading to Eq. (20) which gives the connection between the measured noise amplitude V_{Nrms} and dimensionless noise D. In the idealized case Eq. (16) finds γ ranges from $\pi/2$ to $\pi/4$ in Eq. (20), with $\pi/2$ being appropriate for the noise levels used in Fig. 6. Adjusting the value of γ can eliminate the relative shift, but there is nothing to be gained since the appearance of an optimal noise window for LSR at an appropriate value of I_{ON} is already apparent.

One can also *reconfigure* the system to another set of logic functions, namely the fundamental OR/NOR logic, by simply including a non-zero value for g_2. The parameter g_2 effectively changes the relative position and depth of the wells of the bistable system, allowing the response to morph from AND/NAND to OR/NOR. For instance changing g_2 from 0 to 0.062 (with all other parameters unchanged) changes the bifurcation diagram from that in Fig. 1 toFig. 8 showing that the system is now bistable for $\beta_1=0.9$. The result is that the system displays a clear OR and the complementary NOR response as shown by the simulation and circuit results in Fig. 9. The low noise case Fig. 9a shows that at this low noise level the system usually fails to respond. Figure 9a also shows the resting states are reversed from the $g_2=0$ case in Fig. 6a. Figure 9b shows the OR/NOR response at a noise value within the window for LSR, and Fig. 6c shows errors when the noise is too large.

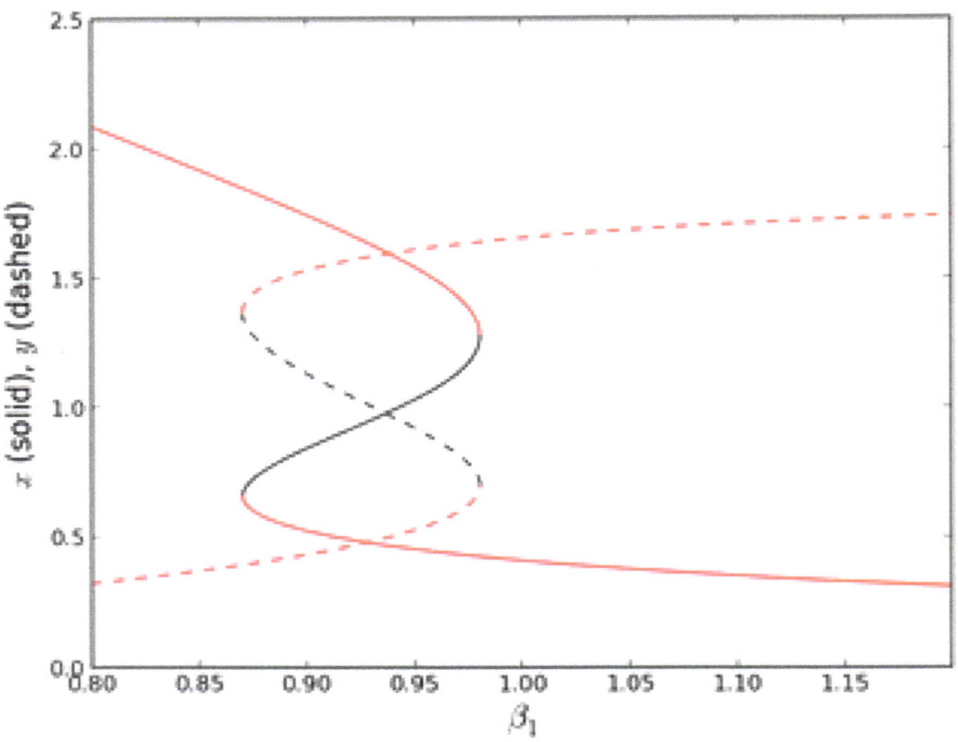

Figure 8: Bifurcation diagram for state-variables x **(solid) and** y **(dashed) configured for OR/NOR.**

x and y are complementary outputs, when one is high the other is low. Red (black) indicate stable (unstable) fixed points. System is bistable for $0.870<\beta_1<0.980$. Calculated for Eqs. (2)–(3) with $n=2.4$, $\alpha_1=1.78$, $\beta_2=1$, $g_1=0$, $g_2=0.062$, $I_1=0$, and $D=0$.

http://dx.doi.org/10.1371/journal.pone.0076032.g008

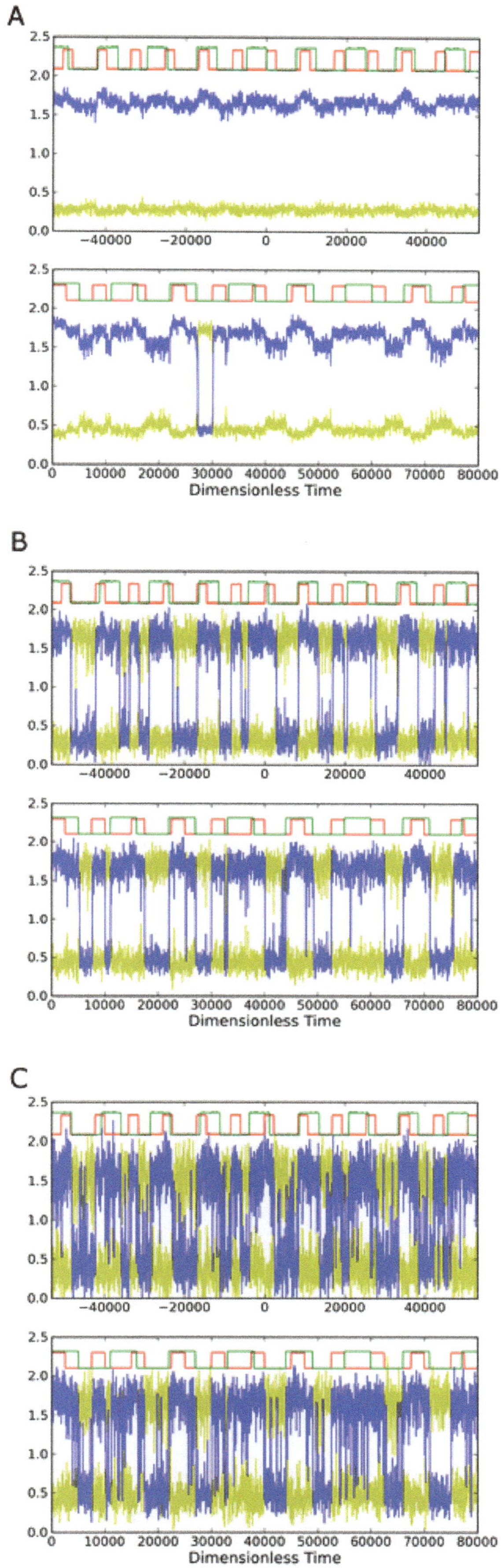

Figure 9: Time series for circuit measurements (upper graph) and simulations (lower) for different noise strengths showing OR/ NOR LSR.

Panel (b) has noise level within the optimal range for displaying OR/NOR characteristics. Noise strengths in simulation and in circuit are: (a) $D=0.043$ and $V_{Nrms}=200mV$, (b) $D=0.107$ and $V_{Nrms}=500mV$, (c) $D=0.15$ and $V_{Nrms}=700mV$. Voltages and times for the circuit measurements have been converted to dimensionless quantities as described in the text.

http://dx.doi.org/10.1371/journal.pone.0076032.g009

In summary, our results show that the dynamics of the two variables x and y with $g_2=0$, mirror AND and the complementary NAND gate characteristics. Further, when $g_2\neq0$, we obtain a clearly defined OR/NOR gate. Since x is low when y is high and vice-versa, the dynamics of the two variables always yield *complementary* logical outputs, simultaneously. That is, if $x(t)$ operates as NAND/NOR, $y(t)$ will give AND/OR.

These results extend the scope and indicate the generality of the recently observed phenomena of Logical Stochastic Resonance through experimental verifications. Further, these observations may provide an understanding of the information processing capacity of synthetic genetic networks, with noise aiding logic patterns. It also may have potential applications in the design of biologically inspired gates with added capacity of reconfigurability of logic operations.

We have also demonstrated that the electronic circuit provides an additional tool for investigating dynamics of proposed genetic networks. The circuit measurements are complementary to numerical simulations, thereby giving indication of the robustness of a particular network design and potential for successful realization in a biological system.

Thus the results presented in this work suggest new directions in biomolecular computing, and indicate how robust computation may be occurring at the scale of regulatory and signalling pathways in individual cells. Design and engineering of such biologically inspired computing systems not only present new paradigms of computation, but can also potentially enhance our ability to study and control biological systems [26].

AUTHOR CONTRIBUTIONS

Conceived and designed the experiments: EHH SKD JK SS. Performed the experiments: EHH EK. Analyzed the data: EHH EK. Wrote the paper: EHH SKD SS.

REFERENCES

1. Mano MM (1993) Computer system architecture (3. ed.). Pearson Education, I-XIII, 1–524 pp.
2. Bartee T (1991) Computer architecture and logic design. Computer science series. McGraw-Hill. URL http:// books. google. com/books?id=b9dQAAAAMAAJ.
3. Bulsara A, Jacobs E, Zhou T, Moss F, Kiss L (1991) Stochastic resonance in a single neuron model: Theory and analog simulation. Journal of Theoretical Biology 152: 531–555. doi: 10. 1016/s0022-5193(05)80396-0
4. Levin JE, Miller JP (1996) Broadband neural encoding in the cricket cereal sensory system enhanced by stochastic resonance. Nature 380: 165–168. doi: 10. 1038/380165a0
5. Gammaitoni L, Hänggi P, Jung P, Marchesoni F (1998) Stochastic resonance. Rev Mod Phys 70: 223–287. doi: 10. 1103/revmodphys. 70. 223
6. Hänggi P (2002) Stochastic resonance in biology how noise can enhance detection of weak signals and help improve biological information processing. ChemPhysChem 3: 285–290. doi: 10. 1002/1439-7641(20020315)3:3<285::aid-cphc285>3. 0. co;2-a
7. Murali K, Sinha S, Ditto WL, Bulsara AR (2009) Reliable logic circuit elements that exploit nonlinearity in the presence of a noise floor. Phys Rev Lett 102: 104101. doi: 10. 1103/physrevlett. 102. 104101
8. Murali K, Rajamohamed I, Sinha S, Ditto WL, Bulsara AR (2009) Realization of reliable and exible logic gates using noisy nonlinear circuits. Applied Physics Letters 95: 194102. doi: 10. 1063/1. 3245318
9. Guerra DN, Bulsara AR, Ditto WL, Sinha S, Murali K, et al. (2010) A noise-assisted reprogrammable nanomechanical logic gate. Nano Letters 10: 1168–1171. doi: 10. 1021/nl9034175
10. Worschech L, Hartmann F, Kim TY, Hoing S, Kamp M, et al. (2010) Universal and reconfigurable logic gates in a compact three-terminal resonant tunneling diode. Applied Physics Letters 96: 042112. doi: 10. 1063/1. 3302457
11. Fierens P, Ibez S, Perazzo R, Patterson G, Grosz D (2010) A memory device sustained by noise. Physics Letters A 374: 2207–2209.
12. Sinha S, Cruz JM, Buhse T, Parmananda P (2009) Exploiting the effect of noise on a chemical system to obtain logic gates. EPL (Europhysics Letters) 86: 60003. doi: 10. 1209/0295-5075/86/60003
13. Singh KP, Sinha S (2011) Enhancement of "logical" responses by noise in a bistable optical system. Phys Rev E 83: 046219. doi: 10. 1103/physreve. 83. 046219
14. Perrone S, Vilaseca R, Masoller C (2012) Stochastic logic gate that exploits noise and polarization bistability in an optically injected vcsel. Opt Express 20: 22692–22699. doi: 10. 1364/oe. 20. 022692
15. Ando H, Sinha S, Storni R, Aihara K (2011) Synthetic gene networks as potential flexible parallel logic gates. EPL (Europhysics Letters) 93: 50001. doi: 10. 1209/0295-5075/93/50001
16. Dari A, Kia B, Bulsara AR, Ditto W (2011) Creating morphable logic gates using logical stochastic resonance in an engineered gene network. EPL (Europhysics Letters) 93: 18001. doi: 10. 1209/0295-5075/93/18001
17. Dari A, Kia B, Wang X, Bulsara AR, Ditto W (2011) Noise-aided computation within a synthetic gene network through morphable and robust logic gates. Phys Rev E 83: 041909. doi: 10. 1103/physreve. 83. 041909
18. Sharma PR, Somani P, Shrimali MD (2013) Bio-inspired computation using synthetic genetic network. Physics Letters A 377: 367–369. doi: 10. 1016/j. physleta. 2012. 12. 002
19. Hasty J, Isaacs F, Dolnik M, McMillen D, Collins JJ (2001) Designer gene networks: Towards fundamental

cellular control. Chaos: An Interdisciplinary Journal of Nonlinear Science 11: 207–220. doi: 10. 1063/1. 1345702

20. Elowitz M, Lim WA (2010) Build life to understand it. Nature 468: 889–890. doi: 10. 1038/468889a

21. Elowitz MB, Leibler S (2000) A synthetic oscillatory network of transcriptional regulators. Nature 403: 335–338. doi: 10. 1038/35002125

22. Gardner TS, Cantor CR, Collins JJ (2000) Construction of a genetic toggle switch in escherichia coli. Nature 403: 339–342.

23. Wang J, Zhang J, Yuan Z, Zhou T (2007) Noise-induced switches in network systems of the genetic toggle switch. BMC Systems Biology 1: 50. doi: 10. 1186/1752-0509-1-50

24. Hellen EH, Volkov E, Kurths J, Dana SK (2011) An electronic analog of synthetic genetic networks. PLoS ONE 6: e23286. doi: 10. 1371/journal. pone. 0023286

25. Hellen EH, Dana SK, Zhurov B, Volkov E (2013) Electronic implementation of a repressilator with quorum sensing feedback. PLoS ONE 8: e62997. doi: 10. 1371/journal. pone. 0062997

26. Benenson Y (2012) Biomolecular computing systems: principles, progress and potential. Nat Rev Genet 13: 455–468. doi: 10. 1038/nrg3197

CHAPTER 10

MULTICELLULAR COMPUTING USING CONJUGATION FOR WIRING

Angel Gon˜ i-Moreno[1], Martyn Amos[2], Fernando de la Cruz[3]

[1]Systems Biology Program, Centro Nacional de Biotecnologı´a CSIC, Cantoblanco-Madrid, Spain

[2]School of Computing, Mathematics and Digital Technology, Manchester Metropolitan University, Manchester, United Kingdom

[3]Departamento de Biologı´a Molecular e Instituto de Biomedicina y Biotecnologı´a de Cantabria (IBBTEC), Universidad de Cantabria-CSIC-SODERCAN, Santander, Spain

ABSTRACT

Recent efforts in synthetic biology have focussed on the implementation of logical functions within living cells. One aim is to facilitate both internal "re-programming" and external control of cells, with potential applications in a wide range of domains. However, fundamental limitations on the degree to which single cells may be re-engineered have led to a growth of interest in multicellular systems, in which a "computation" is distributed over a number of different cell types, in a manner analogous to modern computer networks. Within this model, individual cell type perform specific sub-tasks, the results of which are then communicated to other cell types for further processing. The manner in which outputs are communicated is therefore of great significance to the overall success of such a scheme. Previous experiments in distributed cellular computation have used global communication schemes, such as quorum sensing (QS), to implement the "wiring" between cell types. While useful, this method lacks specificity, and limits the amount of information that may be transferred at any one time. We propose an alternative scheme, based on specific cell-cell conjugation. This mechanism allows for the direct transfer of genetic information between bacteria, via circular DNA strands known as plasmids. We design a multi-cellular population that is able to compute, in a distributed fashion, a Boolean XOR function. Through this, we describe a general scheme for distributed logic that works by mixing different strains in a single population; this constitutes an important advantage of our novel approach. Importantly, the amount of genetic information exchanged through conjugation is significantly higher than the amount possible through QS-based communication. We provide full computational modelling and simulation results, using deterministic, stochastic and spatially-explicit methods. These simulations explore the behaviour of one possible conjugation-wired cellular computing system under different conditions, and provide baseline information for future laboratory implementations.

INTRODUCTION

The growing field of synthetic biology [1]–[4] is concerned with the application of engineering principles, concepts and techniques to the modification and/or construction of biological systems. This (re-)engineering may be motivated by a desire to better understand the underlying biological substrate, or by novel applications of biological "devices". Although the field traces its roots back to early work on genetic engineering, it emerged as a serious research area at the turn of the century, with the simultaneous publication of two significant papers. The first, by Elowitz and Leibler [5],

Citation: Goñi-Moreno A, Amos M, de la Cruz F (2013) Multicellular Computing Using Conjugation for Wiring. PLoS ONE 8(6): e65986. doi:10.1371/journal.pone.0065986.

described a fully-synthetic genetic oscillator engineered into the *E. coli* bacterium. Their "repressilator" relied on mutual repression of a "loop" of connected genes in order to achieve oscillation. The other paper, due to Gardner, Cantor and Collins [6], outlined the design and construction of a synthetic toggle switch (also in *E. coli*), the state of which could be "flipped" from outside by either chemical or thermal induction. Both of these constructions are now standard motifs in the design of synthetic biological systems, and provided inspiration for the construction of a number of genetic devices [7]–[9]. However, just as the pioneers of computer technology quickly incorporated the early transistor into larger circuits in order to build the first solid-state computers, researchers in synthetic biology rapidly sought to build ever larger devices using these gene-based components.

There exists, though, a fundamental limitation on the amount and type of novel genetic "circuitry" that may be introduced into a single living cell. As the authors of [10] argue, "...establishing the wiring of an electronic circuit just requires linking each pair of connected elements by a wire (e.g. a piece of copper). But inside a cell, the cables need to have a different implementation: different proteins must be used for each different pair. Additionally, because of the intrinsic difficulties of implementing them, the resulting constructs are usually specific for the given problem and cannot be reused afterwards". Because of these related issues of cross-talk and lack of modularity, many researchers now seek an alternative approach. By expanding the scope of synthetic biology beyond *single-cell* solutions, and into the domain of *multicellular* systems, we seek to harness the inherent power of biological "nanotechnology", but in a way that readily allows for scalability, noise tolerance and component reusability. With this in mind, attention is turning to the engineering of microbial*consortia* [11]; *multiple* populations of microbes that can interact to perform functions beyond those achievable by *individual* populations. The power of such consortia is derived from both their *robustness* and their ability to perform complex tasks in a *distributed* fashion. These attributes are due to two main features; *communication* between consortium members (that is, the exchange of information between individuals), and *division of labour* (the overall behaviour of the system is the result of combining many sub-tasks, each performed by individuals or sub-populations) [11].

Recently, the notion of distributed, multicellular computation using engineered cells has gained increasing traction [10]. One of the first such systems was demonstrated by Basu *et al.* in 2005 [12], in which the authors demonstrated programmed pattern formation, using populations of engineered "sender" and "receiver" cells. More recently, Tamsir *et al.* [13] showed how simple logic gates may be constructed by programming communication between bacterial colonies, and Regot *et al.* [14] demonstrated a similar system in yeast. Central to the implementation of multicellular computation is controlled *communication* between cells and populations of cells. So far, this has *generally* been implemented using the *global*communication capabilities offered by *quorum sensing* (QS) [13]–[15] (although the use of bacteriophage has also recently been proposed [16]). Within the QS system, one cell (sender) uses small signalling molecules that diffuse over distance and thus reach other cells (receivers) [17]. Apart from the implementation of logical functions, QS has been used for other purposes, such as the synchronization of engineered oscillators [18]. However, we believe that multicellular computation will greatly benefit from a more *varied* range of communication protocols. As Ortiz and Endy highlight [18], QS-based communication is limited by both the *type* and *content* of messages that are possible using chemical signals.

We therefore propose a scheme for multicellular computation based on a *local* communication protocol. The foundation for this is the process of genetic exchange between bacteria known as *bacterial conjugation* [19], which has often been likened to "bacterial sex" [20], [21]. During conjugation, two cells establish a direct, bridge-like connection, called the *pilus*, which brings the cells together. A separate channel is then opened in the respective cell walls, through which a single DNA strand is transferred from the *donor* cell to the *recipient* cell [20]. The importance of this transfer process is that it facilitates the transmission of large, specific genetic messages, which can have arbitrary content. We therefore seek to harness its potential in order to facilitate communication within an engineered cellular population.

In order to achieve this goal, we use *site-specific recombination systems* [22], which allow individual cells to dynamically rewrite their DNA "message". Recently, three bacterial site-specific recombination systems have been used to implement biocomputing devices. The Cre/*lox* system was used to engineer a genetic switch [23], and the *fim* system was used to engineer both a multiplexer [24] and a sequential switch-based memory [25], [26]. The latter example made use of the Hin/*hix* system, which was also used to solve various small instances of mathematical problems in bacterial populations [27], [28]. These implementations demonstrate the power and applicability of such site-specific recombination systems. Futhermore, as this manuscript was under preparation, Siuti *et al.* [29] published their recombinase-based approach to the implementation of logic and memory functions in *E. coli*, underlining the utility of this approach. We present the results of extensive simulation-based experiments, which support the in-principle feasibility of our approach. This work offers a firm foundation for experimental investigations into distributed multicellular computation using conjugation as a core "technology".

RESULTS

We first describe the communication mechanism through conjugation by showing how it may be used, in principle, to implement a single Boolean NOR function. Secondly, we expand the concept by designing a distributed population to implement the exclusive OR (XOR) function, based on mixing three bacterial strains with individual NOR functionality. While using only one site-specific recombination system is sufficient for the NOR-based approach, a combination of two is necessary for the adequate functioning of the XOR computation. For clarity of description, we use abstract labels for the components (except those involved in the recombinase-based logic systems).

Wiring with Conjugative Plasmids: 2-Strain NOR Population

Recall that the two-input NOR function is a negated OR, and thus returns the value 1 if and only if both inputs are zero, and 0 in all other cases. Figure 1A shows how a NOR gate may be constructed using two engineered bacterial strains that communicate via the QS molecule AHL (as in [13]). The evaluation of the NOR function is executed by a *sender* strain, and the output is sent to the *receiver* strain via QS. In this way, QS acts as a *wire* connecting the two components. When no inputs are present (top row of 1A), gene *G1* is *off* (not being transcribed), and the AHL signalling molecules (controlled by a repressible promoter) are expressed by *G2*. These molecules arrive at the *receiver* cell and, after binding to the corresponding transcription factor, induce expression of gene *G3* (the reporter gene). This latter product is read as the output of the NOR logic function (fluorescence detection of the amount of green fluorescent protein, GFP). On the other hand, when one or more inputs are present (bottom row, "A and/or B"), their corresponding promoter is activated (*pA* and/or *pB*), and gene *G1* expresses a repressor, which in turns inhibits the production of AHL. Thus, the*receiver* cells are *off*, and no fluorescence is observed.

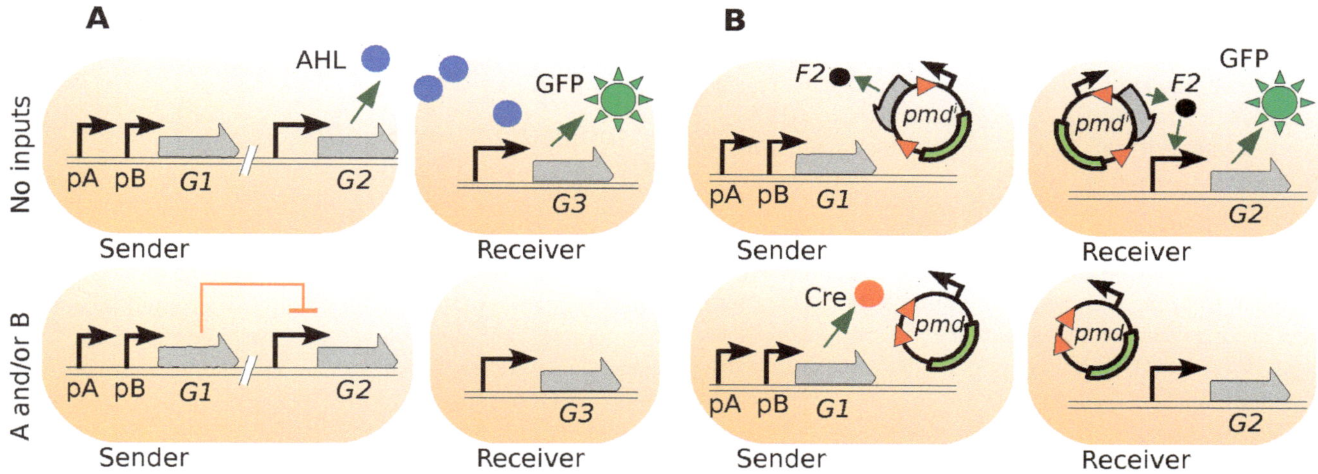

Figure 1: Intercell connection of a NOR logic function.

A. *Traditional* wiring through AHL signalling. When no inputs (*A* and *B*) are present (upper row), the AHL quorum-sensing effectors (expressed by gene *G2* and controlled by a repressible promoter) are produced by the Sender cell and the light (green fluorescent protein, GFP, transcribed by an inducible promoter) is switched on in the Receiver cell. On the other hand, when one or more inputs are present (lower row), AHL production is repressed (via the expression product of gene *G1*) and the fluorescence switched off. B. Wiring by using conjugation. In the 0-0 case (no inputs, upper row) the plasmid pmd^i (i=inducer) *travels* from the Sender to the Receiver cell without modification and induces the expression of GFP. In any other case, the protein Cre is expressed (from gene *G1*) in the Sender, which irreversibly alters the plasmid by deleting the segment between *lox* sites (red triangles). The modified plasmid cannot produce inducers (*F2*), and the Receiver is switched off.

http://dx.doi.org/10.1371/journal.pone.0065986.g001

Our *alternative* design for a distributed NOR gate using conjugation as wiring is shown inFigure 1B. As before, the sender cell computes the NOR function and communicates the result to the receiver cell. The key difference with the previous scheme lies in the communication mechanism; here, we use *conjugative plasmids* [30], rather than QS, to transmit a result. Conjugative plasmids are circular strands of DNA that may be transferred between bacterial cells during the process of conjugation [20], [31]. Although, in the wild, these plasmids allow bacteria to exchange potentially useful genes, here we use them for the transmission of *logical values*.

When no inputs are present (Figure 1B(top row)), promoters *pA* and *pB* are not induced, and the expression product of *G1* (in this case, protein Cre) is not present in the cell. As a result, a constitutive promoter in the plasmid pmd^i (a plasmid that produces inducers) expresses its downstream gene, which results in inducer *F2*.

When the sender cell comes into contact and conjugates with a recipient, thus forming a "wiring" connection, the plasmid is transferred into the receiver cell. Inducers *F2* are then produced inside the receiver by plasmid pmd^i, which are in charge of inducing the expression of the reporter gene (*G2*). If the resulting product is the green fluorescent protein (GFP), the fluorescence (output) is *turned on*. Otherwise (one or more inputs, bottom row), the protein Cre is expressed, which *deletes* a specific DNA segment surrounded by *lox* sites in the plasmid. As a result, the plasmid pmd^i is converted into *pmd* (a plasmid that is unable to produce inducers, due to the lack of the lox-flanked DNA segment). When transferred into the receiver cell, this plasmid will *not* express inducer *F2* , and fluorescence emission will be switched *off*. Thus, the conjugation wire successfully transmits the signal via aplasmid.

It is important to note that when the modified plasmid *pmd* is produced - via deletion - in the sender and transferred to the receiver, it cannot coexist in the latter strain with the previously-introduced unaltered plasmids pmd^i (before the input signals were introduced to the population). Incompatibility is a property of plasmids that contain the same replication genes[32], and is used here to induce "competition". In this scenario, the population of plasmids in the receiver cells gradually shifts towards the "new" (transmitted) plasmid, as the latter gradually replaces the original. Since both plasmids are viewed as identical from the point of view of replication control, the copies of pmd^i and pmd *compete* for an effective resource - the number of plasmids that are allowed in the cell, referred to as the *copy number*. Since copies of pmd are continuously "pumped" to receiver cells, in practical terms they eliminate pmd^i. As we see later, this fact has been incorporated into the model, and results in a progressive change in the plasmid population of the receiver, from pmd^i to pmd.

Single Cell Behaviour

The simulated single-cell behaviour of this design is observed, over time, in Figure 2, where the logic case 1-1 (both inputs present) is applied to a population initially in the 0-0 state. In this idealised set of simulations, we consider the existence of only one cell of each kind (sender and receiver). We also assume a constant conjugation process (as if the wire was permanently *connected*), so that both cells share the plasmid populations (a theoretical ideal state induced in order to test the individual components of the system). As we observe in both deterministic (top) and stochastic (middle) plots, we see a short *flash* from the receivers (GFP initially being expressed), while the expected output should be "0" (corresponding to the NOR 1-1 case). This is due to the fact that from time 0, the senders transfer plasmid pmd^i (in its initial configuration) into the receivers. Thus, inducer $F2$ is initially expressed. As Cre is produced, it transforms the plasmid pmd^i into pmd (the transformation of pmd^i being seen in the bottom graph). When no pmd^i plasmids remain in the ideal system studied in Figure 2, the amount of GFP is controlled only by degradation, and the desired output is reached. Other molecular relations corresponding to the deterministic simulation.

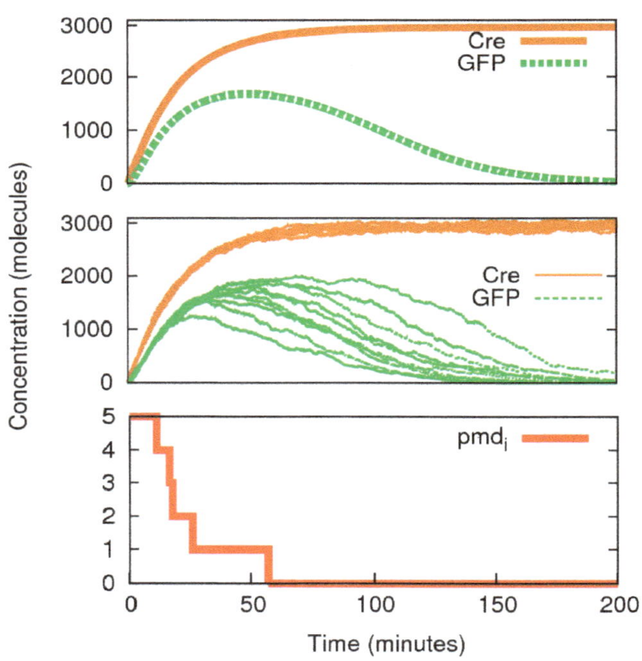

Figure 2: Deterministic and stochastic time evolution of the 1-1 logic case.

These results show the simulation of Figure 1B with input molecules (A and B) set up to 500 (each) (k_{11}=k_{12}=500 molecules min^{-1}) and the copy number of the plasmid to 5. By monitoring (deterministically, upper graph, one run; stochastically, middle graph, ten runs) proteins Cre (in the Sender) and GFP (in the Receiver) we see how GFP production is initially triggered by those plasmids that have not been modified yet (pmd^i). When Cre has been functioning long enough (t≃57 min in this simulation) the remaining GFP is only controlled by degradation rates, as no more fluorescent proteins are being expressed. Lower graph shows the stochastic evolution of pmd^i over time, which determines the delay in displaying the correct output according to the NOR logic function (0 output for the 1-1 case).

http://dx.doi.org/10.1371/journal.pone.0065986.g002

Multicellular Behaviour

A more realistic simulation, which considers physical interactions between cells of a population, is shown in Figure 3. While the cell's logic is simulated via deterministic equations, plasmid numbers (pmd^i and pmd) are discrete (see Section Methods for details on discretization). Importantly, deterministic behaviours are not appropriate for low molecular concentrations, and can result in unrealistic behaviours. Two populations are grown into surfaces with a different distribution of inputs over 24 h. The first (Figure 3A top) has no inputs present and, as a result, recipient cells (receivers) display green light (GFP) corresponding to a logic "1" in output. The second population is grown on a simulated surface containing *both* inputs (A and B, Figure 3A). In the latter simulation, we clearly see the effects that the initial *flash*(explained in Figure 2) has on the output of the circuit. While the plasmids pmd^i are transformed in the donors at an early stage of the process, those that were initially copied into the receivers are constantly being replicated (until they reach their respective copy number) in the population. This *competition* between pmd acquisition and pmd^i replication (in receivers) results in fluorescent cells after 24 h. However, this fluorescence eventually decays (plasmids pmd^i are eventually lost), and we clearly observe the output "0". Figure 3B shows the process of *losing* plasmid pmd^i (due to the proliferation of cells without any copy of pmd^i). At time 930 min, a fluorescent cell has only one copy of pmd^i, due to the lack of further replication of the existing plasmids in the cell (the rest of the copies are pmd). After division (at 950 min) the pmd^i plasmid will go to one of the daughters (all the plasmids are shared randomly). This cell cannot maintain the level of fluorescence after a few min (GFP decay; 990 min). Almost identical behaviour to that seen in input case 1-1 is observed in cases 0-1 and 1-0 (data not shown), as Cre is also expressed in sender cells.

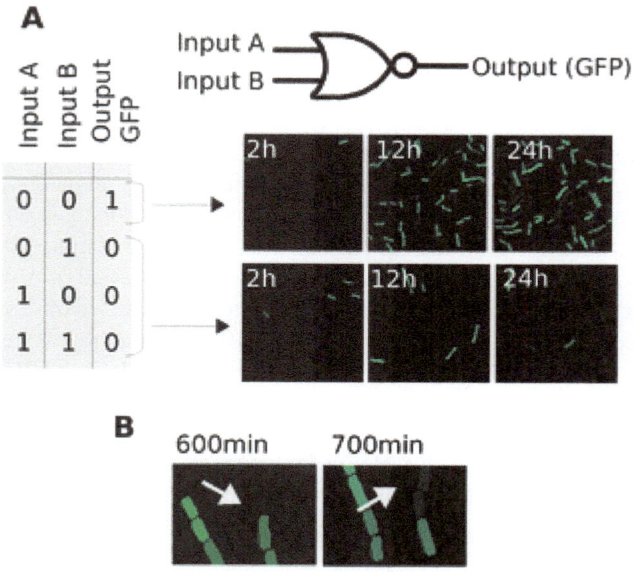

Figure 3: Spatial behaviour of the NOR "wire" over 24 h.

A. Truth table of the NOR "wire" represented in Figure 1B where inputs A and B can be "0" (0 molecules) or "1" (1000 molecules). Two simulated cell populations are monitored at 2, 12 and 24 h, where the intensity of green colour is directly proportional to the concentration level of GFP (green fluorescent protein). The first one (upper row) corresponds to the logic case 0-0, resulting in an output "1" (lights *on*). The second one (lower row) corresponds to the logic case 1-1 (similar functioning for 0-1 and 1-0 cases) resulting in lights *off*. B. Detail of the 1-1 case simulation where the plasmid with inducer (pmd^i) is being replaced by the modified version due to inflow (conjugation) and replication after division. Numbers in cells represent the amount of pmd^i plasmids. After division of a cell with only 1 pmd^i, one of the daughter cells will have no copies of this set. Thus, GFP will decay.

http://dx.doi.org/10.1371/journal.pone.0065986.g003

By varying the conjugation frequency we observe changes in the quantitative information of the system, while qualitative behaviours are still the same. Increasing this frequency causes cells to be overloaded by plasmids (above their copy number), and the replication of plasmids becomes less frequent. For details about cell features, including conjugation frequencies, please see the Methods section.

Extending the Approach: 3-Strain XOR Population

We now show how collections of NOR gates may be connected together in order to compute the exclusive OR (XOR) function, as in [13]. Figure 4 shows the design of a three-strain distributed XOR population, as a proof of principle of the extensibility of the initial approach. The XOR (exclusive OR) function outputs "1" when the inputs are different and "0" when the inputs are the same. Importantly, the inputs are found as *clear* digital values, which are either abundant ("1") or non-existent ("0"). This avoids potential problems with *half*-values [33]. This logic function may be simulated by connecting three NOR gates, as shown in Figure 4A, which is the configuration of our engineered three-strain population. Exactly the same scheme was used in [13], only with QS molecules connecting the gates. The cells named NOR_1 are the "donors" of the community, and their inputs correspond to the inputs of the whole XOR function (molecules *A* and *B*). Their output, *F1* (Figure 4A) encodes the information that travels through the "wire" towards the next logic gates NOR_2 and NOR_3. These latter gates take *F*las one of their inputs, and either A or B (respectively) as the other. The output of NOR_2 and NOR_3 cells, which we call *F2* (the same signal), is the final output of the XOR function. InFigure 4B we show the truth tables of individual strains as well as the emergent XOR logic.

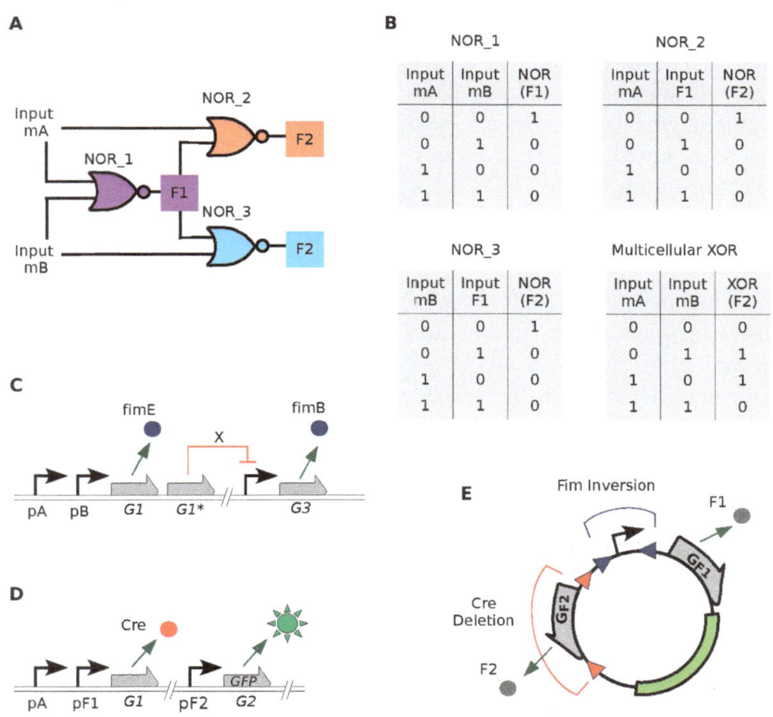

Figure 4: Multicellular design of a distributed XOR circuit.

A. Schema of the three-strain population and its *connections*. Cell NOR_1 (donor) takes its inputs from the concentration of molecules A and B (inputs of the whole circuit) and its output is named "F1" (which is a specific plasmid configuration). The inputs for cell NOR_2 (recipient) are molecule A and F1. Similarly, the inputs for cell NOR_3 (recipient) are B and F1. The output for the XOR function, "F2" is the combination of outputs from NORs 2 and 3 (green fluorescent protein, GFP, in this example). B. Truth tables for each single gate and for the full XOR circuit. C. Inside logic of NOR_1. In the case 0-0 (no inputs present), gene *G3* expresses *FimB* which, in turns, alters the plasmid present in that strain by inverting the promoter in random direction. In any other case, gene *G1* expresses *FimE* which inverts the promoter region of the plasmid pointing towards gene "F2"; While gene $G1^*$ expresses repressor X which stops the production of *FimB* . D. Inside logic of cell NOR_2. Once the plasmid is in the cell (*wire connected*), if no inputs present, GFP is expressed. Otherwise, *Cre* is produced (inducers A bound to inducible promoter pA and/or inducer $F1$ bound to $pF1$) which deletes gene $F2$ from the plasmid. (Cell NOR_3 has the same circuitry but it is sensitive to input B instead of A). E. *Wiring* plasmid. Due to Fim inversion, the promoter can point towards G_{F1} or G_{F2}. Due to Cre deletion, gene G_{F2} can be removed.

http://dx.doi.org/10.1371/journal.pone.0065986.g004

The configuration of NOR_1 cells is depicted in Figure 4C. If no inputs are present, gene $G3$ - controlled by a repressible promoter - produces FimB. This causes random inversions of the promoter in the plasmids (Figure 4E). As a result, about half of the plasmids will have the promoter pointing towards the gene named G_{F1}, while the rest will have the promoter pointing towards gene G_{F2}. Thus, there will be a positive amount of molecule $F1$ in the environment. This output is taken as a logical "1". Otherwise, inducible promoters pA and/or pB switch *on*-genes $G1$ and $G1*$, which produce FimE and repressor X respectively. Repressor X inhibits the expression of $G3$. Thus, FimB will disappear from the sender cells, and the presence of FimE causes directional inversion in the promoter, which will then point towards gene G_{F2}. There will thus be a lack of expression products $F1$ in the environment, which represents a logical "0". Importantly for our design, as noted in [24], the FimE protein causes directional inversions with nearly 100% fidelity, while FimB causes inversions on both directions equally well [34], [35].

Figure 4D shows the inside *program* of NOR_2 cells. Importantly, this configuration is identical to the combination of the senders and receivers of Figure 1B. If no inputs are present, which means that input molecule A is not in the environment and molecule $F1$ is not being produced by the plasmid (all promoters are pointing at gene G_{F2}), the protein Cre is not expressed. Thus, the segment surrounded by *lox* sites in the plasmid is not deleted. Molecules $F2$ are then effectively expressed and, in turn, induce the production of GFP. If either one or both inputs are present (because A is present and/or $F1$ is being expressed by one of the plasmids), protein Cre is expressed. This causes a modification in the plasmid, which loses the G_{F2} gene region by deletion. No inducers of GFP are present and the fluorescence is*switched off*.

Simple Behaviour

A simplified simulation (not considering spatial factors or physical dynamics) of the donor (NOR_1) cells, according to a specific input profile over time, is shown in Figure 5. Input case 1-1 (both inputs present) is induced at 0 min by *introducing* the inputs to the population and maintaining their concentration. At time 200 min we stop introducing inputs, and their concentration is only controlled by degradation rates until case 0-0 is reached. From time 400 min to 500 min we repeat the process. According to that input profile, we observe deterministic oscillation of FimE/FimB (top graph), as well as oscillation between the two possible plasmids in NOR_1 (middle graph) (pmd_{F2}, plasmid where the promoter points towards G_{F2} and pmd_{F1}, plasmid where the promoter points towards G_{F1}). The latter relation is shown stochastically (bottom graph), with a copy number of 5 plasmids, in order to show realistic transitions within discrete numbers. In all graphs we clearly observe the delay produced between the time we stop introducing inputs and the time they effectively disappear from the system. It is important to note that the initial distribution of plasmids (t=0) is no longer repeated in the simulations. That pristine condition is only valid at the very beginning, and is not representative of the working system. From this simulation we can clearly see how the logic is fully reversible, requiring only changes to inputs, with no modification to the cells. This feature is possible precisely because of the reversible nature of the fimE/B system (i.e., it does not delete the segment).

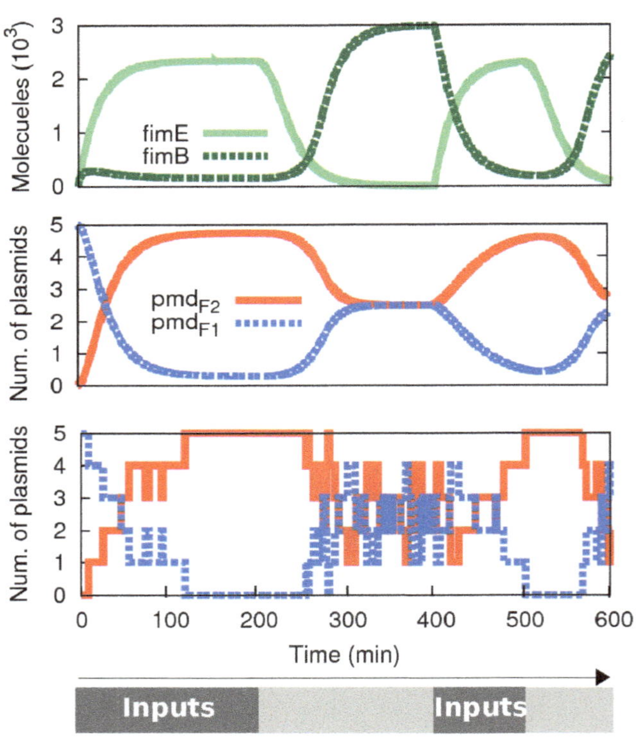

Figure 5: Time evolution of NOR_1 according to a specific input profile.

Logic case 1-1 is induced during the intervals [0...200] min and [400...500] min (A and B=500 molecules -constant entry- during the interval). The case 0-0 domains during the rest of the 600 min. According to that profile we observe the deterministic oscillation of FimE/FimB (top graph) as well as the oscillation between the two possible plasmids in NOR_1 (

pmd_{F2} and pmd_{F1}). The latter relation is shown deterministically (middle graph) and stochastically (bottom graph) (copy number=5). Delays in response are due to input degradation times.

http://dx.doi.org/10.1371/journal.pone.0065986.g005

Spatial Simulation

A full spatial simulation of the XOR population is shown in Figure 6. The three different strains are grown over a surface that contains the corresponding combination of inputs (the four logic cases shown). The visual output is represented by the intensity of green colour (high="1"; low="0") proportional to the concentration of GFP in each cell. All snapshots are taken after 32 h*cultivation* (cells have a doubling time of approximately 100 min). As expected, cases 0-1 and 1-0 are the ones that display a positive output following the XOR function. Below each snapshot, we show a graph that pictures the number of plasmids of each four possibilities (pmd_{F2}, promoter pointing towards G_{F2}; pmd, promoter pointing towards G_{F2} after its deletion; pmd_{F1_F2}, promoter pointing towards G_{F1}; pmd_{F1}, promoter pointing towards G_{F1}after gene G_{F2} deletion). The number of plasmids is averaged per cell within the receiver population (cells NOR_2 and NOR_3) at the same time as the snapshot is taken (32 h).

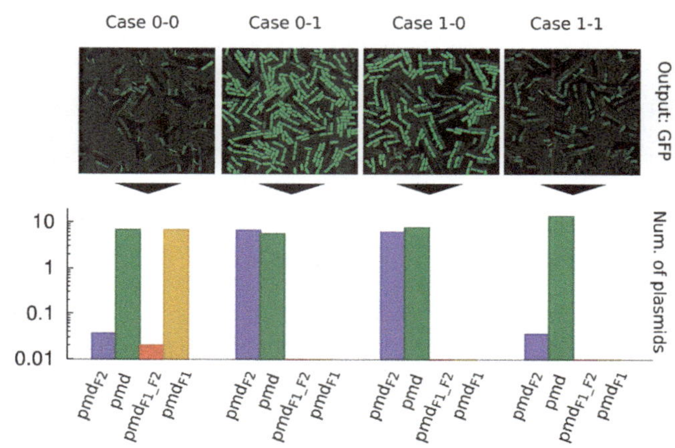

Figure 6: Spatial simulation of the XOR population using the four logic cases.

For each input case (0-0, 0-1, 1-0 and 1-1) a snapshot of the population after 32 hours is shown along a profile (bar chart) of the average number of plasmids of each kind (the four possibilities) in recipients (NOR_2 and NOR_3) (copy number=10). The output, visually identified as green colour cells (from green fluorescent protein, GFP) corresponds to the XOR function. The average number of plasmids pmd_{F2} in a population is directly proportional to the fluorescence of it. The "y" axis in the bar graph is displayed in logarithmic scale. Short bright green lines in spatial figures of the population represent a conjugation process currently happening (those "springs" link donor and recipient).

http://dx.doi.org/10.1371/journal.pone.0065986.g006

The phenomenon of unwanted light *flashes* still occurs at this point in NOR_2 and NOR_3. As described previously, the presence of this phenomenon does not affect the overall functioning of the circuit. Also, conjugation frequency plays an important role in the quantitative behaviour. However, qualitative functioning is still that of an XOR function.

DISCUSSION

It seems increasingly clear that a significant amount of future research in the field of synthetic biology will be concerned with the construction of engineered microbial consortia [10], [11],[14]. The growing complexity of proposed applications (e.g., in energy, biomedical engineering, or environmental monitoring and bio-remediation [36]) will require approaches that allow a task to be broken down into separate sub-tasks, each of which may be addressed by a separately-optimised population of cells. From an historical perspective, this is entirely consistent with the development of computer software, which has moved from a *monolithic* model (in which an application is self-contained and isolated from other programs) to an object-oriented, *distributed*model, in which a system is viewed as a collection of *interacting entities* [37]. Apart from the benefits accrued from a division of labour, mixed populations are also robust to environmental perturbations, which will be particularly significant in "real world" applications [11]. In order to achieve this distributed, engineered cellular computation, we require a reliable *communication*protocol that will allow cells and populations to exchange information, exert control and influence the overall system behaviour. As has already been demonstrated, several such protocols are possible, based on (for example) quorum sensing (QS) [13], [14] or phage infection [16]. In this paper, we propose a third alternative communication scheme, based on bacterial *conjugation*.

A clear advantage of the proposed system is the *amount* of information that may be communicated. Quorum-sensing systems are based on the exchange of small signalling molecules, which are ideal for global synchronization of a system, where the controlling signal may be relatively simple. However, such systems lack the richness and complexity of cell-cell DNA-based methods (such as conjugation or phage), using which we may exchange relatively large "packets" of genetic material [16]. An added advantage of using conjugation as a wiring system is the requirement for *physical contact* between donor and recipient cells. Thus, if cells are constrained in their movement (e.g., because they are growing on a solid surface), conjugation allows fine-grained *spatial resolution* of computational processes (as opposed to the global "broadcast" model of QS). This will allow, for example, the precise targeting of the*location* of a specific input signal within a monolayer of communicating cells.

Another important benefit of engineered microbial consortia is *modularity* and *reusability* of components [10]. The ability to combine basic components together in order to build larger structures is a fundamental engineering principle, and facilitating the reuse of cellular systems is of paramount importance to practitioners in synthetic biology. We illustrate this with reference to our proposed XOR circuit. Our initial two-strain population implementing the NOR function is effectively non-reusable, as it cannot be "reset". This problem is solved in the XOR population, where the system may compute indefinitely due to the reversible nature of the fimE/B system.

Computational simulations allow us to "test" proposed cellular systems in many different scenarios. In this paper, we provide not only deterministic/stochastic single-cell simulations (which provide us with an idealised perspective of the system, impossible to achieve *in vivo*), but also multicellular spatial simulations, which allow us to better understand the *population-level* dynamics of the system. Taken together, these computational studies offer a valuable insight into the proposed system, in order to prepare future wet-lab implementations of our design. Such a framework will allow for relatively easy investigation of implementation-specific issues, which may be expensive or time-consuming to test in the laboratory. For example, in future, we may use our *in silico* approach to explore a wider combination of site-specific recombination tools (only two were combined in this work), and how we may harness their different functionalities (such as insertions, cointegrations, deletions and inversions). Implementation of conjugation-based wiring requires the consideration of various practical issues. Most importantly, there are two types of conjugation channels; those that work with bacteria moving freely in a *liquid* medium (liquid maters, e.g., plasmids F and R64), and those that only work on the surface of a *solid* medium (surface maters, e.g., plasmids RP4, R388 and R46) [38]. The latter allow better control of the experiments, since conjugation only starts when the bacterial population is spread on a solid surface (e.g., an agar plate), and conjugation stops when the cells are taken out of the surface [39]. As a result, the experimenter could divide the computing process in steps, if required. Further control over the experimental course may be achieved by selection for any of the computing strains, for instance by applying antibiotic selection. Periodic selection might be necessary since the various strains used in the computation might have different growth rates, particularly if they are loaded with plasmids (cells containing plasmids typically grow more slowly than plasmid-free ones). Therefore, even if the initial concentrations are 50% donors D and 50% recipients R, this ratio can shift rapidly. Growth rates can be adjusted by applying sub-inhibitory concentrations of antibiotics, and the effects of such changes may be checked using our simulations. As stated in the MethodsSection, a conjugation experiment can be prolonged by mixing the cells of a mating and plating out again on fresh plates. Additional donors (or recipients) may be added at this stage to equilibrate the D/R ratio.

Another advantage of our system, which provides an obvious line of enquiry for future work, is the potential *scalability* of the system. There are many plasmid conjugation systems that work in an essentially orthogonal manner (meaning each plasmid uses its own DNA transport system, and the different systems do not interfere, even if acting on the same cell). This is easily understood, since each conjugative system promotes only the transfer of the plasmid it recognizes as its own [20]. In the *E. coli* bacterium there are at least 12 different conjugative systems, coded by compatible plasmids, that can act *together* in the *same* cell [40]. Many strains of *E. coli* isolated from clinical settings carry three or more plasmids, and these assemblies are stable and evolutionarilyy successful [41]. Moreover, a given conjugative plasmid can bring about the mobilization of several independent mobilizable plasmids [42]. Thus, the numbers of potential orthogonal systems quickly escalate.

We fully expect future work in synthetic biology to adhere closely to this model, in which computational simulations and laboratory investigations are inextricably linked in a tight cycle of feedback. Our proposed system offers one possible addition to the ever-growing "toolbox" available to biological engineers, and we hope that experimental validation of its basic principles will be quickly forthcoming.

METHODS

Modelling Genetic Logic

The model used for all simulations is supplied in this Section. We make the assumption that modelling the translation processes and the slow transcriptions (basal rates) make no qualitative difference. Thus, we adopt the following two simplifications in our system: 1) transcription and translation are joined into one single process called "transcription", and 2) basal transcription rates are considered to be null in the system.

Ordinary differential equations (ODEs) from 1 to 13 explain the deterministic dynamics of the first approach (2-strain NOR population) using an *idealized* two-cell environment where physical contact is constant:

$$dG_1/dt = -k_1AG_1 + K_{-1}G_1^a - k_2BG_1 + k_{-2}G_1^b \quad (1)$$

$$dG_1^a/dt = k_1AG_1 - k_{-1}G_1^a - k_3BG_1^a + k_{-3}G_1^{ab} \quad (2)$$

$$dG_1^b/dt = k_2BG_1 - k_{-2}G_1^b - k_4AG_1^b + k_{-4}G_1^{ab} \quad (3)$$

$$dG_1^{ab}/dt = k_3BG_1^a - k_{-3}G_1^{ab} + k_4AG_1^b - k_{-4}G_1^{ab} \quad (4)$$

$$dCre/dt = k_5G_1^a + k_6G_1^b + k_7G_1^{ab} - k_{13}Cre \quad pmd_i - k_8Cre \quad (5)$$

$$dA/dt = k_{11} - k_1AG_1 - k_4AG_1^b - k_9A \quad (6)$$

$$dB/dt = k_{12} - k_2BG_1 - k_3BG_1^a - k_{10}B \quad (7)$$

$$dpmd_i/dt = -k_{13}Cre \quad pmd_i \tag{8}$$

$$dpmd/dt = k_{13}Cre \quad pmd_i \tag{9}$$

$$dG_2/dt = -k_{15}G_2F2 + k_{-15}G_2^{F2} \tag{10}$$

$$dG_2^{F2} = k_{15}G_2F2 - k_{-15}G_2^{F2} \tag{11}$$

$$dF2/dt = k_{14}pmd_i - k_{15}G_2F2 + k_{-15}G_2^{F2} - k_{17}F2 \tag{12}$$

$$dGFP/dt = k_{16}G_2^{F2} - k_{19}GFP \tag{13}$$

where G_1 denotes the gene of the sender cell without inputs bound; G_1^a, G_1^b and G_1^{ab} denotes the same gene with inputs A, B and both bound (respectively) to its promoters; Cre is the Cre protein; A and B are the inputs; pmd_i and pmd the plasmid with the segment that contains the gene in charge of expressing inducers or without it (respectively); G_2 denotes the gene in the receiver cell without inducer bound; G_2^{F2} is the same gene with inducers bound to its promoter; $F2$ are the inducers expressed by the *wiring* plasmid; and *GFP* are the green fluorescent proteins.

Biochemical reactions and kinetic rate values (k_i) are detailed. All rates are specified within standard ranges found in the literature [43]–[48]. Kinetic rate *k13* has been fixed according to the experimental results of [23]. Importantly, similar parameters have the similar values, following the objective of defining as general a system as possible (free from parameter constraints that could compromise future validation). A perturbation analysis is performed by adding Gaussian noise to every rate (being the mean the original value and the standard deviation 20% of the mean) and running the simulations several times. This analysis reinforces our conclusion about the qualitative robustness of our model. The quantitative behaviour may change without altering the overall goal of the computational process.

Stochastic simulations, that take into account the randomness of the chemical reactions, used the Gillespie algorithm [49] . In this approach, time is discrete and the rates (k_i) indicate the propensity of each reaction to happen.

The deterministic behaviour of cells NOR_1 (component of the 3-strain XOR population) are described by equations 14–20:

$$dfimE/dt = k_5G_1^a + k_6G_1^b + k_7G_1^{ab} - k_8fimE - k_{25}fimE \quad pmd_{F1} - k_{26}fimE \quad pmd_{F2} \tag{14}$$

$$dX/dt = k_5G_1^a + k_6G_1^b + k_7G_1^{ab} - k_{22}XG_3 + k_{-22}G_3^X - k_{20}X \tag{15}$$

$$dG_3/dt = -k_{22}XG_3 + k_{-22}G_3^X \tag{16}$$

$$dG_3^X/dt = k_{22}XG_3 - k_{-22}G_3^X \tag{17}$$

$$dfimB/dt = k_{21}G_3 - k_8fimB - k_{23}fimB \quad pmd_{F1} - k_{24}fimB \quad pmd_{F2} \tag{18}$$

$$dpmd_{F1}/dt = -k_{23}fimB \quad pmd_{F1} + k_{24}fimB \quad pmd_{F2} - k_{25}fimE \quad pmd_{F1} \tag{19}$$

$$dpmd_{F2}/dt = k_{23}fimB \quad pmd_{F1} - k_{24}fimB \quad pmd_{F2} + k_{25}fimE \quad pmd_{F1} \tag{20}$$

where $fimE$ and $fimB$ denote the inversion proteins; X represents the repressor; G_3 is the gene that expresses $fimB$; G_3^X denotes the same gene when the repressor X is bound to its promoter; and pmd_{F1} and pmd_{F2} are the possible plasmid configurations inside NOR_1 cells. Equations 1, 2, 3, 4, 6 and 7 are also used in this example as they define dynamics for existing components in NOR_1 cells. Reactions and rates are, as before. Importantly, k_{-22} is a low value in order to simulate a strong repression done by X.

Separately, NOR_2 cells are simulated by ODEs 21 to 24. As stated in Section Results, the inside *program* of these cells are similar to the combination of both senders and receivers of the 2-strain NOR design. The only difference relies on the four plasmid configurations (instead of two) that can be present in the cell. Thus, their equations are the same as in the NOR example but with some modifications:

$$dCre/dt = k_5 G_1^a + k_6 G_1^b + k_7 G_1^{ab} - k_{13} Cre$$

$$pmd_{F2} - k_{27} pmd_{F2}^{F1} - k_8 Cre \tag{21}$$

$$dF1/dt = k_{28} pmd^{F1} + k_{29} pmd_{F2}^{F1} - k_2 F1 G_1 - k_3 F1 G_1^a - k_{10} F1 \tag{22}$$

$$dpmd_{F2}^{F1}/dt = -k_{27} Cre \quad pmd_{F2}^{F1} \tag{23}$$

$$dpmd^{F1}/dt = k_{27} Cre \quad pmd_{F2}^{F1} \tag{24}$$

where $F1$ is the expression product that induces the production of Cre; pmd_{F2}^{F1} and $pmd^{F1}$are the plasmid with or without (respectively) gene $F2$ when the promoter is pointing towards gene F1. The plasmids pmd_i and pmd (equations 8 and 9 also used in cells NOR_2) make reference to plasmids pmd_{F2} and pmd. The gene G_1 inside the NOR_1 cell is not the same than the gene G_1 inside the NOR_2 cell, altough they share the label. Stochastic simulations are done (via Gillespie algorithm) by using the biochemical reactions with the kinetic rates.

Population Dynamics

A simulation tool for population dynamics is needed in order to get realistic spatial results (Figures 3, 6) for the whole system behaviour. We present a simulation framework for bacterial growth, movement and horizontal gene transfer called DiSCUS (DIscrete Simulation of Conjugation Using Springs - http://code.google.com/p/discus/). Using an agent-based model (AbM) approach, our software combines the management of intercellular interactions with the simulation of intracellular genetic networks. AbMs are widely used to study microbial growth and biofilm formation [50]–[53] and conjugation has also been included in several simulations [31], [54], [55], but these (unrealistically) consider cells as abstract circular objects. Having rod-shaped cells (as in [56]–[59]) is essential to obtain accurate conjugation dynamics. Thus, DiSCUS is the first platform, to our knowledge, to offer the possibility of simulating conjugation dynamics in rod-shape bacteria.

ODEs are introduced *inside* every cell independently, so each bacterium runs its own copy of the circuit according to its design. While the circuit is simulated deterministically, the plasmid number is always discretized using several build-in functions. The discrete number corresponding to the plasmid concentration is used when: a) the cell divides, when a random selection of the plasmids will be part of each daughter (without losing plasmids); b) the plasmids have lower copy number than the limit, in which case they increase their copies; and c) the cell transfers (through conjugation) a small number of plasmids (1–3) to the recipient. The discretized values are always used to update deterministic concentrations. This process is essential to assure the numbers are kept consistent and realistic.

Conjugation events take place while the cells carry out their normal growing activity. DiSCUS handles conjugation processes through a probability distribution that can be easily tuned to correspond to the behaviour of different cell types. The *low* and *high* conjugation frequencies used in this paper are obtained via visual validation against real data taken from a recent study[60] where a difference is established between original donors (*low* conjugation frequency) and new donors or *transconjugants* (*high* conjugation probability). Biomechanical validation of DiSCUS against real data [59] has also been performed. In our design there is only one conjugation frequency, as transconjugant cells are unable to conjugate between them or back to donors (due to the fact that they do not carry a complete genetic transfer system). However, we run duplicate simulations considering both alternative frequencies (*low*: Figures 3 and; *high*: Figures 6) in order to explore the range of correct functioning of the system. We observe quantitative changes, but not significant alterations in the qualitative logic behaviour of the circuits. In order to allow conjugation to achieve complete infection of the recipient population during the experiment, the cells were mixed (shuffled) every 400 min so that new pairs donor-receiver can arise in the population.

AUTHOR CONTRIBUTIONS

Conceived and designed the experiments: AG-M MA FdlC. Performed the experiments: AG-M. Analyzed the data: AG-M. Contributed reagents/materials/analysis tools: AG-M. Wrote the paper: AG-M MA FdlC.

REFERENCES

1. Oldham P, Hall S, Burton G (2012) Synthetic biology: mapping the scientific landscape. PLoS One 7: e34368. doi: 10. 1371/journal. pone. 0034368
2. Heinemann M, Panke S (2006) Synthetic biology{putting engineering into biology. Bioinformatics 22: 2790–9. doi: 10. 1093/bioinformatics/btl469
3. Andrianantoandro E, Basu S, Karig DK, Weiss R (2006) Synthetic biology: new engineering rules for an emerging discipline. Mol Syst Biol 2: 2006. 0028. doi: 10. 1038/msb4100073
4. De Lorenzo V, Danchin A (2008) Synthetic biology: discovering new worlds and new words. EMBO reports 9: 822–827. doi: 10. 1038/embor. 2008. 159
5. Elowitz MB, Leibler S (2000) A synthetic oscillatory network of transcriptional regulators. Nature 403: 335–338. doi: 10. 1038/35002125
6. Gardner TS, Cantor CR, Collins JJ (2000) Construction of a genetic toggle switch in Escherichia coli. Nature 403: 339–342.
7. Lou C, Liu X, Ni M, Huang Y, Huang Q, et al. (2010) Synthesizing a novel genetic sequential logic circuit: a push-on push-off switch. Molecular systems biology 6. doi: 10. 1038/msb. 2010. 2
8. Purcell O, Savery NJ, Grierson CS, di Bernardo M (2010) A comparative analysis of synthetic genetic oscillators. J R Soc Interface 7: 1503–24. doi: 10. 1098/rsif. 2010. 0183
9. Purcell O, di Bernardo M, Grierson CS, Savery NJ (2011) A multi-functional synthetic gene network: a frequency

multiplier, oscillator and switch. PLoS One 6: e16140. doi: 10. 1371/journal. pone. 0016140

10. Macía J, Posas F, Solé RV (2012) Distributed computation: the new wave of synthetic biology devices. Trends Biotechnol 30: 342–9. doi: 10. 1016/j. tibtech. 2012. 03. 006
11. Brenner K, You L, Arnold FH (2008) Engineering microbial consortia: a new frontier in synthetic biology. Trends Biotechnol 26: 483–9. doi: 10. 1016/j. tibtech. 2008. 05. 004
12. Basu S, Gerchman Y, Collins CH, Arnold FH, Weiss R (2005) A synthetic multicellular system for programmed pattern formation. Nature 434: 1130–4. doi: 10. 1038/ nature03461
13. Tamsir A, Tabor JJ, Voigt Ca (2011) Robust multicellular computing using genetically encoded NOR gates and chemical 'wires'. Nature 469: 212–5. doi: 10. 1038/ nature09565
14. Regot S, Macia J, Conde N, Furukawa K, Kjellen J, et al. (2011) Distributed biological computation with multicellular engineered networks. Nature 469: 207–211. doi: 10. 1038/nature09679
15. Goñi-Moreno A, Redondo-Nieto M, Arroyo F, Castellanos J (2011) Biocircuit design through engineering bacterial logic gates. Natural Computing 10: 119–127. doi: 10. 1007/s11047-010-9184-2
16. Ortiz M, Endy D (2012) Engineered cell-cell communication via dna messaging. Journal of Biological Engineering 6: 16. doi: 10. 1186/1754-1611-6-16
17. Atkinson S, Williams P (2009) Quorum sensing and social networking in the microbial world. J R Soc Interface 6: 959–78. doi: 10. 1098/rsif. 2009. 0203
18. Danino T, Mondragón-Palomino O, Tsimring L, Hasty J (2011) A synchronized quorum of genetic clocks. Nature 463: 326–330. doi: 10. 1038/nature08753
19. Tatum EL, Lederberg J (1947) Gene recombination in the bacterium escherichia coli. Journal of Bacteriology 53: 673–684.
20. De La Cruz F, Frost LS, Meyer RJ, Zechner EL (2010) Conjugative dna metabolism in gramnegative bacteria. FEMS Microbiology Reviews 34: 18–40. doi: 10. 1111/j. 1574-6976. 2009. 00195. x
21. Llosa M, Gomis-Rth FX, Coll M, Cruz Fdl (2002) Bacterial conjugation: a two-step mechanism for dna transport. Molecular Microbiology 45: 1–8. doi: 10. 1046/j. 1365-2958. 2002. 03014. x
22. Grindley ND, Whiteson KL, Rice PA (2006) Mechanisms of site-specific recombination*. Annual Review of Biochemistry 75: 567–605. doi: 10. 1146/annurev. biochem. 73. 011303. 073908
23. Yamanishi M, Matsuyama T (2012) A modified cre-lox genetic switch to dynamically control metabolic ow in saccharomyces cerevisiae. ACS Synthetic Biology 1: 172–180. doi: 10. 1021/sb200017p
24. Moon TS, Clarke EJ, Groban ES, Tamsir A, Clark RM, et al. (2011) Construction of a genetic multiplexer to toggle between chemosensory pathways in escherichia coli. Journal of Molecular Biology 406: 215–227. doi: 10. 1016/j. jmb. 2010. 12. 019
25. Ham TS, Lee SK, Keasling JD, Arkin AP (2008) Design and construction of a double inversion recombination switch for heritable sequential genetic memory. PLoS ONE 3: e2815. doi: 10. 1371/journal. pone. 0002815
26. Ham TS, Lee SK, Keasling JD, Arkin AP (2006) A tightly regulated inducible expression system utilizing the fim inversion recombination switch. Biotechnology and Bioengineering 94: 1–4. doi: 10. 1002/bit. 20916
27. Haynes K, Broderick M, Brown A, Butner T, Dickson J, et al. (2008) Engineering bacteria to solve the burnt pancake problem. Journal of Biological Engineering 2: 8. doi: 10. 1186/1754-1611-2-8
28. Baumgardner J, Acker K, Adefuye O, Crowley S, DeLoache W, et al. (2009) Solving a Hamiltonian path problem with a bacterial computer. Journal of Biological Engineering 3: 11. doi: 10. 1186/1754-1611-3-11
29. Siuti P, Yazbek J, Lu TK (2013) Synthetic circuits integrating logic and memory in living cells. Nature Biotechnology doi: 10. 1038/nbt. 2510
30. Garcillán-Barcia MP, de la Cruz F (2008) Why is entry exclusion an essential feature of conjugative plasmids? Plasmid 60: 1–18. doi: 10. 1016/j. plasmid. 2008. 03. 002
31. Gregory R, Saunders JR, Saunders VA (2008) Rule-based modelling of conjugative plasmid transfer and incompatibility. Biosystems 91: 201–15. doi: 10. 1016/j. biosystems. 2007. 09. 003
32. Novick RP (1987) Plasmid incompatibility. Microbiological Reviews 51: 381–395.
33. Goñi-Moreno A, Amos M (2012) A reconfigurable nand/ nor genetic logic gate. BMC Systems Biology 6: 126. doi: 10. 1186/1752-0509-6-126
34. Gally DL, Bogan JA, Eisenstein BI, Blomfield IC (1993) Environmental regulation of the fim switch controlling type 1 fimbrial phase variation in escherichia coli k-12: effects of temperature and media. Journal of Bacteriology 175: 6186–6193.
35. McClain MS, Blomfield IC, Eisenstein BI (1991) Roles of fimb and fime in site-specific dna inversion associated with phase variation of type 1 fimbriae in escherichia coli. Journal of Bacteriology 173: 5308–5314. doi: 10. 1046/j. 1365-2958. 1997. 2241615. x
36. Khalil AS, Collins JJ (2010) Synthetic biology: applications come of age. Nature Reviews Genetics 11: 367–379. doi: 10. 1038/nrg2775
37. Meyer B (1988) Object-Oriented Software Construction, volume 2. Prentice Hall, New York.
38. Smillie C, Garcilln-Barcia MP, Francia MV, Rocha EPC, de la Cruz F (2010) Mobility of plasmids. Microbiology and Molecular Biology Reviews 74: 434–452. doi: 10. 1128/mmbr. 00020-10
39. del Campo I, Ruiz R, Cuevas A, Revilla C, Vielva L, et al. (2012) Determination of conjugation rates on solid surfaces. Plasmid 67: 174–182. doi: 10. 1016/j. plasmid. 2012. 01. 008
40. Garcilln-Barcia MP, Francia MV, De La Cruz F (2009)

The diversity of conjugative relaxases and its application in plasmid classification. FEMS Microbiology Reviews 33: 657–687. doi: 10. 1111/j. 1574-6976. 2009. 00168. x

41. Grad YH, Godfrey P, Cerquiera GC, Mariani-Kurkdjian P, Gouali M, et al. (2013) Comparative genomics of recent shiga toxin-producing escherichia coli o104:h4: Short-term evolution of an emerging pathogen. mBio 4. doi: 10. 1128/mbio. 00452-12

42. Cabezn E, Sastre J, de la Cruz F (1997) Genetic evidence of a coupling role for the trag protein family in bacterial conjugation. Mol Gen Genet 254: 400–6. doi: 10. 1007/s004380050432

43. Dublanche Y, Michalodimitrakis K, Kummerer N, Foglierini M, Serrano L (2006) Noise in transcription negative feedback loops: simulation and experimental analysis. Molecular systems biology 2. doi: 10. 1038/msb4100081

44. Goñi-Moreno A, Amos M (2012) Continuous computation in engineered gene circuits. Biosystems 109: 52–56. doi: 10. 1016/j. biosystems. 2012. 02. 001

45. Balagaddé FK, Song H, Ozaki J, Collins CH, Barnet M, et al. (2008) A synthetic Escherichia coli predator-prey ecosystem. Mol Syst Biol 4: 187. doi: 10. 1038/msb. 2008. 24

46. Andersen JB, Sternberg C, Poulsen LK, Bjorn SP, Givskov M, et al. (1998) New unstable variants of green uorescent protein for studies of transient gene expression in bacteria. Appl Environ Microbiol 6: 2240–2246.

47. Bolouri H, Davidson EH (2003) Transcriptional regulatory cascades in development: Initial rates, not steady state, determine network kinetics. Proceedings of the National Academy of Sciences 100: 9371–9376. doi: 10. 1073/pnas. 1533293100

48. de Leon SBT, Davidson EH (2009) Modeling the dynamics of transcriptional gene regulatory networks for animal development. Developmental Biology 325: 317–328. doi: 10. 1016/j. ydbio. 2008. 10. 043

49. Gillespie DT (1977) Exact stochastic simulation of coupled chemical reactions. The Journal of Physical Chemistry 81: 2340–2361. doi: 10. 1021/j100540a008

50. Gorochowski TE, Matyjaszkiewicz A, Todd T, Oak N, Kowalska K, et al. (2012) Bsim: an agentbased tool for modeling bacterial populations in systems and synthetic biology. PLoS One 7: e42790. doi: 10. 1371/journal. pone. 0042790

51. Kreft JU, Booth G, Wimpenny JWT (1998) Bacsim, a simulator for individual-based modelling of bacterial colony growth. Microbiology 144: 3275–3287. doi: 10. 1099/00221287-144-12-3275

52. Xavier JB, Picioreanu C, van Loosdrecht MCM (2005) A framework for multidimensional modeling of activity and structure of multispecies biofilms. Environ Microbiol 7: 1085–103. doi: 10. 1111/j. 1462-2920. 2005. 00787. x

53. Lardon LA, Merkey BV, Martins S, Dotsch A, Picioreanu C, et al. (2011) idynomics: nextgeneration individual-based modelling of biofilms. Environ Microbiol 13: 2416–34. doi: 10. 1111/j. 1462-2920. 2011. 02414. x

54. Krone SM, Lu R, Fox R, Suzuki H, Top EM (2007) Modelling the spatial dynamics of plasmid transfer and persistence. Microbiology 153: 2803–16. doi: 10. 1099/mic. 0. 2006/004531-0

55. Merkey BV, Lardon LA, Seoane JM, Kreft JUU, Smets BF (2011) Growth dependence of conjugation explains limited plasmid invasion in biofilms: an individual-based modelling study. Environ Microbiol 13: 2435–52. doi: 10. 1111/j. 1462-2920. 2011. 02535. x

56. Melke P, Sahlin P, Levchenko A, Jönsson H (2010) A cell-based model for quorum sensing in heterogeneous bacterial colonies. PLoS Comput Biol 6: e1000819. doi: 10. 1371/journal. pcbi. 1000819

57. Rudge TJ, Steiner PJ, Phillips A, Haseloff J (2012) Computational modeling of synthetic microbial biofilms. ACS Synthetic Biology 1: 345–352. doi: 10. 1021/sb300031n

58. Cho H, Jönsson H, Campbell K, Melke P, Williams JW, et al. (2007) Self-organization in highdensity bacterial colonies: efficient crowd control. PLoS Biol 5: e302. doi: 10. 1371/journal. pbio. 0050302

59. Volfson D, Cookson S, Hasty J, Tsimring LS (2008) Biomechanical ordering of dense cell populations. Proceedings of the National Academy of Sciences 105: 15346–15351. doi: 10. 1073/pnas. 0706805105

60. Seoane J, Yankelevich T, Dechesne A, Merkey B, Sternberg C, et al. (2011) An individual-based approach to explain plasmid invasion in bacterial populations. FEMS Microbiology Ecology 75: 17–27. doi: 10. 1111/j. 1574-6941. 2010. 00994. x

CHAPTER 11 OSCILLATORY THRESHOLD LOGIC

Jon Borresen and Stephen Lynch

School of Computing, Mathematics and Digital Technology, Manchester Metropolitan University, Manchester, United Kingdom

ABSTRACT

In the 1940s, the first generation of modern computers used vacuum tube oscillators as their principle components, however, with the development of the transistor, such oscillator based computers quickly became obsolete. As the demand for faster and lower power computers continues, transistors are themselves approaching their theoretical limit and emerging technologies must eventually supersede them. With the development of optical oscillators and Josephson junction technology, we are again presented with the possibility of using oscillators as the basic components of computers, and it is possible that the next generation of computers will be composed almost entirely of oscillatory devices. Here, we demonstrate how coupled threshold oscillators may be used to perform binary logic in a manner entirely consistent with modern computer architectures. We describe a variety of computational circuitry and demonstrate working oscillator models of both computation and memory.

INTRODUCTION

In 2005, the US National Security Agency [1] concluded that transistors were rapidly approaching the limits of functionality and that new technologies needed to be developed in order to overcome this.

Recent advances in novel computation suggest a variety of new computing technologies which may be applicable and have far reaching applications. Of particular note are: quantum computing [2]–[6], all-optical computing [7]–[11], spin computing [12]–[15], chaos computing[16]–[18], and DNA computing [19]–[23]. Each of these suggests a technology which is in many respects fundamentally different from current computing systems and in order for their implementation to be successful, new programming structures and techniques would be required, in addition to any development of reliable working prototypes.

One technology which is considerably more advanced from an implementation perspective is superconductive computing [24], [25], based on Rapid Single Flux Quantum (RSFQ) technology [26]–[30] which effectively uses Josephson Junctions (JJs) to replace transistors as the fundamental active element in any circuit. This technology is more mature than the others mentioned above and has already produced practical digital and mixed-signal circuits with world record breaking processing speeds at exceptionally low power [31]–[36]. The RSFQ-based digital receivers were demonstrated in the field, converting high-frequency wide-band analog

Citation: Borresen J, Lynch S (2012) Oscillatory Threshold Logic. PLoS ONE 7(11): e48498. doi:10.1371/journal.pone.0048498.

communications signals to digital domain taking advantage of extreme sampling speeds of RSFQ circuits [37]–[41]. The JJ RSFQ circuits are fabricated using a relatively simple thin-film process developed in several places [42]–[46]. Recently, a new energy-efficient generation of RSFQ circuits has been introduced [47] complemented with energy-efficient cryogenic memory [48] leading to the implementation of energy-efficient computing.

Neurons can also be simulated using simple JJ circuits [49]–[53]. These circuits could then be connected to form logic gates similar to those appearing in this paper. The NSA report [1] also concluded that the most likely successor to transistor technology would be JJ circuitry. The authors are currently investigating implementation of binary logic using neuronal JJ circuitry and the results will be published at a later date.

One further technology which is somewhat less advanced from an implementation perspective is neuronal computing (or wetware), which uses artificially grown neurons as the processing units. Again such technology would require a considerable shift in how computers were designed and programmed and is unlikely to prove to be a successor to Complementary Metal-Oxide-Semiconductors (CMOS) at any time in the foreseeable future. What neuronal computing does provide is a deeper understanding of brain functionality and possible associated medical benefits that would therefore follow. Recently, it has been reported that scientists were able to grow brain nerve cells affected by Parkinson›s disease using human skin cell samples [54]. These neurons could be connected to form logic circuits similar to those reported in this paper. Degradation of logical functionality could then be used as an assay to determine the effect of drugs or physical damage on neuronal circuitry.

Zanin et al. [55], show that computation can emerge from collective dynamics of an ensemble of networking neurons. Synchronization and desynchronization of neurons using a dynamic weighted network is used to codify binary information. Additionally, neural encoding using conjugate symmetries [56] and the concept of 'winnerless competition' in coupled oscillator networks [57] have also been demonstrated as viable methods of oscillatory computation.

Threshold logic [58], [59] and null convention logic [60] have also been considered, particularly with respect to neural network computing. Beiu et al. [61] provide a review of commercial VLSI hardware implementation of threshold logic up to 2003. With the emergence of nanotechnologies; resonant tunneling, single electron and memristor implementation are also looking promising, especially with respect to energy delay efficiency and reconfigurable circuits[62]–[67]. Additionally, Modified Variable Threshold Logic (MVTL) using JJs has demonstrated high speed, low power processing [68].

In this paper, we demonstrate a novel computational concept using both inhibitory and excitatory connections between threshold oscillators. The use of inhibitory connections in this way draws on ideas from biological neural encoding and has not previously considered as a tool in implementing binary computation. The use of threshold logic in oscillator circuits presents numerous opportunities for novel circuit design. The inherent richness of oscillator dynamics allows for excitatory and inhibitory connections which may be in many respects different from standard threshold weights used in traditional threshold logic: Neuronal circuits are known to display phase and anti-phase synchronization [69] which may have applications in clocking and error correction. Bursting type behavior [70] could be utilized to prevent signal degradation and improve noise resilience. Connections may induce a variety of responses [71]- all or nothing, additive, amplitude or frequency, which may have specific uses in circuit design.

The method by which inhibition may be caused to occur would be specific to the type of oscillator or oscillatory circuit in question. Neural inhibition has been widely studied [72]. In biological neural circuits, excitation and inhibition occur via diffusion of neurotransmitters across a synaptic gap. For inhibition, this has a temporal effect which permits suppression of post synaptic neural activity for specific time intervals (known as the *refractory period*). In this paper, we use a method of inhibition using negative connection weights similar to that employed in standard threshold logic design, however, this is to demonstrate proof of principle and is not the only possible method by which logic gate connections could be made. Electrical circuit oscillators can likewise demonstrate inhibition [73]. Optical oscillator circuits offer the possibility of interference based inhibition [74] and perhaps most importantly, given recent advances in JJ technology, JJ oscillator circuits have also been designed with inhibitory characteristics very similar to those in neural circuitry [52].

The underlying concept is not technology specific and, given recent advances in JJ circuitry and neuronal computing, could be readily implemented [75], [76]. It is also considerably less disruptive in that the underlying binary logic is identical to that employed in CMOS and implementation in the JJ form would be compatible with traditional computing architectures.

We present schematics for simple arithmetic and memory operations and describe these operations as the solution set of a system of linear inequalities. Simulations using a neural oscillator model demonstrate how such a model may be implemented.

METHODS

Computing using oscillators is not a new concept, indeed the first modern computers were made using vacuum tube oscillators, and oscillators in a variety of forms are integral components in many devices. The use of neural oscillators has also been widely studied, however, in all cases the method of computation is derived from concepts of biological neural encoding. Current research into encoding using neural oscillators is therefore spatio-temporal, rate, or more usually synchronization based [77]–[79].

What has not previously been considered is using oscillators as the fundamental components of computing devices (with all the inherent dynamical richness that this provides) and designing them in such a way as to perform binary logic in an equivalent manner to standard transistor logic - that is the oscillator will provide a binary output (1 equivalent to an oscillator firing or 0 where the oscillator does not fire) and the output from a single oscillator can be interpreted in exactly the same way as that of a transistor.

Threshold Logic

Threshold logic has been studied as an alternative to Boolean logic for some time. For many implementations this is advantageous, allowing for reduced component counts and/or number of logic levels, as the implementation of complex logical operations may be achieved using a single gate [80].

Threshold logic gates [81] have a set of inputs $\{I_1, I_2, \ldots I_n\}$, weights $\{w_1, w_2, \ldots w_n\}$ and a binary output y. The output y is typically described by: $y = \phi\left(\sum_{i=1}^{n} I_i w_i\right)$, where the function ϕ is an activation function (eg Heaviside, tanh, sigmoid, piecewise linear, low gain saturation [82]) and the binary output 1 is defined at some threshold $y > T$, say.

Threshold logic implementation has not supplanted standard logic implementation in CMOS due to sensitivity to parameter changes and variable connection weights requiring very low tolerance engineering. Recent advances in nanotechnology, in particular, Resonant Tunneling Devices (RTD) [83] and memristor devices [84] have the potential to overcome such concerns.

Generic Threshold Oscillator Model

A threshold oscillator is an oscillatory device that will begin oscillating when the input to the device is above a certain threshold. Below this level the oscillator remains in a resting state and gives no output. It is possible to use the output of one threshold oscillator as the input of another oscillator to cause the second oscillator to operate (excitation) and under certain circumstances, it is also possible to cause the input of one oscillator to suppress the output of another oscillator (inhibition).

There are numerous viable methods for implementing binary computation using threshold oscillators. In order to perform the logical operations it is necessary that either oscillators with differing thresholds be used or the connections to the oscillators be of differing weights. In our modeling we shall use the latter method as this mimics more closely biological neural systems, from where the idea originated.

Logical operations

Logical operations can be performed in a similar manner to standard logic circuits, however, due to the threshold nature it is possible to formulate logical operations as solutions of sets of linear inequalities. For instance, the AND function can be replicated by a threshold oscillator with two inputs, where the input strengths are scaled such that the total input is only above threshold if both the inputs are on. For a single input or for no input the total input would be below threshold. Defining the inputs to the logical circuits in vector form and scaling the input strength to a binary 1 or 0, we write $\sum I = I_{1,1} + I_{2,1}$ as the total input to the circuit. The threshold equations may be thus written as:

$$\text{for } I = \begin{pmatrix} 0 \\ 0 \end{pmatrix}, \begin{pmatrix} 1 \\ 0 \end{pmatrix}, \begin{pmatrix} 0 \\ 1 \end{pmatrix} \quad \sum Iw < T$$
$$\text{for } I = \begin{pmatrix} 1 \\ 1 \end{pmatrix} \quad \sum Iw > T, \tag{1}$$

where T is the oscillator threshold and w the coupling weight between the inputs and the oscillator performing the AND operation. Clearly the solution to the above system Eq. (1) is $\frac{T}{2} < w < T$. For the logical OR operation, the solution $w > T$ would suffice.

Using threshold oscillators in this manner it is straightforward to implement the logical NOT operation using a negative coupling strength, however, as the logical NOT is effectively redundant in more complex logically complete circuit design where NAND and XOR operations are used, we will present all models using the latter formulations.

Binary Half Adder

One of the simplest computing circuits is the binary half adder. The binary half adder gives the sum of two binary inputs as a two bit binary output. The truth table for the binary half adder is given in Fig. 1A.

A

Input		Output	
I_1	I_2	O_1	O_2
0	0	0	0
0	1	1	0
1	0	1	0
1	1	0	1

B

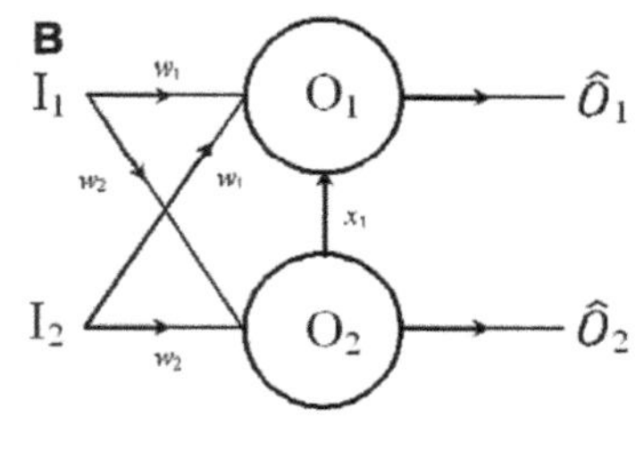

Figure 1: Binary half adder.

(A) Truth table for a binary half adder. (B) Oscillator circuit diagram for a binary half adder comprising two inputs I_1 and I_2 and two oscillators O_1 and O_2. The sum oscillator O_1 will oscillate if either I_1 or I_2 are active. The carry oscillator O_2 will oscillate if both I_1 and I_2 are active. An inhibitory connection from O_2 to O_1 suppresses oscillator O_1 if O_2 is active.

http://dx.doi.org/10.1371/journal.pone.0048498.g001

Standard transistor implementation of a binary half adder uses one XOR gate (to give the sum) and one AND gate (to give the carry). Implementation of this circuit using threshold oscillators can be achieved via a similar design, with two oscillators replicating the logical functions. The AND operation is implemented as described above and the XOR operation can be achieved using an OR operation (as above) with an additional connection from the AND oscillator, which in some way inhibits the operation of the OR oscillator if the AND oscillator is active. The method by which inhibition occurs would be dependent upon the oscillators being used to form the circuitry.

Fig. 1B demonstrates a viable circuit schematic for half adder implementation using two oscillators O_1 and O_2 and two inputs I_1 and I_2, which may themselves be the output from other oscillators in a more complex circuit. Schematically, the circuit design is not dissimilar to standard threshold logic half adders [58], however, due to the nature of the connections between oscillators, implementation may be markedly different. If we consider oscillators with identical thresholds we will require that the coupling strength, w_1, say, from I_1 and I_2 to O_1 be sufficient to cause O_1 to oscillate for only one input and for the coupling strength, w_2, say, from I_1 and I_2 to O_2 to be sufficient for it to oscillate for two inputs. The additional connection x_1, say, from O_2 to O_1 is inhibitory such that if O_2 is oscillating it suppresses O_1. Denoting the output from O_2 as $\hat{O}_2$, the total input to O_1 and O_2 are thus given by:

$$O_1 = \sum Iw_1 - \hat{O}_2 x_1$$
$$O_2 = \sum Iw_2. \tag{2}$$

We can consider such a system as a set of linear inequalities with normalized input vectors Iand threshold T requiring solutions of the form:

$$\text{for } I = \begin{pmatrix} 0 \\ 0 \end{pmatrix} \quad \begin{cases} \sum Iw_1 - \hat{O}_2 x_1 < T \\ \sum Iw_2 < T \end{cases}$$
$$\text{for } I = \begin{pmatrix} 1 \\ 0 \end{pmatrix}, \begin{pmatrix} 0 \\ 1 \end{pmatrix} \quad \begin{cases} \sum Iw_1 - \hat{O}_2 x_1 > T \\ \sum Iw_2 < T \end{cases}$$
$$\text{for } I = \begin{pmatrix} 1 \\ 1 \end{pmatrix} \quad \begin{cases} \sum Iw_1 - \hat{O}_2 x_1 < T \\ \sum Iw_2 > T. \end{cases} \tag{3}$$

Thus, for instance, for a total input of $\sum I = 1$, only O_1 will be above threshold causing oscillation giving a binary equivalent output of 1. If both I_1 and I_2 are active, O_2 will oscillate but O_1 is suppressed if $\hat{O}_2 x_1 > T/2 + w_1$, giving a binary output $1 + 1 = 10$, as required.

Two-Oscillator Full Adder

In order to demonstrate how oscillatory threshold logic scales for operations on larger numbers of bits we will consider the next simplest arithmetic circuit - the binary adder (or full adder). The binary adder has a truth table as in Fig. 2A.

A

Input			Output	
I_1	I_2	I_3	O_1	O_2
0	0	0	0	0
0	0	1	1	0
0	1	0	1	0
0	1	1	0	1
1	0	0	1	0
1	0	1	0	1
1	1	0	0	1
1	1	1	1	1

B

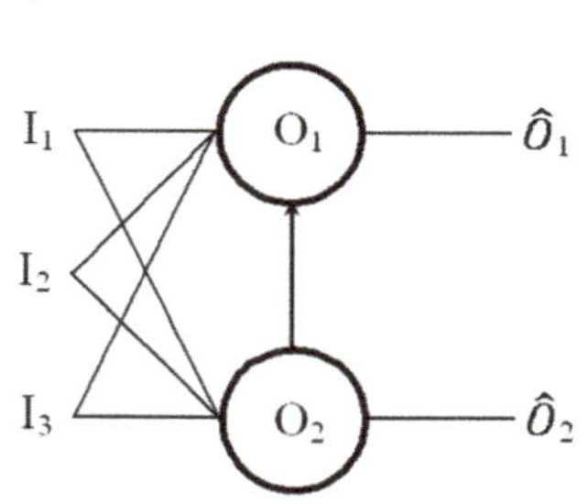

Figure 2: Binary full adder.

(A) Truth table for a binary full adder. (B) Oscillator circuit diagram for a binary full adder comprising three inputs I_1, I_2 and I_3 and two oscillators O_1 and O_2. Oscillator O_1 will oscillate if either I_1, I_2 or I_3 are active. Oscillator O_2 will oscillate if any two of I_1, $I_2$and I_3 are active. An inhibitory connection from O_2 to O_1 suppresses oscillator O_1 if $O_2$is active, however, the inhibition is only sufficient to suppress O_1 for $\sum I = 2$. For inputs of $\sum I = 3$ the total input to O_1 is still sufficient to induce oscillation.

http://dx.doi.org/10.1371/journal.pone.0048498.g002

Using traditional circuitry the binary adder requires five logic gates to perform such an operation. It is possible to formulate the adder circuit using oscillators as a solution set of two linear inequalities and as such only two oscillators are required to perform the operation. The oscillator equations are the same as Eq. (??) and the threshold inequalities are given by:

$$\text{for } I = \begin{pmatrix} 0 \\ 0 \\ 0 \end{pmatrix} \quad \begin{cases} \sum Iw_1 - \hat{O}_2 x_1 < T \\ \sum Iw_2 < T \end{cases}$$
$$\text{for } I = \begin{pmatrix} 1 \\ 0 \\ 0 \end{pmatrix}, \begin{pmatrix} 0 \\ 1 \\ 0 \end{pmatrix}, \begin{pmatrix} 0 \\ 0 \\ 1 \end{pmatrix} \quad \begin{cases} \sum Iw_1 - \hat{O}_2 x_1 > T \\ \sum Iw_2 < T \end{cases}$$
$$\text{for } I = \begin{pmatrix} 1 \\ 1 \\ 0 \end{pmatrix}, \begin{pmatrix} 1 \\ 0 \\ 1 \end{pmatrix}, \begin{pmatrix} 0 \\ 1 \\ 1 \end{pmatrix} \quad \begin{cases} \sum Iw_1 - \hat{O}_2 x_1 < T \\ \sum Iw_2 > T \end{cases}$$
$$\text{for } I = \begin{pmatrix} 1 \\ 1 \\ 1 \end{pmatrix} \quad \begin{cases} \sum Iw_1 - \hat{O}_2 x_1 > T \\ \sum Iw_2 > T. \end{cases} \tag{4}$$

In keeping with other full adder threshold logic designs [58], [85], the full adder schematic (Fig. 2B) is essentially identical to those of the half adder except for the additional input I_3.

It is possible to use additional oscillators to give the the binary sum for $2^n - 1$ inputs using only n oscillators. Each additional oscillator acts as the next binary digit of the required output and the respective input weights are adjusted thusly. Inhibitory connections from the additional oscillator to all other oscillators are required such that the inhibition strength from each new oscillator is scaled accordingly. Thus an exponential increase in computational power could (theoretically) be provided by a linear increase in the number of fundamental components. For larger circuits, the number of interconnections within the circuit would increase considerably, as the computation is effectively encoded in the connections rather than the switches themselves. Given sufficient engineering capability to provide the necessary connectivity, the number of components in any circuit and the time required to perform calculations - as a function of the required switching time of the oscillators - could be considerably reduced in this way.

RESULTS

The Fitzhugh-Nagumo Model

The Fitzhugh-Nagumo system [86], [87] is one of the more well known oscillator models. It is essentially a reduction of the Hodgkin-Huxley equations [88] which describe the action potential of a spiking neuron. The model is fairly straightforward to implement as an electrical circuit, as demonstrated by Binczak et al. [73].

The describing equations are:

$$\frac{dv}{dt} = C + v(v-\alpha)(1-v) - w$$
$$\frac{dw}{dt} = \epsilon(v - \gamma w), \quad (5)$$

where v is a fast variable (in biological terms - the action potential) and w represents a slow variable (biologically - the sodium gating variable). The parameters α, γ and ϵ dictate the threshold, oscillatory frequency and the location of the fixed points for v and w. The model will begin to oscillate when the input current C is above a critical threshold C_c. For all the following simulations, the threshold $T \approx 0.5$.

It is possible to couple the oscillators together via various methods. For biological neural systems, where there is synaptic coupling between neurons the coupling function is complex, relying on diffusion of neurotransmitters across a synaptic gap. The connections between neurons may either depolarize (excite) or hyperpolarize (inhibit) the post synaptic neuron.

Crucially, the hyperpolarizing inhibitory effect has a temporal component such that if inhibition occurs, the post synaptic neuron remains inhibited for some period of time after the pre-synaptic neuron fires. It is not straightforward to simulate such a system using the Fitzhugh-Nagumo model without either integration of the signal pulse or introducing arbitrary conditions on oscillators receiving an inhibitory pulse - which would not be viable from an implementation perspective. As such we will employ a method which is phenomenologically similar to neural hyperpolarization but is not necessarily consistent with any biological process.

Implementation by coupling through either the fast v variable or the slow w variable are equally viable. As can be seen from Fig. 3 and Fig. 4 for varying inputs C, the fast voltage oscillates with a fairly constant maximum and minimum whilst the slow variable w oscillates around a fixed point at approximately C. Any coupling function to be used must take into account the specific dynamics of whichever variable is used.

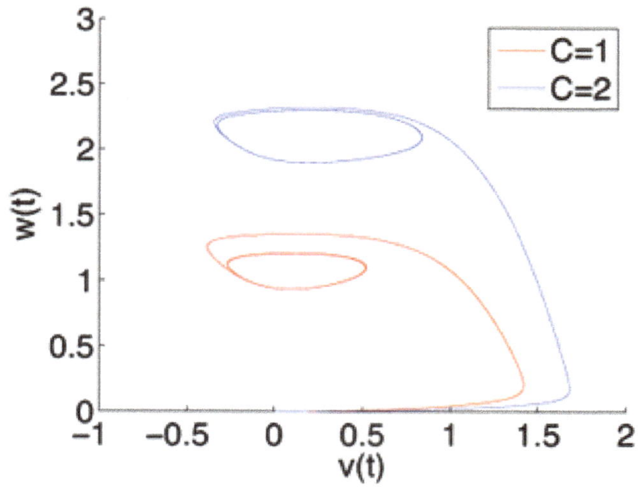

Figure 3: Phase portrait for Fitzhugh-Nagumo model with $\alpha = 0.1, \epsilon = 0.1$ **and** $\gamma = 0.1$ **shown for two input strengths** $C = 1$ **and** $C = 2$.

From an initial resting state at $(0,0)$, the transient dynamics is of a wide trajectory before settling onto a limit cycle.

http://dx.doi.org/10.1371/journal.pone.0048498.g003

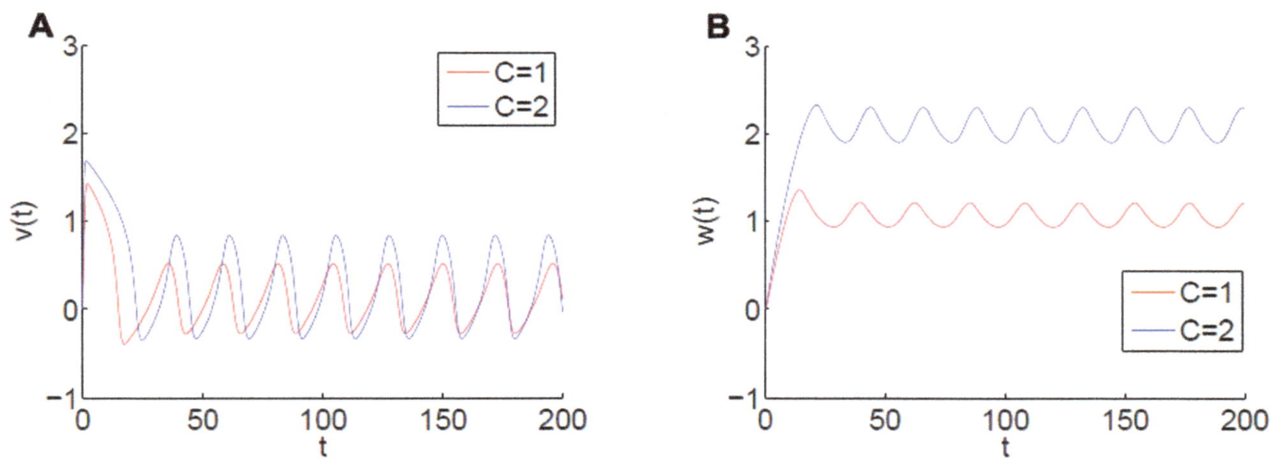

Figure 4: Associated time series of v **and** w **for corresponding Fig. 3.**

Note that after transients, (A) $v(t) \in [-1,1]$ for $C = 1$ and $C = 2$, (B) whereas $w(t) \approx C$.

http://dx.doi.org/10.1371/journal.pone.0048498.g004

As is common in such biologically inspired models we will use a sigmoidal transfer function between oscillators of the form:

$$S(x) = \frac{1}{1 + e^{m(-x+c)}}, \quad (6)$$

where c is the threshold at which the output begins to rise and m denotes the steepness of the curve of the function $S(x)$. In biological systems, neural connections can exhibit plastic responses and become 'tuned' (via some Hebbian learning rule [89]) allowing for more reliable excitation and inhibition. Choosing suitable values of m and c would in many respects replicate such a process.

Fitzhugh Nagumo Half Adder and Full Adders

Numerical simulations for systems of Fitzhugh Nagumo oscillators coupled as in Figs. 1B and2B will now be discussed. It is also shown that by adding one more oscillator to the two-oscillator full adder it is possible to construct a three-oscillator seven-input full adder.

The inputs to the logical circuits are oscillatory, being provided by Fitzhugh-Nagumo oscillators with similar coupling and parameter values to the computational oscillators. Oscillatory inputs of this form have been chosen over continuous inputs, as this demonstrates the necessary robustness of signal integrity which would be required for larger computational circuits. Continuous inputs to the computational oscillators would be equally viable and present no difficulties in implementation. As such the matrix form for the input weights for each oscillator is 4×4 rather than 2×2 as two additional oscillators are used as inputs.

One solution, in matrix form, to the inequalities Eq. (2) and Eq. (3) for the binary half adder, would be:

$$W = \begin{pmatrix} 0 & 0 & 0.8 & 0.45 \\ 0 & 0 & 0.8 & 0.45 \\ 0 & 0 & 0 & 0 \\ 0 & 0 & -1.5 & 0 \end{pmatrix}, \quad (7)$$

where $c = 0.5$ in Eq. (5) for the inputs I_1 and I_2. This would give the parameter values in Fig. 1B as $w_1 = 0.8, w_2 = 0.45$ and $x_1 = 1.5$. The time series for such is shown in Fig. 5.

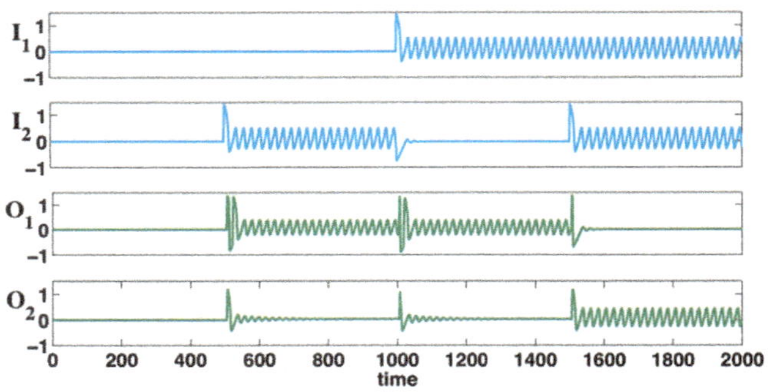

Figure 5: Time series for Fitzhugh-Nagumo binary half adder.

The parameter values taken for the transfer function $S(x)$, see Eq. (5), are $m = 100$and $c = 0.6$. All binary combinations of oscillatory inputs I_1 and I_2 give the required binary outputs O_1 and O_2 (see Fig. 1A). Here O_1 represents the sum and O_2 the carry in terms of standard logical circuitry. The observed pulse like behavior after switching is due to the wide trajectories taken by the input oscillators after being perturbed from the resting state.

http://dx.doi.org/10.1371/journal.pone.0048498.g005

A two-oscillator binary full adder can be constructed by simply introducing another input, I_3, say, as in Fig. 2B. Fig. 6 shows the time series for the Fitzhugh-Nagumo two-oscillator full adder.

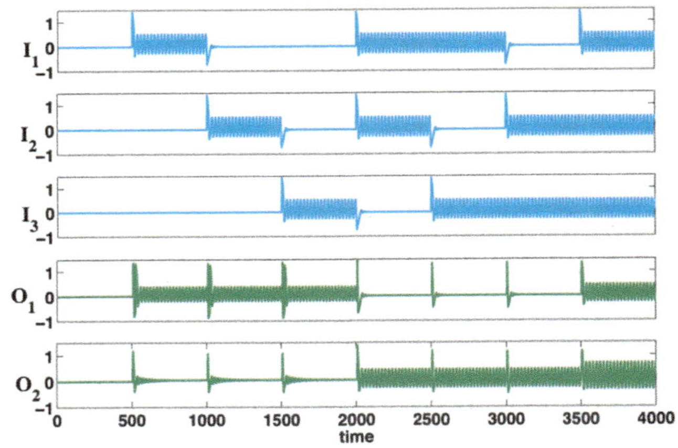

Figure 6: Time series for a Fitzhugh-Nagumo two oscillator full adder.

All binary combinations of oscillatory inputs I_1, I_2 and I_3 give the required binary outputs for O_1 and O_2 (see Fig. 2A). The weights used in this example were again $w_1 = 0.8$, $w_2 = 0.4$ and $x_1 = 1.5$ (see Eq. (4)).

http://dx.doi.org/10.1371/journal.pone.0048498.g006

To conclude this section, we consider a three-oscillator seven input full adder (see Fig. 7). Fig. 7A shows a schematic of the three-oscillator seven input adder comprising twenty one excitatory connectors and three inhibitory connectors. Fig. 7B shows the time series of the seven input three-oscillator full adder.

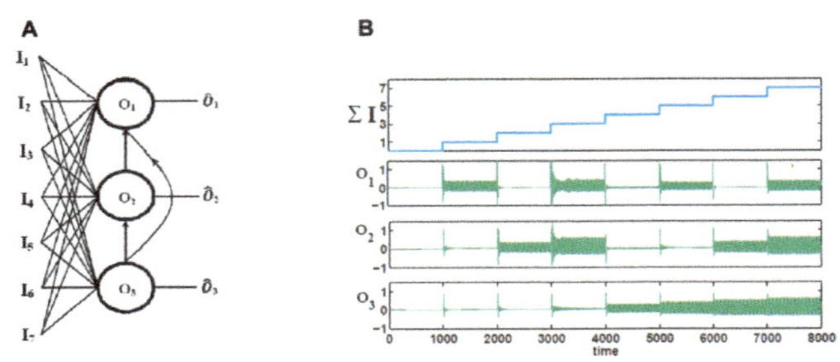

Figure 7: Seven input, three oscillator full adder.

(A) Block diagram for seven input, three oscillator full adder. (B) Time series for a Fitzhugh-Nagumo three-oscillator full adder. The total input strength ΣI is shown in the first row. All binary combinations of oscillatory inputs I_1 to I_7 give the required binary outputs for O_1, O_2 and O_3. The weights used in this example were $w_1 = 0.8$, $w_2 = 0.385$, $w_3 = 0.175$, $x_1 = 1.5$, $x_2 = 1.5$ and $x_3 = 3.3$, where x_1 is the inhibitory connection from O_2 to O_1, x_2 is from O_3 to O_2 and x_3 is from O_3 to O_1. Note that the amplitude of oscillation increases when the sum of the input to any oscillator is greatly above threshold, however, the binary on/off response is as required.

http://dx.doi.org/10.1371/journal.pone.0048498.g007

Fitzhugh Nagumo 2×2 Bit Binary Multiplier

In order to more fully demonstrate the applicability of such a model we have also simulated a more complex circuit, namely a 2×2 bit binary multiplier. Such a circuit outputs the binary multiple of two, 2 bit binary inputs. Although it is possible to perform such a calculation using only four oscillators we have used a standard circuit implementation of a binary multiplier to demonstrate how oscillators could be used to replace transistors as the fundamental units of computing devices without the need for architectural redesign.

Fig. 8A shows the logic table for a 2×2 bit binary multiplier and Fig. 8B shows the corresponding schematic of the oscillator circuit. The time series of the 2×2 bit binary multiplier is displayed in Fig. 9.

A

Input				Output			
I_1	I_2	I_3	I_4	O_1	O_5	O_7	O_8
0	0	0	0	0	0	0	0
0	0	0	1	0	0	0	0
0	0	1	0	0	0	0	0
0	0	1	1	0	0	0	0
0	1	0	0	0	0	0	0
0	1	0	1	0	0	1	0
0	1	1	0	0	1	0	0
0	1	1	1	0	1	1	0
1	0	0	0	0	0	0	0
1	0	0	1	0	1	0	0
1	0	1	0	1	0	0	0
1	0	1	1	1	1	0	0
1	1	0	0	0	0	0	0
1	1	0	1	0	1	1	0
1	1	1	0	1	1	0	0
1	1	1	1	1	0	0	1

B

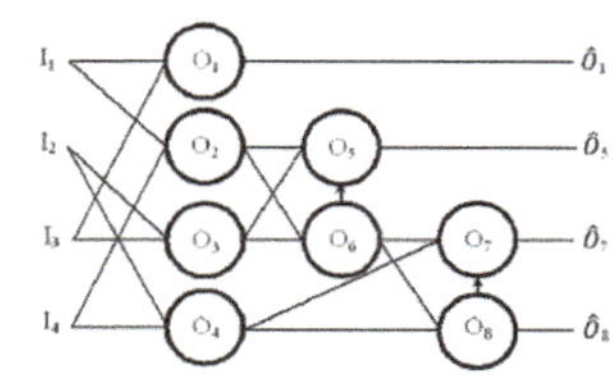

Figure 8: A 2×2 **bit multiplier.**

(A) Truth table for 2×2 bit binary multiplier. (B) Schematic of the binary oscillator 2×2 bit multiplier using oscillators based on standard circuitry.

http://dx.doi.org/10.1371/journal.pone.0048498.g008

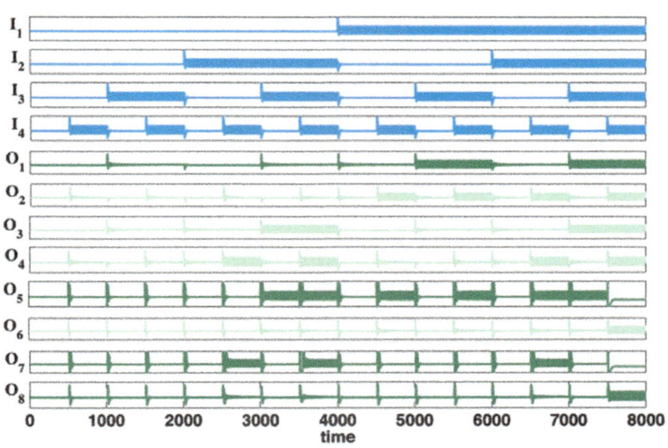

Figure 9: Time series of a 2×2 **bit multiplier using oscillators based on standard circuitry.**

All possible binary combinations for inputs I_1, I_2, I_3 and I_4 are shown. The oscillators O_1, O_5, O_7 and O_8 represent binary digit representation of the powers of 2. Oscillators O_2, O_3, O_4 and O_6 are used in the internal circuitry only and their time traces have been made fainter to emphasize the binary output oscillators. For example, at time $t\in[3500,4000]$ the inputs $I_1=0, I_2=1$ representing the decimal 2 and $I_3=1, I_4=1$, representing the decimal 3, giving an output where O_5 and O_7 are oscillating, being the binary equivalent $(10+100=110)$ of the decimal 6. Note O_1 and O_8 are not oscillating. This is equivalent to column eight above and row eight in the truth table Fig. 8A.

http://dx.doi.org/10.1371/journal.pone.0048498.g009

Fitzhugh Nagumo Set Reset Flip-Flop

Fig. 10 shows a schematic of a Set-Reset (SR) flip-flop circuit, the input I_1 is commonly referred to as the Set and input I_2 is referred to as the Reset. Output $\hat{O}_2$ is the complement of output $\hat{O}_1$. Note that both oscillators require a constant input I_C, say, for the circuit to function properly. This circuit acts as a memory, storing a bit and presenting it on its output $\hat{O}_1$, as can be seen in Fig. 11.

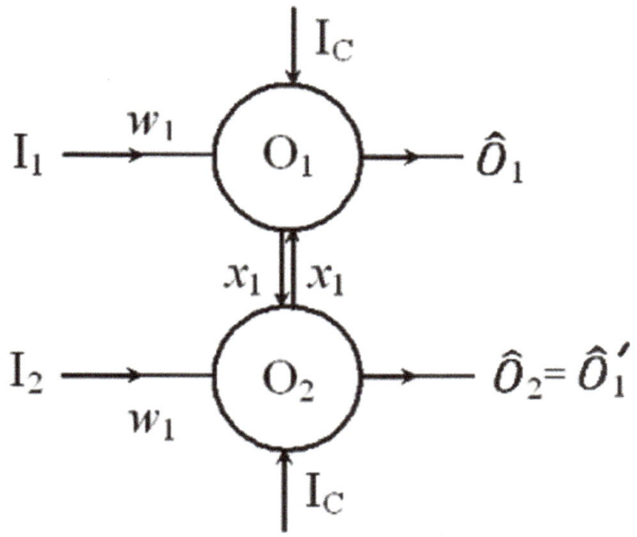

Figure 10: Schematic of an SR flip flop using oscillators.

Both oscillators O_1 and O_2 receive an external input current I_c and there are inhibitory connections x_1 between the oscillators. From an initial state where one oscillator is active the other will remain suppressed. External inputs I_1 or I_2 to the inactive oscillator will induce a switch. The oscillators are effectively performing NOR logical operations as in a conventional SR NOR latch.

http://dx.doi.org/10.1371/journal.pone.0048498.g010

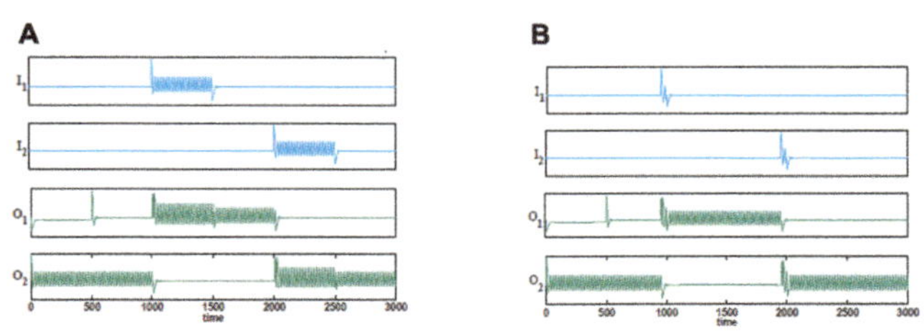

Figure 11: Time series of an SR flip-flop using oscillations to switch.

(A) Simulation with $w_1=0.5$ and $x_1=-1$. The simulation is initialized using a single external current to O_2 for $t\in[0,500]$. At $t=500$, oscillator O_1 also receives an external input current, however, it is suppressed by the output from O_2. The initial state has O_2active and O_1 inactive. A continuous switching pulse is provided by I_1 at $t=1000$. At $t=1500$, this switching pulse is turned off, but O_1 and O_2 remain in the switched state (as required). A further switch is performed at $t=2000$ using a continuous pulse to O_2. (B) Time series of an SR flip-flop using single input pulses to switch, with $w_1=0.5$ and$x_1=-1$. The switching is performed as for case A, however, only one pulse cycle (ballistic propagation) is used. Note the switching occurs as required and the system remains switched once the pulse has been received.

http://dx.doi.org/10.1371/journal.pone.0048498.g011

The SR flip-flop described here is an application of the 'winnerless competition' principle [57]. In the absence of coupling between the oscillators, both will remain active.

However, a symmetric inhibitory coupling between them ensures that from an initial state, where only one oscillator is active, the other will remain suppressed in the absence of any external perturbation. When an input is given to the inactive oscillator this is switched on, simultaneously suppressing the previously active oscillator. When the external input is turned off, the system remains in the switched state. Note that for a switch to occur, an input pulse of only one period is required (see Fig. 11). Switching using a single pulse in this way can open an opportunity to use ballistic propagation of signals between gates and memory cells, which could significantly reduce the energy required to operate memory circuits, where currently power intensive line charging is required to initiate memory switches.

One important consideration, particularly with respect to flip-flop circuits is the ability to switch accurately in the presence of noise [90]. A detailed examination of error rates arising from such noise would be specific to the oscillator and the underlying circuit implementation. For the general case considered here (a Fitzhugh-Nagumo implementation) repeated simulations have demonstrated that the system is particularly resistant to such noise (see Fig. 12) as is often the case with coupled oscillator dynamics [91], [92]. As the oscillators in the flip-flop circuit will only 'switch' if provided with a pulse above threshold, for a noise induced switch to occur (an error) it would be required that the noise amplitude itself be above this threshold. Although such an error is conceivable, no such error has been observed for Gaussian white noise with a standard deviation below 5% of the oscillator threshold. It should also be noted that in any JJ implementation of an oscillatory SR flip-flop, noise is effectively reduced to zero due to the required supercooling.

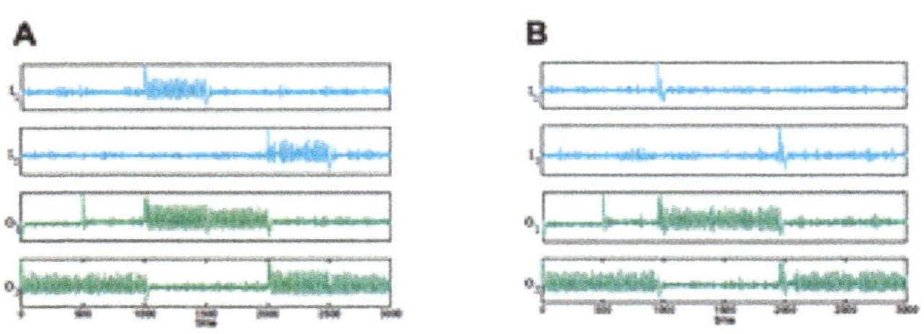

Figure 12: Controlled switching in the presence of noise.

(A) and (B) Time series of SR flip-flop as in Fig. 11 with additional Gaussian noise, mean=0, standard deviation=0.05, added to all oscillators. In both cases the switching occurs as required.

http://dx.doi.org/10.1371/journal.pone.0048498.g012

DISCUSSION

Here we have demonstrated how coupled threshold oscillators may be used as the principle components of the next generation of computers. Such implementation using binary logic is not disruptive, in that, from a programming and architectural perspective no significant changes are required.

There are clearly considerations concerning the accurate functioning of any such highly connected circuits, particularly with respect to multiple inputs to a single gate [93], [94], however, it should not be overly problematic to reduce the component numbers significantly given modern engineering capabilities.

The Fitzhugh-Nagumo simulations of the half adder, two-oscillator full adder, three-oscillator full adder, and 2×2 bit multiplier demonstrate threshold oscillators performing all the necessary components of arithmetic logic, while the SR flip-flop demonstrates the potential for very low power memory, particularly when ballistic propagation is considered.

Although the Fitzhugh-Nagumo models demonstrated here are in many ways phenomenologically similar to neural dynamics, we are not attempting to make any inference as to neural dynamics themselves. Moreover, we use the Fitzhugh-Nagumo system as an exemplification of the idea. In practice, implementation via Fitzhugh-Nagumo circuitry would be unlikely as this would offer very little in terms of speed or power consumption, however, any implementation via JJs or optical oscillators could be achieved in a very similar manner to that described for the Fitzhugh-Nagumo model whilst providing exceptional processing speed for minimal power usage.

There is currently a drive to low power exascale supercomputing (a computer which performs more than 10^{18} floating point operations per second (FLOPS)). The previous world's fastest supercomputer, the K computer, operates at a maximum performance of 10.51×10^{15} FLOPS and requires approximately $12MW$ of power (excluding the power for the cooling system which is typically in excess of 25% of power). This has recently been overtaken by the Sequoia IBM BlueGene/Q operating at 16.32×10^{15} FLOPS and using $7.89MW$ [95]. Even with such continued improvements in power consumption it is clear that without a significant technological breakthrough beyond that offered using standard CMOS transistor technology the prospect of an exascale computer is currently unviable. The use of oscillatory threshold logic presents a plausible avenue for implementation for which the engineering capability currently exists and which could be readily implemented.

ACKNOWLEDGMENTS

The authors would like to gratefully acknowledge the useful comments of Teresa Serrano-Gotarredona (Academic Editor) and the anonymous reviewers. We would also like to acknowledge input and advice from Professor Steve Furber and Dr Oleg Mukhanov.

AUTHOR CONTRIBUTIONS

Conceived and designed the experiments: JB SL. Performed the experiments: JB SL. Analyzed the data: JB SL. Contributed reagents/materials/analysis tools: JB SL. Wrote the paper: JB SL.

REFERENCES

1. National Security Agency (2005) Superconducting Technology Assessment report. Available: www. nitrd. gov/PUBS/nsa/sta. pdf. Accessed 2012 June 27.
2. Deutsch D (1985) Quantum-theory, the Church-Turing principle and the universal quantum com-puter. Proc Roy Soc A 400: 97–117. doi: 10. 1098/rspa. 1985. 0070
3. Lloyd S (1993) A potentially realizable quantum computer. Science 261: 1569–1571. doi: 10. 1126/ science. 261. 5128. 1569

4. Majer J, Chow JM, Gambetta JM, Koch J, Johnson BR, et al. (2007) Coupling superconducting qubits via a cavity bus. Nature 449: 443–447. doi: 10. 1038/nature06184
5. Morton JJL, McCamey DR, Erikkson MA, Lyon SA (2011) Embracing the quantum limit in silicon computing. Nature 479: 345–353. doi: 10. 1038/nature10681
6. Weber B, Mahapatra S, Ryu H, Lee S, Fuhrer A, et al. (2012) Ohm's law survives to the atomic scale. Science 335: 64–67. doi: 10. 1126/science. 1214319
7. Szöke A, Daneu V, Goldhar J, Kirnit NA (1969) Bistable optical element and its applications. Appl Phys Lett 15: 376.
8. Smith SD (1984) Towards the optical computer. Nature 307: 315–316. doi: 10. 1038/307315a0
9. Arsenault A, Fournier-Bidoz SB, Hatton B, Miguez H, Tétrault N, et al. (2004) Towards the synthetic all-optical computer: science fiction or reality? Journal of Materials Chemistry 14: 781–794. doi: 10. 1039/b314305h
10. Pahari N, Das DN, Mukhopahadhyay S (2004) All optical methods for the addition of binary data by nonlinear materials. Applied Optics 43: 6147–6150. doi: 10. 1364/ao. 43. 006147
11. Lynch S, Steele AL (2011) Nonlinear optical fibre resonators with applications in electrical en-gineering and computing, Applications of Nonlinear Dynamics and Chaos in Engineering. Santo Banerjee, Mala Mitra, Lamberto Rondoni (Eds.), Springer 1: 65–84. doi: 10. 1007/978-3-642-21922-1
12. Julliere M (1975) Tunneling between ferromagnetic-films. Phys Lett A 54: 225–226. doi: 10. 1016/0375-9601(75)90174-7
13. Datta S, Das B (1990) Electronic analog of the electro-optic modulator. Appl Phys Lett 56: 665–667. doi: 10. 1063/1. 102730
14. Meservey R, Tedrow PM (1994) Spin-polarized electron tunneling. Physics Reports 238: 173–243. doi: 10. 1016/0370-1573(94)90105-8
15. Dery H, Dalal P, Cywinski L, Sham LJ (2007) Spin-based logic in semiconductors for reconfigurable large-scale circuits. Nature 447: 573–576. doi: 10. 1038/nature05833
16. Kia B, Dari A, Ditto WL, Spano ML (2011) Unstable periodic orbits and noise in chaos computing. Chaos 21: 047520. doi: 10. 1063/1. 3664349
17. Kia B, Spano ML, Ditto WL (2011) Chaos computing in terms of periodic orbits. Phys Rev E 84: 036207. doi: 10. 1103/physreve. 84. 036207
18. Zanin M, Papo D, Sendina-Nadal I, Boccaletti S (2011) Computation as an emergent feature of adaptive synchronization. Phys Rev E 84: 060102. doi: 10. 1103/physreve. 84. 060102
19. Adleman LM (1994) Molecular computation of solutions to combinatorial problems. Science 266: 1021–1024. doi: 10. 1126/science. 7973651
20. Boneh D, Dunworth C, Lipton RJ, Jiri S (1996) On the computational power of DNA. Discrete Applied Mathematics 71: 79–94. doi: 10. 1016/s0166-218x(96)00058-3
21. Amos M, Paun G, Rozenberg G, Salomaa A (2002) Topics in the theory of DNA computing. Theoretical Computer Science 287: 1, 3–38. doi: 10. 1016/s0304-3975(02)00134-2
22. Kari L, Gloor G, Yu S (2000) Using DNA to solve the Bounded Post Correspondence Problem. Theoretical Computer Science 231: 192–203. doi: 10. 1016/s0304-3975(99)00100-0
23. Xu J, Tan GJ (2007) A review on DNA computing models. Journal of Computational and Theo-retical Nanoscience 4: 1219–1230. doi: 10. 1166/jctn. 2007. 001
24. Dorojevets M, Bunyk P, Zonoviev D, Likharev K (1999) COOL-0: Design of an RSFQ subsystem for petaflops computing. IEEE Trans Appl Supercond 9: 3606–3614. doi: 10. 1109/77. 783810
25. Takagi N, Murakami K, Fujimaki A, Yoshikawa N, Inoue K, et al. (2008) Proposal of a desk-side supercomputer with reconfigurable data-paths using Rapid Single-Flux-Quantum circuits. IEICE Trans Electron E91-C: 350–355. doi: 10. 1093/ietele/e91-c. 3. 350
26. Mukhanov OA, Semenov VK, Likharev K (1987) Ultimate performance of the RSFQ Logic Circuits. IEEE Trans Mag 23: 759–762. doi: 10. 1109/tmag. 1987. 1064951
27. Kaplunenko VK, Khabipov MI, Koshelets VP, Likharev KK, Mukhanov OA, et al. (1989) Experi-mental study of the RSFQ logic elements. IEEE Trans Mag 25: 861–864. doi: 10. 1109/20. 92422
28. Likharev K, Semenov V (1991) RSFQ logic/memory family: A new Josephson-junction technology for sub-terahertz clock-frequency digital systems. IEEE Trans Appl Supercond 1: 3–28. doi: 10. 1109/77. 80745
29. Mukhanov OA (1994) Superconductive single flux quantum technology. Solid-State Circuits Con-ference, Digest of Technical Papers 126–127 doi:10. 1109/ISSCC. 1994. 344703.
30. Bunyk P, Likharev K, Zinoviev D (2001) RSFQ technology: Physics and devices. International Journal of High Speed Electronics and Systems 11: 257–306. doi: 10. 1142/s012915640100085x
31. Mukhanov OA, Rylov SV (1997) Time-to-Digital converters based on RSFQ digital counters. IEEE Trans Appl Supercond 7: 2669–2672. doi: 10. 1109/77. 621788
32. Kirichenko AF, Sarwana S, Gupta D, Rochwarger I, Mukhanov O (2003) Multi-channel time digi-tizing system. IEEE Trans Appl Supercond 13: 454–458. doi: 10. 1109/tasc. 2003. 813898
33. Mukhanov O, Gupta D, Kadin AM, Semenov VK (2004) Superconductor analog-to-digital con- verters. Proc of the IEEE 92: 1564–1584. doi: 10. 1109/jproc. 2004. 833660
34. Filippov T, Dorojevets M, Sahu A, Kirichenko A, Ayala C, et al. (2011) 8-bit asynchronous wave-pipelined RSFQ arithmetic-logic unit,. IEEE Trans on Applied Superconductivity 21: 847–851. doi: 10. 1109/tasc. 2010. 2103918
35. Filippov T, Sahu A, Kirichenko AF, Vernik IV, Dorojevits

M, et al. (2012) 20 GHz operation of an asynchronous wave-pipelined RSFQ arithmetic-logic unit,. Physics Procedia 36: 59–65. doi: 10. 1016/j. phpro. 2012. 06. 130

36. Ortlepp T, Wetzstein O, Engert S, Kunert J, Toepfer H (2011) Reduced power consumption in superconducting electronics, IEEE Trans. on Applied Superconductivity 21: 770–775. doi: 10. 1109/tasc. 2011. 2117410

37. De Escobar LA, Mukhanov O, Hitt R, Littlefield W (2006) High performance HF-UHF all digital RF receiver tested at 20 GHz clock frequencies. Military Communications Conference, MILCOM 2006. IEEE 1–6 doi:10. 1109/MILCOM. 2006. 302479.

38. Vernik V, Kirichenko D, Dotsenko VV, Miller R, Webber RJ, et al. (2007) Cryocooled wideband digital channelizing RF receiver based on low-pass ADC. Superconductor Science and Technology 20: S323–S327. doi: 10. 1088/0953-2048/20/11/s05

39. Wong J, Dunnegan R, Deepnarayan G, Kirichenko D, Dotsenko V, et al. (2007) High performance, all digital RF receiver tested at 7. 5 GigaHertz. Military Communications Conference, MILCOM 2007. IEEE 1–5 doi:10. 1109/MILCOM. 2007. 4455052.

40. Mukhanov OA, Kirichenko D, Vernik IV, Filippov TV, Kirichenko A, et al. (2008) Superconductor digital-RF receiver systems,. IEICE Trans Electron E91-C: 306–317. doi: 10. 1093/ietele/e91-c. 3. 306

41. Littlefield W, Mukhanov O, Gupta D, Hitt D (2009) Ultra high speed ADCs and DSP brings direct digital RF beam forming to MILSATCOM phased array apertures. Military Communications Conference, 2009. MILCOM 2009. IEEE 1–7 doi:10. 1109/MILCOM. 2009. 5379946.

42. Brock DK, Kadin AM, Kirichenko AF, Mukhanov OA, Sarwana S, et al. (2001) Retargeting RSFQ cells to a submicron fabrication process. IEEE Trans Appl Supercond 11: 369–372. doi: 10. 1109/77. 919359

43. Tolpygo SK, Yohannes D, Hunt RT, Vivalda JA, Donnelly D, et al. (2007) 20 kA/cm2 process development for superconducting integrated circuits with 80 GHz clock frequency. IEEE Trans Appl Supercond 17: 946–951. doi: 10. 1109/tasc. 2007. 898571

44. Abelson LA, Kerber GL (2004) Superconductor integrated, circuit fabrication technology,. Proc of the IEEE 92: 1517–1533. doi: 10. 1109/jproc. 2004. 833652

45. Satoh T, Hinode K, Akaike H, Nagasawa S, Kitagawa Y, et al. (2005) Fabrication process of planarized multi-layer Nb integrated circuits. IEEE Trans Appl Supercond 15: 78–81. doi: 10. 1109/tasc. 2005. 849698

46. Andres S, Blamire MG, Buchholzet FI, Crété DG, Cristiano R, et al. (2010) European roadmap on superconductive electronics - status and perspectives. Physica C 470: 2079–2126. doi: 10. 1016/j. physc. 2010. 07. 005

47. Mukhanov O (2011) Energy-efficient Single Flux Quantum technology. IEEE Trans Appl Super-cond 21: 760–769. doi: 10. 1109/tasc. 2010. 2096792

48. Ryazanov VV, Bol'ginov VV, Sobanin DS, Vernik IV, Tolpygo SK, et al. (2012) Magnetic Josephson junction technology for digital and memory applications,. Physics Procedia 36: 35–41. doi: 10. 1016/j. phpro. 2012. 06. 126

49. Mukhanov OA, Semenov VK (1988) Reproduction of the Single Flux Quantum pulses in Josephson junction systems. II. Neuristor lines and logic elements. Mikroelektronika [Sov. Microelectronics] 17: 96–102.

50. Mizugaki Y, Nakajima K, Sawada Y, Yamashita T (1994) Implementation of new superconducting neural circuits using coupled SQUIDs. IEEE Trans Appl Supercond 4: 1–8. doi: 10. 1109/77. 273058

51. Dana SK, Sengupta DC, Hu CK (2006) Spiking and bursting in Josephson junction,. IEEE Transactions on Circuits and Systems 11-Express Briefs 10: 1031–1034. doi: 10. 1109/tcsii. 2006. 882183

52. Crotty P, Schult D, Segall K (2010) Josephson junction simulation of neurons,. Phys Rev E 82: 011914. doi: 10. 1103/physreve. 82. 011914

53. Onomi T, Maenami Y, Nakajima K (2011) Superconducting neural network for solving a combinatorial optimization problem,. IEEE Trans Appl Supercond 21: 701–704. doi: 10. 1109/tasc. 2010. 2092397

54. Jiang H, Ren Y, Zhong P, Ghaedi M, Hu Z, et al. (2012) Parkin controls dopamine utilization in human midbrain dopaminergic neurons derived from induced pluripotent stem cells,. Nature Communications 3: 688. doi: 10. 1038/ncomms1669

55. Zanin M, Del Pozo F, Boccaletti S (2011) Computation emerges from adaptive synchronization of networking neurons,. PloS ONE 11: e26467. doi: 10. 1371/journal. pone. 0026467

56. Ashwin P, Borresen J (2004) Encoding via conjugate symmetries of slow oscillations for globally coupled oscillators. Phys Rev E 70: 026203. doi: 10. 1103/physreve. 70. 026203

57. Afraimovich V, Rabinovich M, Varona P (2004) Heteroclinic contours in neural ensembles and the winnerless competition principle. Int. Journal of Bifurcation and Chaos 14: 1195–1208. doi: 10. 1142/s0218127404009806

58. Muroga S (1971) Threshold Logic and Its Applications Wiley.

59. Siu KY, Roychowdhury VP (1994) On optimal depth threshold circuits for multiplication and related problems. SIAM Journal on Discrete Mathematics 7: 284–292. doi: 10. 1137/s0895480192228619

60. Fant KM, Brandt SA (1997) Null Convention Logic Theseus Logic Inc.

61. Beiu V, Quintana JM, Avedillo MJ (2003) VLSI implementation of threshold logic: a survey. IEEE Trans on Neural Networks, Special Issue on Hardware Implementations 14: 1217–1243. doi: 10. 1109/tnn. 2003. 816365

62. Goser KF, Pacha C, Kanstein A, Rossmann ML (1997) Aspects of system and circuits for nano-electronics. Proc IEEE 85: 558–573. doi: 10. 1109/5. 573741

63. Likharev K (2003) Electronics below 10 nm. in Greer J, Korkin A, Labanowsky J (eds.): Nano and Giga Challenges in Microelectronics. Elsevier 27–68. doi: 10.

1016/b978-044451494-3/50002-0

64. Wei YI, Shen J (2011) Novel universal threshold logic gate based on RTD and its application. Microelectronics Journal 42: 851–854. doi: 10. 1016/j. mejo. 2011. 04. 005
65. Chua L (1971) Memristor-the missing circuit element. IEEE Trans Circuit Theory 18: 507–519. doi: 10. 1109/ tct. 1971. 1083337
66. Strukov DB, Snider GS, Stewart DR, Stanley W (2008) The missing memristor found. Nature 453: 80–83. doi: 10. 1038/nature06932
67. Rajendran J, Manem H, Karri R, Rose GS (2012) An energy-efficient memristive threshold logic circuit. IEEE Trans on Computers 61: 474–487. doi: 10. 1109/tc. 2011. 26
68. Inoue A, Kotani S, Imamura T, Hasuo S (1993) Niobium based Josephson circuit technology. Superconductivity 1: 1863–1877. doi: 10. 1016/0964-1807(93)90335-y
69. Han SK, Kurrer C, Kuramoto Y (1995) Dephasing and bursting in coupled neural oscillators. Phys Rev Lett 75: 3190–3193. doi: 10. 1103/physrevlett. 75. 3190
70. Hindmarsh JL, Rose RM (1984) A model of neuronal bursting using three coupled first order differential equations. Proc Roy Soc B 221: 87–102. doi: 10. 1098/ rspb. 1984. 0024
71. Boucsein C, Tetzlaff T, Meier R, Aertsen A, Naundorf B (2009) Dynamical response properties of neocortical neuron ensembles: Multiplicative versus additive noise. J Neuroscience 29: 1006–1010. doi: 10. 1523/jneurosci. 3424-08. 2009
72. Selverston AL, Moulins M (1985) Oscillatory neural networks. Annual Review of Physiology 47: 29–48. doi: 10. 1146/annurev. ph. 47. 030185. 000333
73. Binczak S, Jacquir S, Bibault J-M, Kazantsev VB, Nekorkin VI (2006) Experimental study of electrical Fitzhugh-Nagumo neurons with modified excitability. Neural Networks 19: 684–693. doi: 10. 1016/j. neunet. 2005. 07. 011
74. Kiesel N, Schmid C, Weber U, Ursin R, Weinfurter H (2005) Linear optics controlled-phase gate made simple. Phys Rev Letts 95: 210505. doi: 10. 1103/physrevlett. 95. 210505
75. Borresen J, Lynch S (2012) Binary half-adder using oscillators. International Patent Number WO 2012/001372 A1 (PDR/CLR/PM328208WO) 1–57.
76. Borresen J, Lynch S (2012) Binary half-adder and other logic circuits. UK Patent Number GB 2481717 A, Priority Data 1011110. 2 1–57.
77. Smeal RM, Ermentrout GB, White JA (2010) Phase response curves and synchronized neural networks. Phil Trans R Soc B. 365 (1551) 2402–2407. doi: 10. 1098/ rstb. 2009. 0292
78. Barreiro AK, Shea-Brown E, Thilo EL (2010) Time scales of spike-train correlation for neural oscillators with common drive. Phys Rev E 81: 011916. doi: 10. 1103/physreve. 81. 011916
79. Zhang X, Wang R, Zhang Z, Qu J, Cao J, et al. (2010) Dynamic phase synchronization charac-teristics of variable high-order coupled neuronal oscillator population. Neurocomputing 73 (13–15) 2665–2670. doi: 10. 1016/j. neucom. 2010. 05. 001
80. Waser R (2012) Nanoelectronics and Information Technology. Wiley
81. McCulloch W, Pitts W (1943) A logical calculus of the ideas imminent in nervous activity. Bull Math Biophys 5: 115–133. doi: 10. 1007/bf02478259
82. Lynch S (2010) Dynamical Systems with Applications using Maple. Springer doi: 10. 1007/978-0-8176-4605-9
83. Wei Y, Shen J (2011) Novel universal threshold logic gate based on RTD and its application. Microelectronics Journal 42: 851–854. doi: 10. 1016/j. mejo. 2011. 04. 005
84. Rajendran J, Manem H, Karri R, Rose GS (2012) An energy efficient memristive threshold logic circuit. IEEE Trans on Computers 61: 474–487. doi: 10. 1109/tc. 2011. 26
85. Current KW (1994) Current-hyphen mode CMOS multiple-valued logic-circuits. IEEE J Solid-State Circuits 29: 95–107. doi: 10. 1109/4. 272112
86. Fitzhugh R (1961) Impulses and physiological states in theoretical models of nerve membranes. Biophys 1182: 445–466. doi: 10. 1016/s0006-3495(61)86902-6
87. Nagumo J, Arimoto S, Yoshizawa S (1970) An active pulse transmission line simulating 1214-nerve axons. Proc. IRL 50: 2061–2070. doi: 10. 1109/jrproc. 1962. 288235
88. Hodgkin AL, Huxley AF (1952) A qualitative description of membrane current and its application to conduction and excitation in nerve. J Physiol 117: 500–544. Reproduced in Bull Math Biol (1990) 52: 25–71.
89. Hebb DO (1949) The organization of behaviour: A Neurophysiological Theory. Wiley
90. Elgamel MA, Faisal MI, Bayoumi MA (2005) Noise metrics in flip-flop designs. IEICE Trans Inf & Syst E88-D 7: 1501–1505. doi: 10. 1093/ietisy/e88-d. 7. 1501
91. Ashwin P, Orosz G, Wordsworth J, Townley S (2007) Dynamics on networks of cluster states for globally coupled phase oscillators. SIAM J on Applied dynamical systems 6: 728–758. doi: 10. 1137/070683969
92. Ashwin P, Borresen J (2005) Discrete computation using a perturbed heteroclinic network. Phys Letts A 347: 208–214. doi: 10. 1016/j. physleta. 2005. 08. 013
93. Reischuk R (2000) Can large fanin circuits perform reliable computations in the presence of faults? Theoretical Computer Science 240: 319–335. doi: 10. 1016/s0304-3975(99)00237-6
94. Kuo PY, Wang CY, Huang CY (2011) On rewiring and simplification for canon-icity in threshold logic circuits. IEEE/ACM International Conference on Computer-Aided Design (ICCAD) 396–403. doi: 10. 1109/iccad. 2011. 6105360
95. Top 500 supercomputer sites. Available www. top500. org. Accessed 2012 June 27.

CHAPTER 12

ELECTRICAL MAXWELL DEMON AND SZILARD ENGINE UTILIZING JOHNSON NOISE, MEASUREMENT, LOGIC AND CONTROL

Laszlo Bela Kish[1], Claes-Go¨ ran Granqvist[2]

[1]Department of Electrical and Computer Engineering, Texas A&M University, College Station, Texas, United States of America

[2]Department of Engineering Sciences, The A˚ngstro¨m Laboratory, Uppsala University, Uppsala, Sweden

ABSTRACT

We introduce a purely electrical version of Maxwell's demon which does not involve mechanically moving parts such as trapdoors, etc. It consists of a capacitor, resistors, amplifiers, logic circuitry and electronically controlled switches and uses thermal noise in resistors (Johnson noise) to pump heat. The only types of energy of importance in this demon are electrical energy and heat. We also demonstrate an entirely electrical version of Szilard's engine, i.e., an information-controlled device that can produce work by employing thermal fluctuations. The only moving part is a piston that executes work, and the engine has purely electronic controls and it is free of the major weakness of the original Szilard engine in not requiring removal and repositioning the piston at the end of the cycle. For both devices, the energy dissipation in the memory and other binary informatics components are insignificant compared to the exponentially large energy dissipation in the analog part responsible for creating new information by measurement and decision. This result contradicts the view that the energy dissipation in the memory during erasure is the most essential dissipation process in a demon. Nevertheless the dissipation in the memory and information processing parts is sufficient to secure the Second Law of Thermodynamics.

INTRODUCTION

Heat engines [1] utilize temperature differences to produce work while "heat demons" [2]–[8]employ information about instantaneous amplitudes of thermal fluctuations and execute control to produce a temperature difference and/or work. There has been an upsurge of interest in "heat demons", as evidenced from an extensive recent literature [9]–[18]. Furthermore there is an old debate [8]–[14], [19]–[21] on the question whether the energy dissipation due to erasure of information in the memory is the fundamental process to save the Second Law of Thermodynamics or if the generation of new information via measurement and decision [2],[13] and control requirements [9] is more important.

This paper introduces new and improved types of demons: they are purely electrical systems which employ the thermal voltage noise (Johnson noise) of resistors in thermal equilibrium. One of the demons pumps heat (similarly to Maxwell›s demon) and the other produces work (similarly to Szilard›s engine). Their common characteristics is that nothing but electrical energies appear in the fluctuations, measurement, decision, control and memory parts of the demons, and that all of the energy loss is due to dissipative

Citation: Kish LB, Granqvist C-G (2012) Electrical Maxwell Demon and Szilard Engine Utilizing Johnson Noise, Measurement, Logic and Control. PLoS ONE 7(10): e46800. doi:10.1371/journal.pone.0046800.

electrical transport. This situation is very different from that of earlier versions of Maxwell›s demon and Szilard›s engine wherein various forms of energies are (implicitly) employed in the various components. In the case of Maxwell›s demon, for example, the detection of an approaching molecule and its velocity require photoelectronics-based sensing that includes laser light (photon energy), photodiodes, electrical signals and energy, and an electromechanical system to drive the trapdoor, as well as related kinetic and potential mechanical energies. Our demon, which is a linear system, is also different in nature from ratchet-based demons [5] and Brillouin's diode-based ones [6], [7]because those utilize nonlinearity instead of measurement, decision and control. Similarly, in Szilard›s engine, detecting the location of the molecule, controlling the lever and repositioning the piston require photonic, electrical and mechanical energies and imply related forms of energy dissipation.

Complete evaluations of the various energy dissipation channels have not been performed in earlier work on Maxwell›s demon and Szilard›s engine, which may be associated with difficulties to assess the transfer of energies between their different forms. However our purely electrical demons, which are introduced in this paper, offer fundamental advantages and provide a convenient methodology for such analyses.

DISCUSSION AND RESULTS

Electrical Maxwell Demon Utilizing Johnson noise and Control

General Description

The electrical Maxwell demon described below employs the Johnson noise of resistors as well as measurement, decision, logic operations and control for pumping heat. One should note that control systems in general contain units for making measurements, decision and executing control. Figure 1 shows an electrical Maxwell demon in its starting stage 1 extracting energy from the resistor to the left. The switch is in position 1, and the Johnson noise current of the resistor yields voltage fluctuations $U(t)$ in the capacitor with variance

$$U_{eff}^2 = \langle U^2(t) \rangle = \frac{kT}{C} \quad (1)$$

in accordance with Boltzmann's energy equipartition theorem. It leads to a mean energy of $kT/2$ in the capacitor [1], where U_{eff} is the effective noise voltage (root mean square, RMS, voltage) on the capacitor C. This voltage fluctuation is a Gaussian process with exponential relaxation and correlation time (relaxation time) τ expressed as

$$\tau = RC, \quad (2)$$

where R is the resistance, and with an amplitude distribution function $g(U)$ given by

$$g(U) = \frac{1}{U_{eff}\sqrt{2\pi}} \exp\left[-\frac{U^2}{2U_{eff}^2}\right]. \quad (3)$$

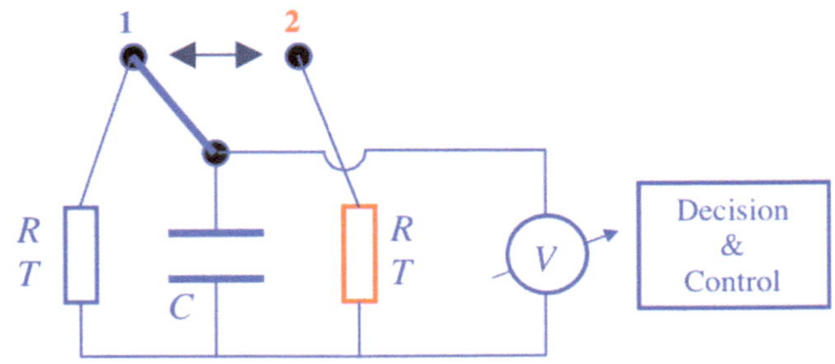

Figure 1: Electrical Maxwell demon with Johnson noise in stage 1.

The temperature T is the same in both resistors. The thick line indicates the moving part of the switch and the double-ended arrow the switching positions.

http://dx.doi.org/10.1371/journal.pone.0046800.g001

The voltage $U(t)$ is monitored as illustrated in Figure 2, and the demon stays in stage 1 until the voltage reaches a chosen arbitrary threshold value U_{th}, which corresponds to the energy

$$E_{th} = 0.5CU_{th}^2 \quad (4)$$

on the capacitor. The switch is then flipped to position 2, see Figure 3, and the demon gets into its stage 2 wherein the capacitor is discharged and the energy E_{th} is pumped into the resistor on the right. In stage 2, $U(t)$ is monitored on the capacitor, as shown in Figure 4, and when it reaches the zero threshold level with $U(t)=0$ the switch is flipped into its position 1 shown in Figure 1. Thus the demon gets into stage 1 again, and the full cycle of the demon is completed through this step.

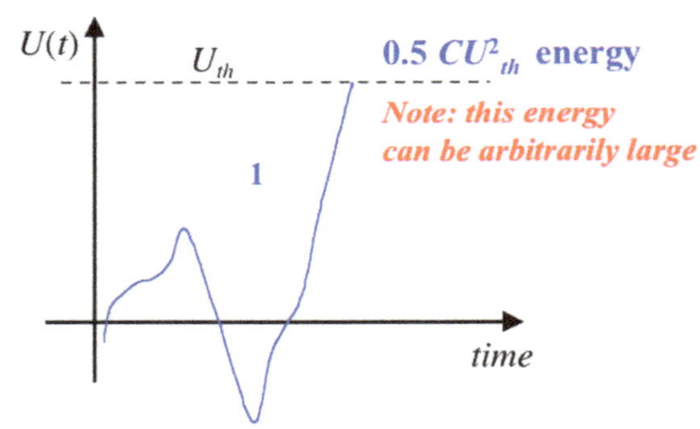

Figure 2: Voltage fluctuations in the capacitor during stage 1 for an electrical Maxwell demon with Johnson noise.

(Note that the real voltage time function typically exhibits more random fluctuation events, as discussed later.)

http://dx.doi.org/10.1371/journal.pone.0046800.g002

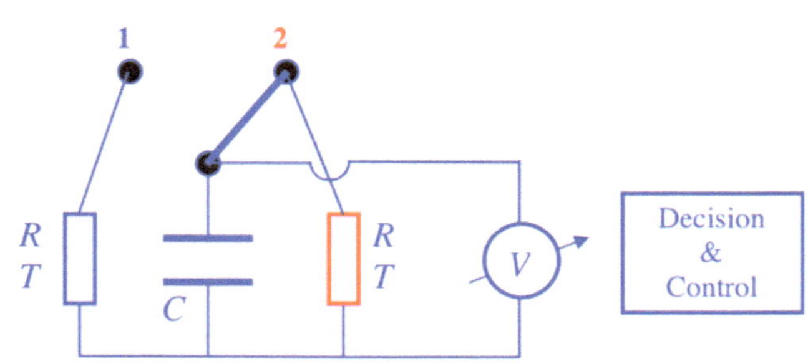

Figure 3: Electrical Maxwell demon with Johnson noise in stage 2.

http://dx.doi.org/10.1371/journal.pone.0046800.g003

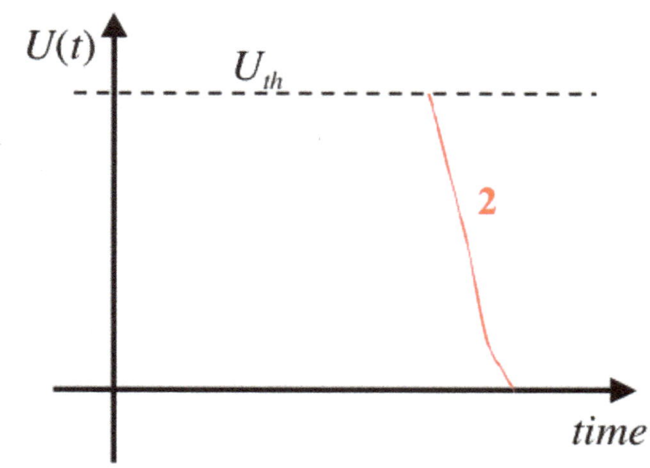

Figure 4: Voltage fluctuations on the capacitor during stage 2 for an electrical Maxwell demon with Johnson noise.

(Note that real voltage fluctuations typically exhibit more random events, as discussed later.)

http://dx.doi.org/10.1371/journal.pone.0046800.g004

By the end of stage 1, the capacitor has extracted the energy $E_{th}=0.5CU_{th}^2$ from the resistor on the left, and this energy is fully dissipated in the resistor on the right by the end of stage 2. When the demon gets back to stage 1 to start a new cycle, the energy in the capacitor is zero. Thus by the end of the new stage 1 process, the capacitor again extracts $E_{th}=0.5CU_{th}^2$ from the resistor on the left, and this energy will once more be fully dissipated in the resistor on the right by the end of the new stage 2. Figure 5 illustrates voltage fluctuations during the whole cycle.

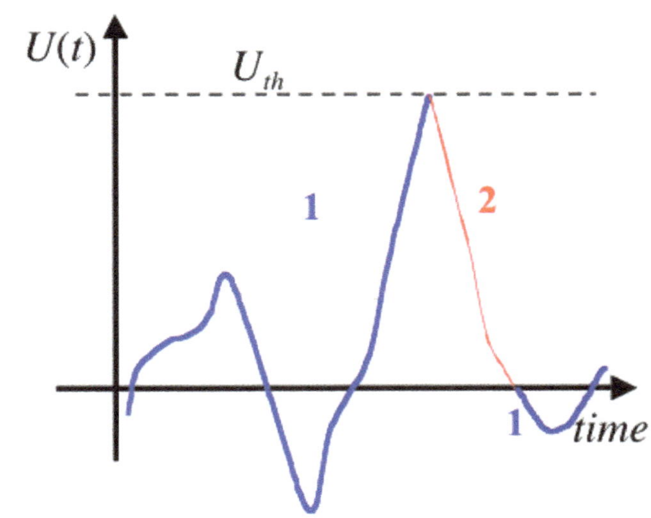

Figure 5: Voltage fluctuations in the capacitor, indicating the role of the two threshold levels U_{th} **and 0 during the whole cycle, and the beginning of a new cycle.**

http://dx.doi.org/10.1371/journal.pone.0046800.g005

It is important to note that, in an ideal system, the heat pumped during a single cycle does not have an upper limit because, according to Eq. 2, the threshold amplitude U_{th} (and thus the energy E_{th}) can be chosen to be arbitrarily large.

Cycle Duration versus Extracted Heat/Work during a Cycle

The cycle frequency of Maxwell's demon is usually not part of its energy balance, but we note that an increase of the threshold U_{th} will cause an exponential slowdown of the demon as a result of the increased cycle time needed to reach U_{th}, as further discussed later. The duration of the exponential discharge process in stage 2 scales logarithmically with the pumped energy (threshold energy) E_{th}, whereas the duration of stage 1 scales exponentially in the limit of large E_{th}, as discussed below. Thus the main question is the average duration of stage 1, i.e., the mean first-passage time T_1 of the voltage fluctuation of the capacitor between zero and U_{th} when this energy is much greater than the average thermal energy in the capacitor; the threshold is given by

$$E_{th}=0.5CU_{th}^2>>0.5kT. \tag{5}$$

This expectation agrees with an earlier, analytic treatment of Gaussian noise with exponential relaxation, wherein it was derived in an asymptotic limit [22]. Note that first passage times belong mostly to the field of "unsolved problems of noise", and analytic results (e.g. [22]–[24]) exist only for the asymptotic limit indicated by Eq. 5. In the present analysis we use results from the analytic treatment in [22].

A dimensionless first passage time τ_1 from zero amplitude to the level U_{th} is obtained [22]according to

$$\tau_1(\eta)=\frac{\sqrt{\pi}}{2\eta}\exp(\eta^2), \tag{6}$$

where $\tau_1(\eta)=T_1(\eta)/\tau$. Here η is the dimensionless threshold T_1 and τ is the relaxation time of the Lorentzian. In our system the dimensionless threshold is $\eta=\frac{1}{\sqrt{2}}\frac{U_{th}}{U_{eff}}$, where $U_{eff}=\sqrt{kT/C}$ is the effective value of the noise. Thus the real first passage time T_1 from zero amplitude to level U_{th} is

$$\begin{aligned}T_1(\zeta)=\tau_1(\eta)\tau&=RC\frac{\sqrt{\pi}}{2\frac{1}{\sqrt{2}}\frac{U_{th}}{U_{eff}}}\exp\left(\frac{U_{th}^2}{2U_{eff}^2}\right)\\&=RC\sqrt{\frac{\pi}{2}\frac{U_{eff}^2}{U_{th}^2}}\exp\left(\frac{U_{th}^2}{2U_{eff}^2}\right)=\\&=RC\sqrt{\frac{\pi}{2}\frac{kT}{CU_{th}^2}}\exp\left(\frac{CU_{th}^2/2}{kT}\right)\\&=RC\sqrt{\pi\frac{kT}{4E_{th}}}\exp\left(\frac{E_{th}}{kT}\right)\propto RC\sqrt{\frac{kT}{E_{th}}}\exp\left(\frac{E_{th}}{kT}\right).\end{aligned} \tag{7}$$

In conclusion, even though an arbitrarily large energy E_{th} can be transferred by the Maxwell demon type heat pump within a cycle, the cycle duration scales exponentially with this energy, which means that increasing E_{th} results in an exponential scaling down of the mean power.

On Energy Requirement and Dissipation

In this section we discuss the energy dissipation in the limit of high threshold, $E_{th}>>kT$(where Eq. 7 holds), and small error probability. A full assessment of the energy dissipation needed to run the demon requires consideration of all of its building elements. Such analyses have not been common in prior work because the demon›s functional characteristics have usually not been described in sufficient detail. To avoid mistakes caused by such deficiencies, Figure 6 shows the complete functional block scheme required to realize Maxwell's demon. A possible physical realization of this scheme is given in Figure 7.

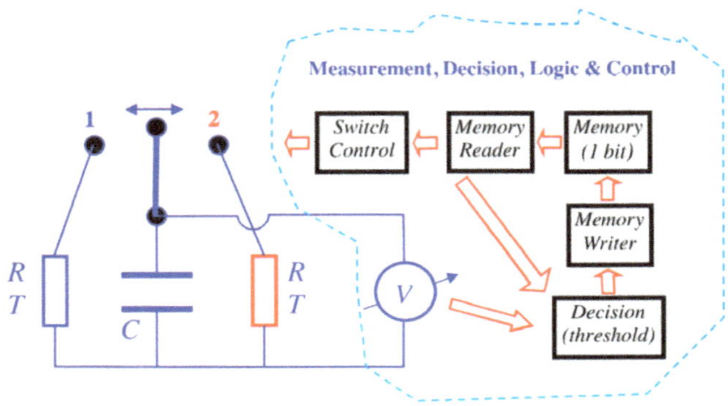

Figure 6: Electrical Maxwell demon showing a block scheme of the building elements for measurement, decision and control.

The voltmeter and the threshold device create new information, and the rest of the system only processes or utilizes this binary information.

http://dx.doi.org/10.1371/journal.pone.0046800.g006

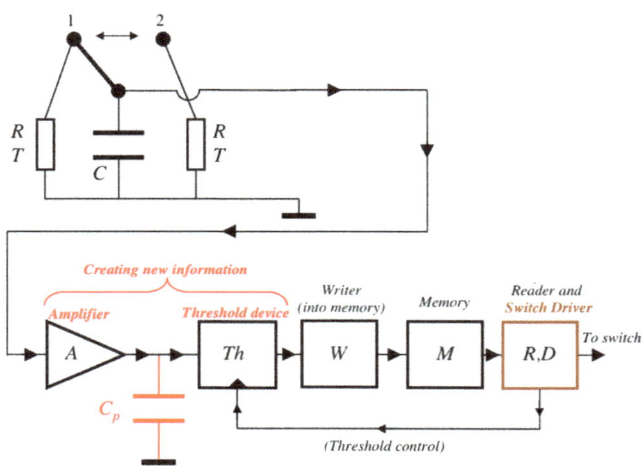

Figure 7: Possible realization of an electrical Maxwell demon.

Some of the elements can be joined in other realizations, but the energy dissipation sources described here remain.

http://dx.doi.org/10.1371/journal.pone.0046800.g007

It is important to identify the dominant causes of energy dissipation. The first two stages of the measurement, decision, logic and control unit in Figure 6 are the amplifier and the threshold device in Fig. 7, which represent the measurement and decision. For example, the output of the threshold device can be a positive or negative pulse representing the moment when the switch state must be changed from 1 to 2 or from 2 to 1 and triggering the memory to alter its state accordingly. The memory stores binary, single-bit information—a high or low value—and these two-bit states signify the actual state (1 or 2) of the switch. During a single cycle of the demon, the single-bit information cycles through its two states low-high-low or high-low-high depending on the choice of the designer of the system: whether the low-bit represents the switch-state 1 and the high-bit the switch-state-2, or vice versa.

The reader and switch driver device is basically an amplifier system that reads out the memory status and checks the voltage controlled switches, which either make a connection between the capacitor and point 1 while breaking the former connection to point 2, or vice versa. Thus these are two electronically controlled switches—between the capacitor and 1 and between the capacitor and 2, respectively—driven in alternate fashion.

We now consider the elements for handling the binary information and note that according to Brillouin›s negentropy principle [6], [7] the processing of each bit of information will dissipate energy of at least $kT\ln(2)$, and processing with acceptable error probability $(<<0.5)$ will require even greater energy dissipation E_d [9], [25]. Within the correlation time (reciprocal bandwidth) of thermal fluctuations and in the small-error probability limit [9], [25] the lower boundary of energy dissipation can be described by the Rice formula [25]. The required energy is logarithmically divergent when the error probability goes to zero according to

$$E_{d,bin} \approx kT\ln\left(\frac{1}{\varepsilon}\right) \quad [\text{Joule/bit}], \tag{8}$$

where $\varepsilon << 0.5$ is the bit error probability during the correlation time τ. Demanding the same numerical value of the error probability [9] for a longer cycle time (or clock period) $T_c >> \tau$makes the energy dissipation increase logarithmically as a result of the additivity of independent errors generated in different non-overlapping correlation time intervals. It follows that

$$E_{d,bin} \approx kT\ln\left(\frac{1}{\varepsilon_c}\frac{T_c}{\tau}\right) \quad [\text{Joule/bit}], \tag{9}$$

where $\varepsilon_c < 0.5$ is the bit error probability during T_c. Using the result of Eq. 7 we obtain the following relation for the lowest possible energy dissipation of each binary switching element (transistor, etc) in the demon:

$$E_{d,bin} \geq kT\ln\left[\frac{1}{\varepsilon_c}\sqrt{\frac{kT}{E_{th}}}\exp\left(\frac{E_{th}}{kT}\right)\right]$$
$$= E_{th} + kT\left[\ln\left(\frac{1}{\varepsilon_c}\sqrt{\frac{kT}{E_{th}}}\right)\right]. \tag{10}$$

We conclude that the Second Law of Thermodynamics is satisfied by even a single binary unit (such as any logic gate or the memory) for the case of a fixed E_{th}/kT ratio and error probability approaching zero. The energy dissipation scales roughly with the energy threshold E_{th}, i.e., in the small error limit it scales with the pumped energy.

Turning now to the major sources of energy dissipation, we consider the amplifier and the threshold device, which are the analog circuitry elements where new information is created. Their energy dissipation scale exponentially with E_{th} and is much larger than the amount required by the Second Law of Thermodynamics. Theoretically, if the chosen threshold level U_{th} is large enough, the amplifier is not needed and it is sufficient with a threshold device. However, any such device can be modeled as an amplifier with either saturation and/or positive feedback, and thus we keep the amplifier stage because it helps illustrate the point where the energy dissipation scales exponentially with E_{th}.

In order to elucidate the origin of the amplifier's energy dissipation it should be realized that any amplifier has a component of inertia. To choose a simple illustrative example, a hydraulic amplifier has inertia due to the mass of its components. In the present case of an electrical voltage amplifier, inertia ensues from parasitic capacitances and inductances. We demonstrate this issue of electrical inertia by considering the final capacitor, positioned at the point where the threshold device makes the decision about the threshold crossing, denoted C_p in Figure 7. For the sake of simplicity we assume that the amplification is unity in the amplifier. Under thermal equilibrium, thermal voltage fluctuations (Johnson noise) appear on the capacitor with an RMS value $\sqrt{kT/C_P}$, which corresponds to a mean thermal energy of $kT/2$ in the capacitor. If this Johnson noise is not actively damped it must be kept at a much lower level than the main Johnson noise of the demon we are monitoring, i.e., $\sqrt{kT/C} >> \sqrt{kT/C_P}$. This means that $C_P >> C$ in order to reduce erroneous threshold crossings caused by the parasitic noise. We note in passing that it can be shown that active damping of the parasitic Johnson noise would cause greater energy dissipation than the value discussed below [26].

Figure 8 illustrates three possible trajectories of the voltage within stage 1 as its amplitude goes from zero to U_{th}. Curves A and B represent monotonically increasing functions which reach U_{th} with relatively modest energy investment, i.e., $0.5C_pU_{th}^2 >> kT/2$ plus the thermalization losses due to the resistances, etc (not shown in the circuitry). However curve $C$is qualitatively different and non-monotonic. This feature causes serious additional energy dissipation and all of the energy invested in the trajectory from points a to b is completely lost. For example, in the case of a symmetric power supply (i.e., with electrodes: +, −, and zero (ground)) the parts of the trajectory with positive velocity (increasing amplitude) are supplied by the positive electrode of the power source, while the decreasing parts of the trajectory are supplied by the negative electrode, when in both cases the current flows into the ground. Both of these currents represent positive power extracted from the positive and negative parts of the supply, respectively. Even though a net energy was dissipated in the voltage-path from a to b, the whole charging process is back to zero at b, and all of the earlier invested energy was lost.

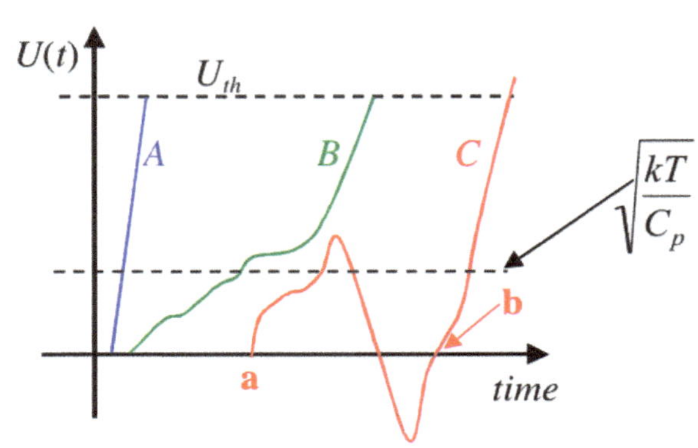

Figure 8: Illustration of the main source of energy dissipation in the electrical Maxwell demon.

An exponential number of random fluctuation events as a function of the threshold energy take place before reaching the threshold. The energy dissipation between the points **a** and **b** represent complete energy loss and do not contribute to the process of attaining the threshold.

http://dx.doi.org/10.1371/journal.pone.0046800.g008

How many such dissipative random fluctuation events will the voltage on the capacitors exhibit, and how large are these fluctuations? During the correlation time $\tau = RC$ of the noise in the main capacitor the noise typically goes through at least one large wave ranging though the +/− RMS amplitude interval. This means that the related energy dissipation E_{d1} during the correlation time of the main noise is of the order

$$E_{d1,an} \approx \frac{C_P}{C} kT. \tag{11}$$

According to Eq. 7 the expected number of correlation time (τ) periods during the dominant stage 1 part of the clock cycle is $\sqrt{\frac{kT}{E_{th}}} \exp\left(\frac{E_{th}}{kT}\right)$ so that the total energy dissipation is of the order

$$E_{d,an} \approx \frac{C_P}{C} kT \sqrt{\frac{kT}{E_{th}}} \exp\left(\frac{E_{th}}{kT}\right) \propto kT \exp\left(\frac{E_{th}}{kT}\right). \tag{12}$$

Thus the dominant energy dissipation during a single cycle in the large-threshold energy limit is

$$E_{d,tot} > E_{d,an} + E_{d,bin}$$
$$\approx \frac{C_P}{C} kT \sqrt{\frac{kT}{E_{th}}} \exp\left(\frac{E_{th}}{kT}\right) + E_{th} + kT\left[\ln\left(\frac{1}{\varepsilon_c}\sqrt{\frac{kT}{E_{th}}}\right)\right], \tag{13}$$

and hence the approximate scaling in this limit is

$$E_{d,tot} \propto kT \exp\left(\frac{E_{th}}{kT}\right). \tag{14}$$

In conclusion, the dominant energy dissipation in the electrical Maxwell demon takes place in the part which generates the new information and is due its analog processing yielding an exponential dependence of the number of random voltage oscillations during a single cycle.

Electrical Szilard Engine Utilizing Johnson noise and Control

The principles of the Johnson noise driven electrical Szilard engine are the same as those of the electrical Maxwell demon though heat is not pumped but work is executed by a moving capacitor plate used as a piston [1], [27]. The control and all other considerations are inherently the same as for Maxwell's demon except the fact that the Szilard engine requires a twice as large system as the Maxwell demon. Thus two single-bit memories are needed and the related voltage monitoring and switch control must be performed over two capacitors.

Figure 9 shows the main elements and circuitry of the electrical Szilard engine except the decision, logic and control components which are identical to the corresponding parts in Maxwell›s demon. The electrical Szilard engine extracts heat from the resistor and executes work on the piston, which is the moving joint capacitor plate of the two capacitors at the middle.

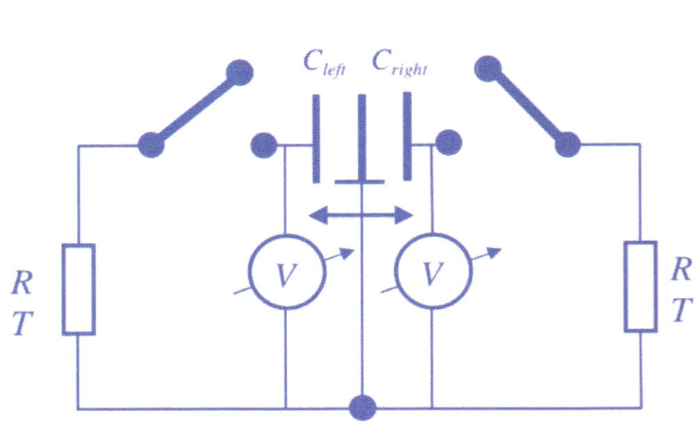

Figure 9: Electrical Szilard engine extracting heat from the resistor and executing work on the piston, i.e., the movable capacitor plate.

http://dx.doi.org/10.1371/journal.pone.0046800.g009

The left capacitor has its maximum capacitance $C_{left} = C_{max}$ when the piston is at the left end while the right capacitor then attains its minimum capacitance $C_{right} = C_{min}$. In the opposite case, when the piston is at the right end, the left capacitor reaches its minimum value$C_{left} = C_{min}$ while the right capacitor has its maximum capacitance $C_{right} = C_{max}$. The capacitor with the minimum capacitance will be charged up to the work charge Q_w at different points of the cycle; they correspond to a threshold voltage $U_{th} = Q_w / C_{min}$ as shown inFigure 2. The corresponding electrical charging energy—analogously to the case of Maxwell›s demon—is much greater than the thermal equilibrium value, i.e.,

$$E_w = 0.5Q_{min}^2 / C >> kT/2. \tag{15}$$

Figure 10A shows an arbitrarily chosen starting point of the cycle. The piston is at the right end and the capacitor to the right has the work charge Q_w. The voltage on the right capacitor is Q_w / C_{max}. The energy difference, compared to the situation with the opposite position of the piston at the same charge, is

$$-\Delta E = 0.5Q_w \left(\frac{1}{C_{max}} - \frac{1}{C_{min}} \right) < 0 \tag{16}$$

implying that the piston has a negative potential energy and sits in a potential well. We suppose that

$$\Delta E = 0.5Q_w \left(\frac{1}{C_{min}} - \frac{1}{C_{max}} \right) >> kT/2, \tag{17}$$

i.e., the piston remains at the bottom of this potential barrier, which is in the vicinity of the right end, while it executes its mechanical Brownian motion.

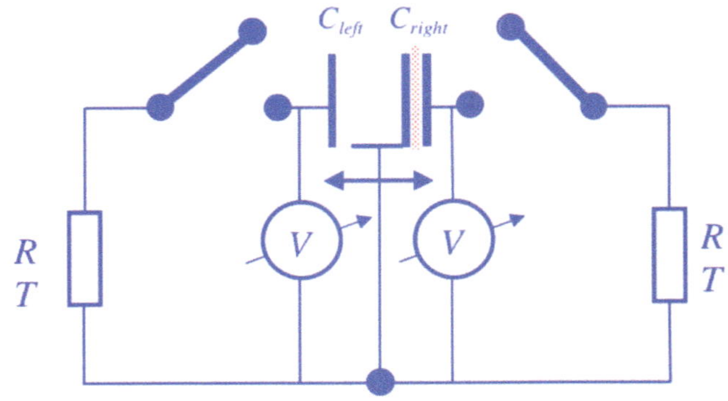

Figure 10: Arbitrarily chosen starting point of the cycle for the electrical Szilard engine.

The piston is at the right end and the capacitor at the right has the work charge Q_w. The corresponding electrical energy is $E_w \gg kT/2$.

http://dx.doi.org/10.1371/journal.pone.0046800.g010

Figure 11 illustrates the operation of the first half-cycle for the electrical Szilard engine. InFigure 11A the left switch is closed in order to charge the left capacitor and the Johnson noise current of the left resistor yields charge fluctuations in the left capacitor, which is a Gaussian noise of the charge; see Figure 10B. After an exponentially long waiting time characterized byEq. 7 the left capacitor, which originally had zero charge, is charged up to the working charge Q_w and energy E_w given by Eq. 15. At that moment the left switch is abruptly opened to keep the charge in the left capacitor and the right switch is closed to discharge the capacitor on the right as shown in Figure 10C. The charge in the right capacitor decays exponentially and, within a time interval of a few RC, thermalizes and crosses the zero charge (zero voltage) level; see Figure 10D. At that moment, as illustrated in Figure 10E, the right switch is abruptly opened and the zero charge (zero voltage) state is preserved in the right capacitor. Then the electrostatic force due to the working charge in the left capacitor moves the piston to the left end, as shown in Figure 10F, while, in accordance with the energy conservation law, the piston executes positive work amounting to ΔE. This is the end of the half-cycle. The piston is at the left end and the capacitor at the left has the work charge Q_w.

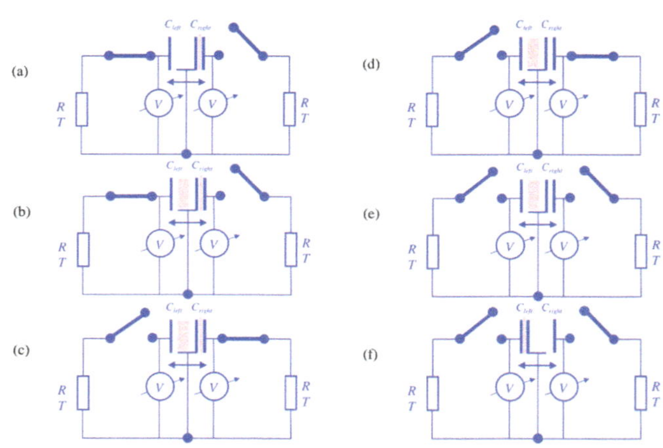

Figure 11: Operation of the first half cycle for the electrical Szilard engine.

(a): The left switch is closed to charge the left capacitor. (b): Johnson noise current yields charge fluctuations in the left capacitor, i.e., a Gaussian noise in the charge. (c): As soon as the left capacitor reaches Q_w, the left switch is abruptly opened to keep the charge in the left capacitor and the right switch is closed to discharge the capacitor on the right. (d): The charge in the right capacitor decays exponentially and, within a time interval of a few *RC*, thermalizes and crosses the zero charge (zero voltage) level. (e): The right switch is abruptly opened and the zero charge (zero voltage) state is preserved in the right capacitor. (f): The force due to the working charge in the left capacitor moves the piston to the left end, which is the end of the half-cycle; the piston is at the left end and the capacitor at the left has the work charge Q_w.

http://dx.doi.org/10.1371/journal.pone.0046800.g011

The remainder of the full cycle comprises the same processes as during the first half with the proper change of the indices signifying "left" and "right" of capacitors and switches, as elucidated in Figure 12.

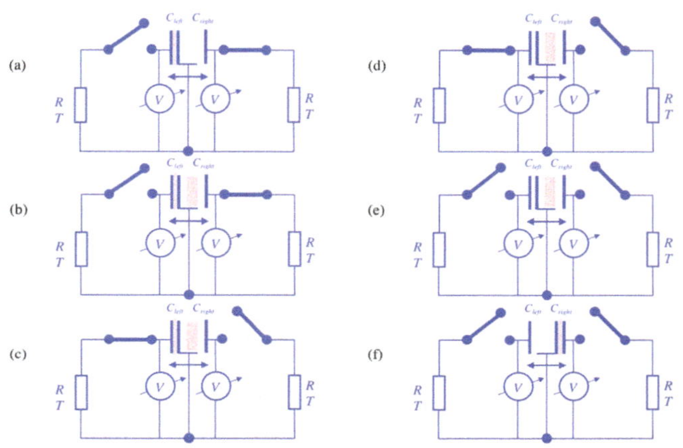

Figure 12: Operation of the second half cycle for the electrical Szilard engine.

(a): The right switch is closed to charge the right capacitor. (b): Johnson noise current yields charge fluctuations in the right capacitor, i.e., a Gaussian noise in the charge. (c): As soon as the right capacitor reaches Q_w the right switch is abruptly opened to keep the charge in the right capacitor and the left switch is closed to discharge the capacitor on the right. (d): The charge in the left capacitor decays exponentially and, within a time interval of a few *RC*, thermalizes and crosses the zero charge (zero voltage) level. (e): The left switch is abruptly opened and the zero charge (zero voltage) state is preserved in the left capacitor. (f): The force due to the working charge in the right capacitor moves the piston to the right end, which is the end of the full cycle and the same state as where the analysis started; see Figure 10. The piston is at the right end and the capacitor in the right has the work charge Q_w.

http://dx.doi.org/10.1371/journal.pone.0046800.g012

Thus the full cycle has been carried out with purely electronic control based on the information from the measurement of charge or voltage in the capacitors. It should be noted that the system is free from the major deficiency of the original Szilard engine, which is that its piston must be artificially relocated into the initial position at the end of the cycle. The two switches represent four possible states and two bits of information, and consequently the unit for measurement, decision, logic and control is basically twice that of the system shown for the electrical Maxwell demon, while the energy dissipation essentially doubles. Thus the energy dissipation in the binary part, including in the memory, is negligible compared to that in the analog part that executes the measurement and creates the new information.

METHODS AND CONCLUSIONS

Two electrical demons utilizing Johnson noise, measurement, decision, logic and control were introduced; they are new versions of Maxwell's demon and Szilard's engine. We showed all of the necessary building elements and analyzed the cycle time versus the energy output. Both demons have an arbitrarily large energy output during a single cycle, and the cycle duration and energy dissipation scale exponentially with this energy output. The exponentially scaling energy dissipation occurs in the analog parts of the demons, where the creation of new information (measurement and decision) takes place. In the binary part, including the memory, the energy dissipation scales only linearly with the energy output and even this dissipation is enough to satisfy the Second Law of Thermodynamics due to the enhanced energy threshold requirements caused by the exponentially long cycle duration and the requirement of keeping the error probability small.

ACKNOWLEDGMENTS

LBK is grateful to Katja Lindenberg for useful discussions about first-passage problems and Eleonora Alfinito and Cecilia Pennetta for valuable remarks during a seminar about this topic. Useful discussions with John Norton, Dave Ferry and Wolfgang Porod about the irreversibility of computation and the Landauer theorem are appreciated.

AUTHOR CONTRIBUTIONS

Analyzed the data: LK CGG. Wrote the paper: LK CGG. Concept: LK CGG. Theory: LK CGG. Analysis: LK CGG. Conclusions: LK CGG.

REFERENCES

1. Kish LB (2011) Thermal noise engines. Chaos Solitons Fractals 44: 114–121. doi: 10. 1016/j. chaos. 2010. 12. 008
2. Szilard L (1929) On the reduction of entropy in a thermodynamic system by the interference of an intelligent being. Z Phys 53: 840–856.
3. Maxwell JC (1871) Theory of Heat. London: Longmans, Green & Co.
4. Leff HS, Rex AS, editors. (2003) Maxwell's Demon 2: Entropy, Classical and Quantum Information, Computing. Bristol: Institute of Physics.
5. Chialvo DR, Millonas MM (1995) Asymmetric unbiased fluctuations are sufficient for the operation of a correlation ratchet. Phys Lett A 209: 26–30. doi: 10. 1016/0375-9601(95)00773-0
6. Brillouin L (1964) Scientific Uncertainty and Information. New York: Academic.
7. Brillouin L (1962) Science and Information Theory. New York: Academic.

8. Zurek WH (1986) Maxwell's demon, Szilard's engine and quantum measurements. In: Moore GT, Scully MO, editors. Proceedings of Frontiers of Non-Equilibrium Quantum Statistical Mechanics. New York: Plenum. pp. 145–150.
9. Kish LB, Granqvist CG (2012) Energy requirement of control: Comments on Szilard's engine and Maxwell's demon. EPL 98: 68001. doi: 10. 1209/0295-5075/98/68001
10. Norton JD (2011) Waiting for Landauer. Stud Hist Philos Mod Phys 42: 184–198.
11. Norton JD (2005) Eaters of the lotus: Landauer's principle and the return of Maxwell's demon. Stud Hist Philos Mod Phys 36: 375–411. doi: 10. 1016/j. shpsb. 2004. 12. 002
12. Norton JD (2012) The end of the thermodynamics of computation: A no go result. (Manuscript to appear in Philos Sci). Available at John Norton's website at Univ. of Pittsburg. Available: http://www. pitt. edu/~jdnorton/papers/No_Go. pdf. Accessed 2012 Sep 21.
13. Granger L, Kantz H (2011) Thermodynamic cost of measurements. Phys Rev E 84: 061110. doi: 10. 1103/physreve. 84. 061110
14. Gea-Banacloche J, Leff HS (2005) Quantum version of the Szilard one-atom engine and the cost of raising energy barriers. Fluct Noise Lett 5: C39–C47. doi: 10. 1142/s0219477505003014
15. Horowitz J, Parrondo JMR (2011) Designing optimal discrete-feedback thermodynamic engines. New J Phys 13: 123019. doi: 10. 1088/1367-2630/13/12/123019
16. Sagawa T, Ueda M (2012) Nonequilibrium thermodynamics of feedback control. Phys Rev E 85: 031114. doi: 10. 1103/physreve. 85. 021104
17. Vaikuntanathan S, Jarzynski C (2011) Modeling Maxwell's demon with a microcanonical Szilard engine. Phys Rev E 83: 061120. doi: 10. 1103/physreve. 83. 061120
18. Van den Broeck C (2010) Thermodynamics of information: Bits for less or more for bits? Nature Phys 6: 937–938. doi: 10. 1038/nphys1834
19. Porod W (1988) Energy requirements in communication – Comment. Appl Phys Lett 52: 2191–2191. doi: 10. 1063/1. 99531
20. Porod W, Grondin RO, Ferry DK (1984) Dissipation in computation. Phys Rev Lett 52: 232–235. doi: 10. 1103/physrevlett. 52. 232
21. Porod W, Grondin RO, Ferry DK, Porod G (1984) Dissipation in computation – Reply. Phys Rev Lett 52: 1206–1206. doi: 10. 1103/physrevlett. 53. 1206
22. Lindenberg K, Seshadri V (1979) Analytic theory of extrema. I. Asymptotic theory for Fokker-Planck processes. J Chem Phys 71: 4075–4084. doi: 10. 1063/1. 438177
23. Palatella L, Pennetta C (2011) Distribution of first-return times in correlated stationary signals. Phys Rev E 83: 041102. doi: 10. 1103/physreve. 83. 041102
24. Pennetta C (2006) Distribution of return intervals of extreme events. Eur Phys J B 50: 95–98. doi: 10. 1140/epjb/e2006-00140-y
25. Kish LB (2004) Moore's law and the energy requirement of computing versus performance. IEE Proc – Circuits Devices Systems 151: 190–194. doi: 10. 1049/ip-cds:20040434
26. Kish LB, Granqvist CG (2011) Energy requirement of control: Comments on Maxwell's demon and Szilard's engine. In: Alfinito E, Leuzzi M, Millithaler J-F (editors). All the Colors of Noise: Essays in Honor of Lino Reggiani. Cormigliano del Brenta: Munari Edizione. pp. 43–50.
27. Davis BR, Abbott D, Parrondo JMR (2001) Thermodynamic energy exchange in a moving plate capacitor. Chaos 11: 747–754. doi: 10. 1063/1. 1394191

CHAPTER 13

ANALYSIS OF FUNCTIONAL GENOMIC SIGNALS USING THE XOR GATE

Mahesh Yaragatti[1] and Qi Wen[2]

[1]Biotechnology Program, University of Pennsylvania, Philadelphia, Pennsylvania, United States of America

[2]Department of Physics, University of Pennsylvania, Philadelphia, Pennsylvania, United States of America

ABSTRACT

Modeling gene regulatory networks requires recognition of active transcriptional sites in the genome. For this reason, we present a novel approach for inferring active transcriptional regulatory modules in a genome using an established systems model of bit encoded DNA sequences. Our analysis showed variations in several properties between random and functional sequences. Cross correlation within random and functional groups uncovered a wave pattern associated with functional sequences. Using the exclusive-OR (XOR) logic gate, we formulated a scheme to threshold signals that may correlate to putative active transcriptional modules from a population of random genomic fragments. It is our intent to use this as a basis for identifying novel regulatory sites in the genome.

INTRODUCTION

A digital transformation of genomic sequences has facilitated the biological discovery of regulatory regions in the genome through computational methods. Since 2003, the ENCODE Consortium has endeavored to identify all the regulatory sites in the human genome; however, there are still 10^5 more regulatory sites than genes in the mammalian genome. This requires the development of creative approaches to identify novel transcriptional regulatory sites in the genome.

There is opulent literature suggesting that one of the primary difficulties in locating active transcriptional sites is due to the majority of the mammalian genome being transcribed and most regulatory sequences being multi-functional [1], [2], [3], [4]. Although we are not presenting results to challenge this idea, it is implicit in our approach that our randomly selected fragments or mini-modules (150–250 bp) from the genome are more likely non-functional by comparison to the experimentally validated active sequences.

It is arguable that our method may only be detecting compositional differences between sequences which may not correlate to functionality. However, regulatory sequences represent a broad array of components including enhancers, repressors, silencers, and other boundary elements. Hence, the genomic data presently available to define specific properties of these individual components is very non-specific since most sites have not yet been discovered in the mammalian genome [5].

Digital delineation of genomic sequences such as DNA sequence barcoding [6] and implementation of boolean logic gates have been widely employed to study

Citation: Yaragatti M, Wen Q (2009) Analysis of Functional Genomic Signals Using the XOR Gate. PLoS ONE 4(5): e5608. doi:10.1371/journal.pone.0005608.

transcription [7],[8], [9]. Unfortunately, there has been limited advancement on the application of boolean logic gates and regulatory module bit coding. In this paper, we used an established systems model of bit encoded sequences from Yaragatti et al. (2009) to identify potentially active modules in the genome through a signal plot. The rationale for our approach was to evaluate the interaction of discrete properties between sequences.

METHODS

Genomic Sequence Database

The random and functional genomic sequences analyzed in this paper were acquired from Yaragatti et al. (2009). As reported, the functional sequences were experimentally validated as enhancers in a mammalian F9 cell line to increase the level of GFP expression when inserted upstream of a promoter site. The average size of these fragments was 149 bp. The random sequences had lengths >150 and <250 and were selected from the mouse genome (UCSC mm9, July 2007 Assembly). As reported, these sequences were encoded as bit vectors based on genomic features (from the UCSC Genome Browser) that were either present (equals 1) or not present (equals 0) in the sequences.

Cross Correlation and XOR Implementation

We implemented the following functions using Matlab scripts to compare a genomic sequence (vector) to other sequences in our dataset. The cross correlation (xcorr) function is a standard built in function that computes raw correlations with no normalization. The xcorr(x,y) function returns the cross-correlation sequence in a length $2*N-1$ vector, where x and y are vectors of length N ($N>1$).

$$\hat{R}_{xy}(m) = \begin{cases} \sum_{n=0}^{N-m-1} x_{n+m} y_n^* & m \geq 0 \\ \hat{R}_{yx}^*(-m) & m < 0 \end{cases}$$

The output vector c has elements given by $c(m)=R_{xy}(m-N)$, $m=1, \ldots, 2N-1$.

A very similar function to xcorr is the exlusive-OR (xor) operation. We used xor(x, y) to perform an operation on the corresponding elements of vectors x and y. The resulting element C(i,j,...) is logically true (i.e., equal to1) if x(i,j,...) or y(i,j,...)–but not both–is nonzero. In both applications, we summed the resultant vector and plotted the normalized scores for each vector.

Averaging the XOR scores for the functional sequences provides a threshold value for isolating putative functional sequences from the random genomic sequence population.

Statistical Analysis

Chi square analysis was performed through a built-in function (with default settings) using Stata v.9E software. All data reported has signficance at the 95% confidence level (i.e., $p<0.05$). The reported t-values were acquired from the online *Bioinformatics* Supplemental Table 1 and were included in our table for comparison purposes.

RESULTS AND DISCUSSION

This paper introduces a novel scheme to isolate candidate functional mini-modules from a random population of genomic fragments based on sequence properties. In an earlier study, it was successfully shown that our genomic fragments can be classified into active and inactive groups by calculating a composite score for each sequence and analyzing it through a support vector machine [10]. We are expanding on this analysis by creating a model that evaluates the interaction of these genomic vectors using the same set of sequence features. In our initial analysis, we presented a t-test that compared the means between vectors to show variation between the groups. To support our current model, a chi square (χ^2) test was performed on the features, by comparing frequency, to justify differences between elements of the vectors. We observed a wider range of χ^2 values, but the results from the chi square analysis are essentially similar to the t-test.

We computed the correlation coefficient to show variation between random and functional sequences with the assumption that the absolute score of a sequence does not directly relate to its functionality. As shown in Figure 1A, the random sequences generated a large range of scores whereas the functional sequences were limited to approximately three values suggesting there was limited variation within the functional group. The rationale for the XOR logic gate (Figure 1B) was to emulate the binary output function of the correlation coefficient with the major exception that similar sequences (i.e., vectors having similar elements) would score equivalently. We implemented a scheme in Figure 1C whereby random sequences with higher scores are more likely to be functional transcriptional modules, which can be selected by calculating thresholds. Although reported in our earlier study, it is important to bear in mind that the random sequences may have contained genic, intergenic, and non-coding regions of the genome.

An important aspect of the support vector machine approach used in Yaragatti et al. (2009) was that each sequence was evaluated by the integration of its discrete features. This paper proposes an analogous model that relates a sequence to other sequences (i.e., the interaction of features) for predicting functionality. This approach may uncover modules active in a specific regulatory network (i.e., transcriptional regulation of a gene) that may otherwise go undetected using current computational methods.

Using the mathematical XOR function, we have introduced a novel approach to approximate loci-specific transcriptional active sites in the genome. This is a simple, but effective, approach that interprets functional DNA as a dynamic bit code of features rather than a static sequence of nucleotide based motifs. Ultimately, we intend to advance this paradigm of analyzing regulatory sequences using bit vectors in more complicated systems with the hope of elucidating the functional landscape of eukaryotic genomes.

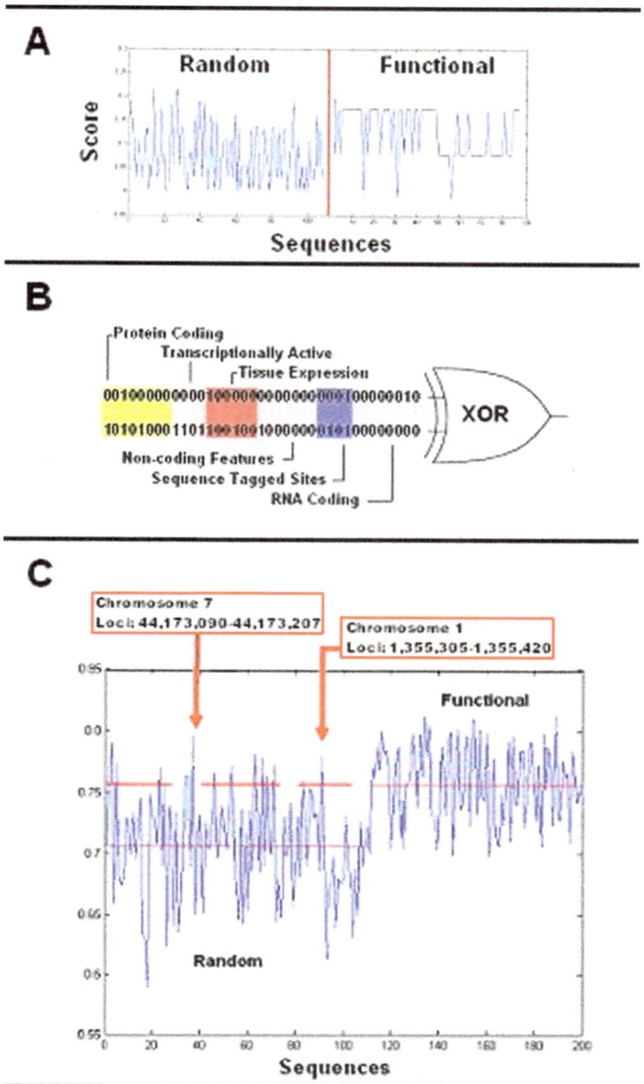

Figure 1: Analysis of bit encoded genomic sequences.

(a) Each sequence (point) is evaluated against every other sequence outputting a binary vector which is summed and normalized. The correlation coefficient was performed on random and functional vectors (t=6.19, f=15.17, df=201, $p<0.05$). The correlation coefficient is a number between 0 and 1 and if there is no relationship between the compared sequences, then a value of 0 is predicted. (b) Schematic of XOR model used for bit encoded genomic sequences. (c) By using the XOR function, we can select sequences from the random population based on a threshold value from the functional modules (t=5.52, f=4.91, df=201, $p<0.05$).

http://dx.doi.org/10.1371/journal.pone.0005608.g001

ACKNOWLEDGMENTS

We are indebted to Claudio Basilico for tipping the iceberg. Special thanks to Hongzhe Li and Isidore Rigoutsos for technical support on this paper. Thanks to Alan Frey, Lisa Dailey, and Nicolas Pelaez for discussions and Talid Sinno for his suggestions. We greatly appreciate Lyle Ungar and Scott Diamond for their support.

Author Contributions

Conceived and designed the experiments: MY. Performed the experiments: MY QW. Analyzed the data: MY QW. Wrote the paper: MY.

REFERENCES

1. Cheng J, Kapranov P, Drenkow J, Dike S, Brubaker S, et al. (2005) Transcriptional maps of 10 human chromosomes at 5 nucleotide resolution. Science 308: 1149–1154.
2. Efroni S, Duttagupta R, Cheng J, Dehghani H, Hoeppner DJ, et al. (2008) Global transcription in pluripotent embryonic stem cells. Cell Stem Cell 5: 437–47.
3. Gingeras TR (2006) The multitasking genome. Nat Genet 6: 608–9.
4. Kapranov P, Willingham AT, Gingeras TR (2007) Genome-wide transcription and the implications for genomic organization. Nat Rev Genet 6: 413–23.
5. Bina M, Wyss P, Lazarus SA, Shah SR, Ren W, et al. (2009) Discovering sequences with potential regulatory characteristics. Genomics 4: 314–22.
6. DasGupta B, Konwar KM, Mandoiu II, Shvartsman AA (2005) DNA-BAR: distinguisher selection for DNA barcoding. Bioinformatics 16: 3424–6.
7. Gingeras TR (2007) Origin of phenotypes: genes and transcripts. Genome Res 6: 682–90.
8. Voelcker NH, Guckian KM, Saghatelian A, Ghadiri MR (2008) Sequence-addressable DNA logic. Small 4: 427–31.
9. Warmflash A, Dinner AR (2008) Signatures of combinatorial regulation in intrinsic biological noise. Proc Natl Acad Sci USA 45: 17262–7.
10. Yaragatti M, Sandler T, Ungar L (2009) A predictive model for identifying mini-regulatory modules in the mouse genome. Bioinformatics 25: 353–7.

CHAPTER 14

INTELLIGENT PHOTOVOLTAIC SYSTEMS BY COMBINING THE IMPROVED PERTURBATION METHOD OF OBSERVATION AND SUN LOCATION TRACKING

Yajie Wang, Yunbo Shi, Xiaoyu Yu, and Yongjie Liu

The Higher Educational Key Laboratory for Measuring & Control Technology and Instrumentations of Heilongjiang Province, School of Measurement–Control Technology and Communications Engineering, Harbin University of Science and Technology, Harbin, China

ABSTRACT

Currently, tracking in photovoltaic (PV) systems suffers from some problems such as high energy consumption, poor anti-interference performance, and large tracking errors. This paper presents a solar PV tracking system on the basis of an improved perturbation and observation method, which maximizes photoelectric conversion efficiency. According to the projection principle, we design a sensor module with a light-intensity-detection module for environmental light-intensity measurement. The effect of environmental factors on the system operation is reduced, and intelligent identification of the weather is realized. This system adopts the discrete-type tracking method to reduce power consumption. A mechanical structure with a level-pitch double-degree-of-freedom is designed, and attitude correction is performed by closed-loop control. A worm-and-gear mechanism is added, and the reliability, stability, and precision of the system are improved. Finally, the perturbation and observation method designed and improved by this study was tested by simulated experiments. The experiments verified that the photoelectric sensor resolution can reach 0.344°, the tracking error is less than 2.5°, the largest improvement in the charge efficiency can reach 44.5%, and the system steadily and reliably works.

INTRODUCTION

Following the development in the world economy, energy consumption has continuously increased. The burning of fossil fuels causes serious pollution to the environment. As a clean energy, solar energy has experienced increased adoption owing to its immense radiation quantity, wide coverage area, and inexhaustible reserves. Therefore, solar photovoltaic (PV) power generation systems have become an inevitable direction in energy development for the future [1–6]. At present, the orientation of solar panels in most solar PV power generation systems is fixed, which leads to low generation efficiency. Solar tracking systems can ensure that solar panels are always oriented toward the sun, which guarantees full photoelectric conversion efficiency. In recent studies, sun-tracking methods have been mainly using photoelectric and are day-numbered [7, 8]. To track the sunlight, a photoelectric tracking system uses photosensitive sensors that can measure the sun angle. Although such a system has high sensitivity, it also has dead zones. It is easily affected by the weather and can lose the target or even track a wrong one [7]. By calculating the movement path of the sun, the day-numbered trajectory tracking method can normally work under clouds. However, its tracking precision is low, and its error accumulates. The tracking trajectory needs to be recalculated for different regions. Unable

Citation: Wang Y, Shi Y, Yu X, Liu Y (2016) Intelligent Photovoltaic Systems by Combining the Improved Perturbation Method of Observation and Sun Location Tracking. PLoS ONE 11(6): e0156858. doi:10.1371/journal.pone.0156858.

to recognize outside light intensity, this system wastes energy by continuously tracking in low or no light [8]. One of the key technonlgies accelerating the growth of solar energy sources is the power management strategy of the electrochemical energy storage system such as the Li-ion battery pack, supercapacitor pack, and dual buffer [9–13].

To make the PV cells produce more electricity, recent research has focused on maximum power point tracking (MPPT) technology. The use of the MPPT technology makes the utilization ratio of PV cells higher, and the output reaches a maximum [14–17]. At present, the main methods to realize the MPPT technology include the perturbation and observation method and the constant-voltage tracking method [18, 19]. The perturbation and observation method compares the current system output with the previous output to determine the next action and thus realizes MPPT. However, it is subject to tracking the step length. It easily oscillates near the maximum power point. In addition, it cannot distinguish the output variation caused by environmental factors, causing the output to deviate from the maximum power point [20]. The constant-voltage tracking method is an approximate tracking method, which fixes the maximum power point as a constant. This method cannot adjust the output along with the change in conditions, leading to low efficiency and large error [21].

The technical contribution of this study is an intelligent control technology for light perception and MPPT. We design and construct an intelligent tracking solar PV power generation system. The core processor of this system is a field-programmable gate array (FPGA). It uses a two-degree-of-freedom (2-DOF) mechanical system and corrects its attitude using closed-loop control. Light-searching sensors are developed using the principle of projection in which a light-intensity sensor unit is combined, making an intelligent light-search perception module. This paper proposes an effective solution to differentiate perturbations, which can avoid oscillation near the maximum power point of the perturbation observation method and malfunction under light-intensity variations. It also maximizes the conversion efficiency of solar PV cells.

SYSTEM STRUCTURE

Overall System Structure

The proposed system is primarily composed of an FPGA core controller, a system power supply unit, a MOSFET drive module, a DC/DC conversion unit, a voltage/current detection module, a sun-positioning recognition module, a liquid crystal display (LCD), a motor drive module, a positioning module, and a mechanical structure. The system general structure is shown in Fig 1.

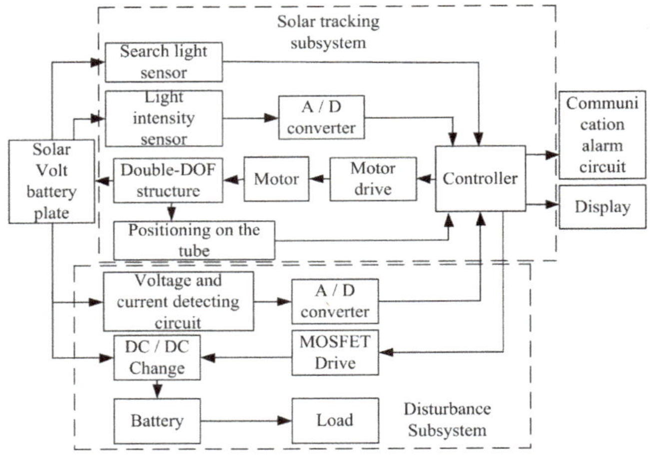

Figure 1: Overall system structure.

http://dx.doi.org/10.1371/journal.pone.0156858.g001

Mechanical Structure

A level-pitch 2-DOF mechanical structure is designed to track the sun, as shown in Fig 2. Items 1 and 9 in the figure are the positioning laser geminate transistors. Item 4 is the photoelectric detection module whose main structure is a light-tracing sensor and whose top is a light-intensity-detection unit. The worm-and-gear structures (Items 2 and 11) are added to the base, which is a DC geared motor structure, to prevent reversal and overturning and to counter any driving moment caused by external stress. In this manner, the problem of counter driving moment is solved, and the control precision and stability of the system are strengthened.

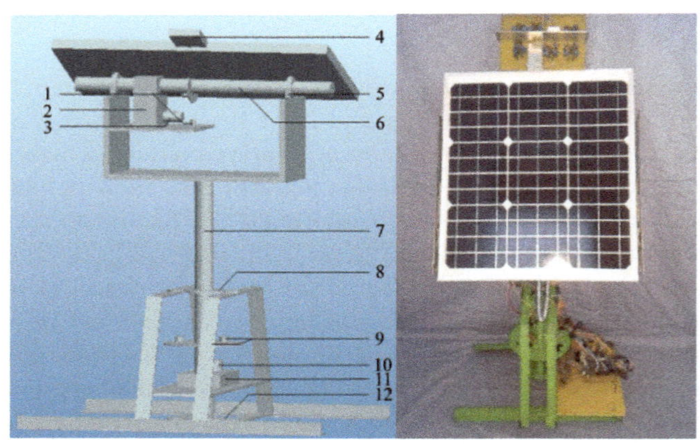

Figure 2: Mechanical structure.

(a) Diagram of the structure. (b) Actual structure.

http://dx.doi.org/10.1371/journal.pone.0156858.g002

System Operation

The system operation can be divided into three main parts: motor, clock, and MPPT controllers. The clock is the main component in the system operation. The system only tracks during a set period of time. During this time, the system corrects its attitude at set intervals. If vision is lost (due to environmental factors such as overcast and rainy weather), it will make a fuzzy adjustment according to the current time, which is the first interval when the environmental light intensity exceeds a photosensitive threshold, and it will find a precise position after it recovers the vision. In the set period, the system adopts a discrete-type tracking method. The system peripherals, which include the motor, motor drive module, light-tracing sensor module, positioning geminate-transistor module, light-intensity-detection module, and display module, are only charged when a set clock signal appears, and the power is interrupted and maintained in standby mode after the required detection and control. Every evening, the system attitude returns to zero, and the control part stays in the standby mode with low energy consumption. To prolong the battery service life, the system adopts a three-stage charging, which uses the MPPT technology during the large-current charging phase to rapidly provide power and to perform real-time monitoring of the environmental light intensity and battery power. During the small-current and trickle-charging stage, all equipment except the controller runs in a time-discrete mode.

DISTURBANCE SUBSYSTEM

MPPT Model

According to the relationgs with the output current and voltage of solar photovoltaic cells, can obtain the relations with the output voltage and power, as shown in Eq 1.

$$P = I_{ph}U - I_0 U[\exp(qU/nkT) - 1] \quad (1)$$

where q is the quantity of electric charge, n is the diode ideal constant, k is Boltzmann constant (1.38×10–23) and T is absolute temperature(°C).

Fig 3 shows that a PV battery can reach maximum output power by regulating the external load in a set external environment. As the output power reaches its peak, the output voltage is defined as maximum power point voltage U_m, the electric current is defined as maximum power point current I_m, and the output power is defined as maximum power point P_m. Because of the influence of external factors such as temperature and light intensity, the maximum power point moves when external conditions change. Thus, a technique is needed to find the maximum power point and to obtain the maximum power output using a control algorithm, namely, the MPPT technology [22].

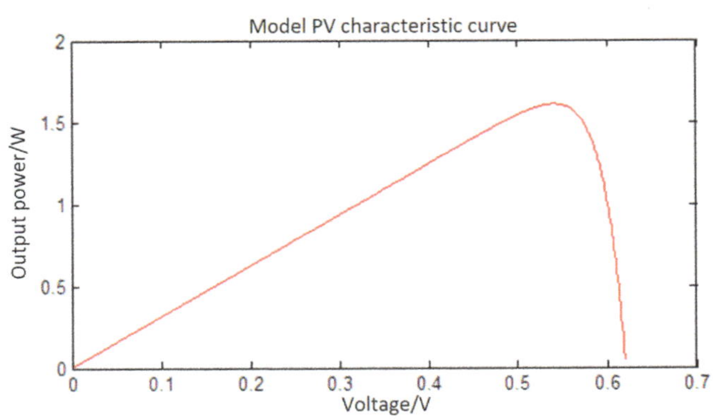

Figure 3: Characteristic P–V curve of PV cells.

http://dx.doi.org/10.1371/journal.pone.0156858.g003

Determination of the MPPT Method

Realization of MPPT is a dynamic optimization process. The system controller monitors and controls the power through real-time monitoring and by regulating the output voltage and current strategy of the solar cell. To ensure that the PV cells are dynamically working at the maximum power point, the present output power should be compared with the previously measured output power, which is memorized in the controller. Then, the interference should be adjusted to increase the output power. The cycle is then repeated. MPPT is implemented as shown in Fig 4.

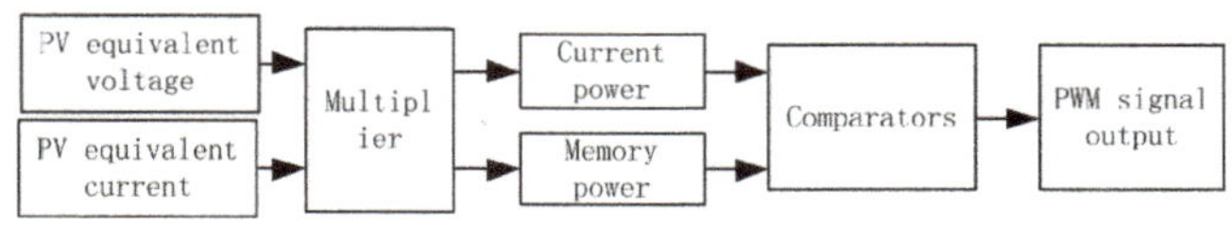

Figure 4: Basic schematic of the MPPT.

http://dx.doi.org/10.1371/journal.pone.0156858.g004

Disturbance Observation

This design incorporates a disturbance observation method to realize the MPPT algorithm. The principle is shown in Fig 5. The basic principle is to disturb the output voltage of the solar panels using pulse-width modulated (PWM) signals output by the software via the MOSFET driver circuit to continuously monitor the output power and voltage of the PV cells and to compute power variation ΔP and voltage variation ΔV. If ΔP is positive, the current disturbance causes the output power of the PV cells to move in the direction of maximum power point, which means that the disturbance is in the right direction. If ΔP is negative, the current disturbance causes the PV cell output power to decrease, which means a change in the disturbance direction is needed and so on until ΔP tends to zero. At this time, the system operates at the maximum power point and maintains the current duty ratio.

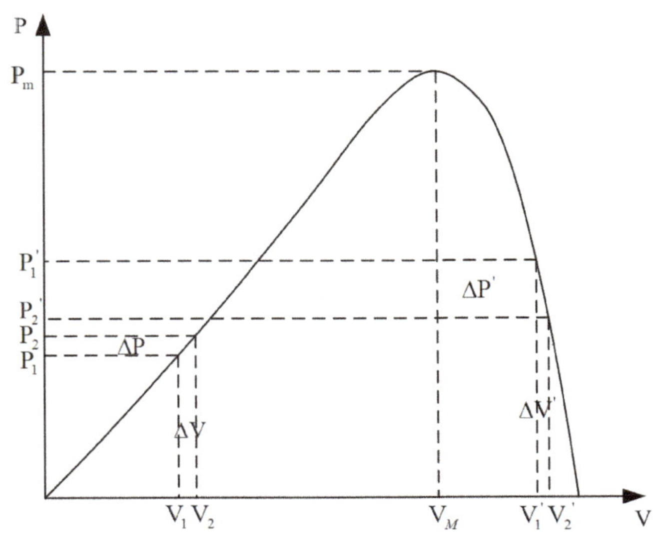

Figure 5: **MPPT scheme.**

http://dx.doi.org/10.1371/journal.pone.0156858.g005

Improved Algorithm Model

The disturbance observation method suffers from defects such as poor anti-interference and a tendency to oscillate near the maximum power point. To solve these problems, this paper proposes an improved disturbance observation method. Fig 6 shows that when the current output is to the left of the maximum power point, the next-state ΔP is negative because decreasing sunshine strength causes the current power to decrease from P1 to P2. According to the discriminant method of perturbation observation, the controller reduces the output voltage V of the PV cells. However, when the maximum power point is to the right of the current power output, the correct adjustment method increases V. The same error also exists when the current power output is to the right of the maximum power point. At this moment, the disturbance observation control would perform a wrong operation in response to the illumination change.

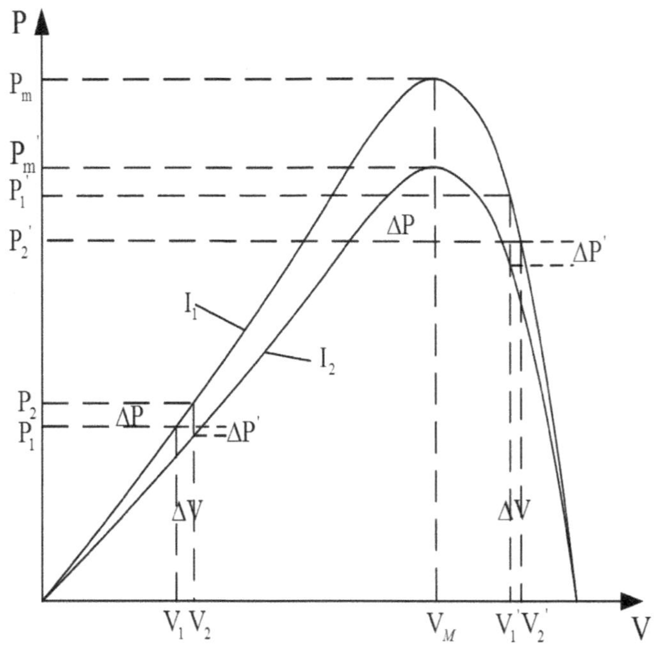

Fig 6; Improved disturbance observation principle.

http://dx.doi.org/10.1371/journal.pone.0156858.g006

To overcome this shortcoming, we improve the algorithm in this study. The environmental light intensity, monitored as ΔP, is classified as positive or negative. When ΔP is less than zero, the variation in the environmental light ΔS is also less than zero, which means that the environmental light intensity interferes with the algorithm. When the duty ratio is kept unchanged, if ΔS is nonnegative (which means that the ambient light does not affect the algorithm), the direction of the disturbance needs to be changed. Similarly, when both ΔP and ΔS is positive, the light-intensity change interferes with the algorithm, and the output voltage maintains its original value. Otherwise, the disturbance should continue. The improved disturbance observation also considers the influence of ΔS on ΔP. It suppresses the interference from the light-intensity change and makes up for the defect in algorithm miscalculation caused by the light-intensity changes. Because the adjusted step length of the disturbance observation method is fixed, the output power characteristic of the PV cells is not linear. Therefore, when the disturbance observation reaches the maximum power point, it would overshoot and reverberate in steady state.

HARDWARE CIRCUIT DESIGN OF THE DISTURBANCE SYSTEM

DC/DC Conversion Circuit

A DC/DC conversion circuit is designed to work together with the MPPT algorithm (to make the output voltage of the PV battery controllable) and to complete the step-charging function of the battery. The DC/DC conversion circuit is directly connected to the two modules, PV cells, and battery. Therefore, security must be integrated into the circuit to make the system safe by preventing component damage caused by problems such as wrong operation and connection error. The DC/DC conversion circuit is shown in Fig 7.

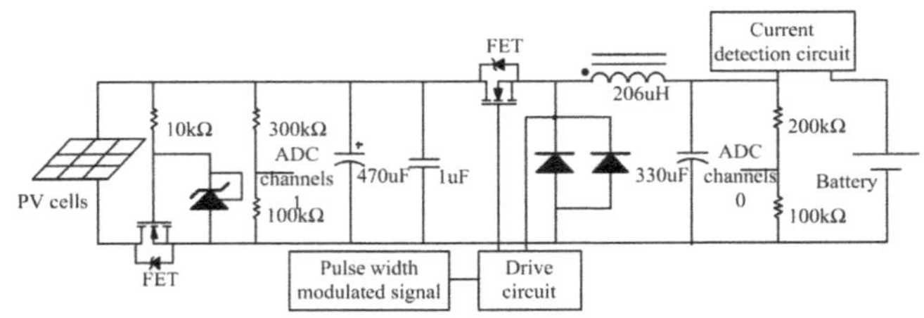

Figure 7: DC/DC conversion circuit.

http://dx.doi.org/10.1371/journal.pone.0156858.g007

The basic parameters of the DC/DC circuit are designed as follows:

1. The input voltage range is determined by the output characteristics of the PV cells; the value ranges from 13.6 to 21.9 V.
2. The maximum dropout voltage value is 200 mV (the output current ranges from 100 mA to 2.5 A within 1 ms).
3. The maximum output current is 2.5 A.
4. The switching frequency is 100 kHz.

Electrical components parameters of the DC/DC circuit were calculated as follows:

The duty ratio ranges from two ratio figures, namely, the ratio of the minimum output to the maximum input and the ratio of the maximum output to the minimum input.

$$D_{\min} = \frac{V_{out,\ \min}}{V_{in,\ \max}} = \frac{13.6}{21.9} = 0.621 \tag{2}$$

$$D_{\max} = \frac{V_{out,\ \max}}{V_{in,\ \min}} = \frac{13.6}{13.6} = 1 \tag{3}$$

The inductance (L) is chosen according to the principle of maximum voltage difference value and minimum duty ratio.

$$u = L\frac{di}{dt}$$

$$L = \frac{(V_{in,\ \max} - V_{out}) \times T_{on}}{\delta \times I_{out,\ \max}} \tag{4}$$

where the compensating coefficient value of δ is 0.1 and T_{on} is the PWM conduction time. Inductance L can be obtained as

$$L = \frac{(21.9 - 13.6) \times 6.21 \times 10^{-6}}{0.1 \times 2.5} = 206.172\mu\text{H}.$$

Choosing the capacitance value should meet the condition of 250 mV (100 mA–2.5 A). The calculation expression is

$$i = C\frac{du}{dt}. \tag{5}$$

Eq 5 can be expressed in vector form as

$$\dot{I} = j\omega C\dot{U}. \tag{6}$$

Eq 6 leads to the capacitance expression

$$C = \frac{\dot{I}}{j\omega\dot{U}} \tag{7}$$

in which the bandwidth of crossover frequency f_c is obtained as

$$f_c = \frac{\Delta I_{out}}{2\pi C_{out}\Delta V_{out}} \tag{8}$$

where ΔI_{out} is the down current, ΔV_{out} is the dropout voltage, and f_c is the crossover frequency.

When the voltage is 10 kHz, the impedance value of the capacitance is obtained by

$$Z_c = \frac{\Delta V_{out}}{\Delta I_{out}} = \frac{250\text{mV}}{2.4\text{A}} = 104.2\text{m}\Omega. \tag{9}$$

According to the calculation results, in the ZL system of RUBYCON Company, we can obtain the 16-V, 330-µF capacitance parameters as follows:

$$I_{c,\,rms} = 760\text{mA}@T_A = 105°\text{C}$$

$$R_{\text{ESR, low}} = 72\text{m}\Omega@T_A = 20°\text{C}$$

$$R_{\text{ESR, low}} = 220\text{m}\Omega@T_A = -10°\text{C}.$$

The capacitive impedance of the equivalent series resistance (ESR) is obtained by

$$\frac{1}{2\pi f_c C_{out}} = \frac{1}{6.28 \times 10k \times 330} = 48\text{m}\Omega$$

$$Zc \le \sqrt{R^2{}_{ESR} + \left(\frac{1}{2\pi f_c C_{out}}\right)^2} = \sqrt{72^2 + 48^2} = 86\text{m}\Omega. \tag{10}$$

The calculation results of the capacitor ESR show that the capacitive impedance is smaller than Zc, which meets the design requirements.

MOSFET Driver Circuit

To realize MOSFET switch control, this design uses the IR2110 driver chip made by IR Company (U.S.). This chip, which has the advantages of both light coupling and electromagnetic separation, is the first choice of the driver circuit required in this design. A typical driving circuit is shown in Fig 8.

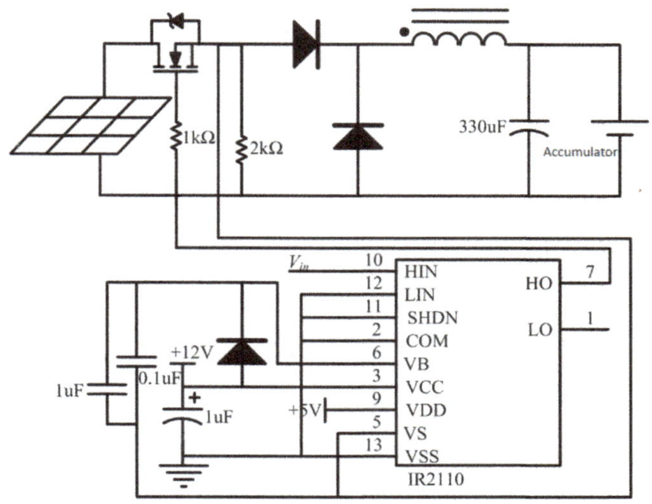

Figure 8: MOSFET typical driving circuit.

http://dx.doi.org/10.1371/journal.pone.0156858.g008

To improve the performance of the driving circuit, the design in this study made some improvements to the traditional driving circuit of the IR2110, such as adopting a high–low end simultaneous reverse-switching mode, decreasing the charge and discharge times, and making the edge of the drive pulse steeper. Such improvements greatly reduced the problem of MOSFET output jitter in the threshold voltage and improved the switching speed and frequency. The improved driving circuit significantly increased the controllable range of the output voltage. After eliminating the jitter, the output voltage value was substantially the same as the theoretical value obtained by the duty-ratio calculation. The improved IR2110 drive circuit is shown in Fig 9, where HO and LO are the high and low drive pulse output ends, respectively, and HIN and LIN are the high and low logic level input ends, respectively.

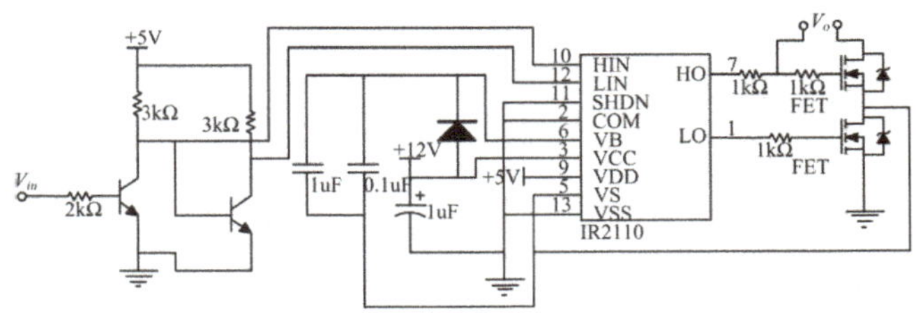

Figure 9: Improved MOSFET drive circuit.

http://dx.doi.org/10.1371/journal.pone.0156858.g009

The improved driving circuit expends less power than the typical one. Compared with the traditional drive circuit, the improved drive circuit eliminates the diode in the DC/DC conversion circuit (which is designed to prevent reverse signal flow) and the resistance for charging. The energy expended by the diode and the resistance cannot be ignored, especially that of the diode. Its energy dissipation is expressed as

$$P = \frac{U^2{}_R}{R} + U_d I_L = 1.212\text{W} \tag{11}$$

where the voltage drop of the diode conduction $U_d = 0.7$ V, $R = 2$ kΩ, load average voltage $U_L = 14$ V, and average current $I_L =$

1.5 A. The voltage of the electric resistance is obtained from the equation $U_R = U_L/D$, where D is the duty ratio, which is equal to 18 V. The improved field-effect tube increases the low energy consumption on the order of milliwatts, which can be ignored. Therefore, the improved circuit can increase the efficiency of the drive circuit in a better manner.

A/D Conversion Circuit

This design incorporates the ADC0809 eight-bit analog/digital (A/D) conversion chip as a modular transition device. The chip has eight input channels. Switching between each channel is realized by an analog switch. The analog signal input in each channel can be successively converted through time-division multiplex access. Because the A/D converter maximum input voltage is limited to a standard voltage, the standard voltage of this design is 5 V, and the highest voltage converted is 21.9 V. As a result, the measured voltage must be processed into a voltage that can be identified by the A/D converter. To be collected by the A/D converter, the measured voltage is proportionally decreased to less than 5 V by adopting a resistance divider. The A/D conversion circuit is shown in Fig 10.

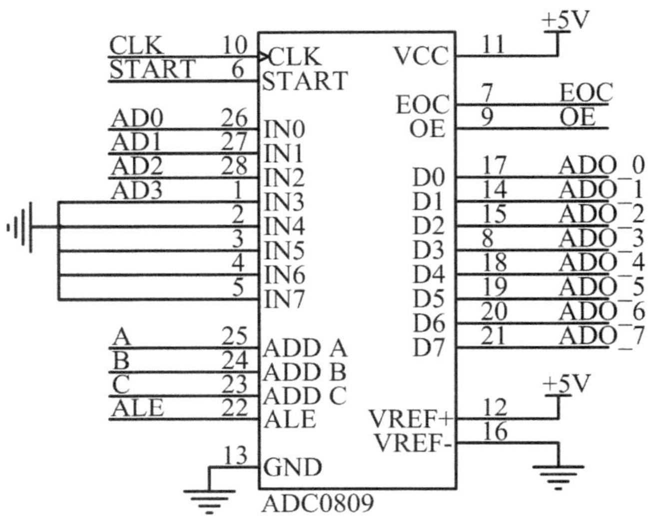

Fig 10: A/D converter circuit.

http://dx.doi.org/10.1371/journal.pone.0156858.g010

Current Detection Circuit

The ACS712 Hall current sensor from Allegro Application Company is used for the current examination. Using a bidirectional DC linear output, it has the characteristics of high sensitivity, high precision, and high cost performance. The ACS712-5 type sensor is used in this design, whose detection ranges from −5 to +5 A. Its accuracy is ±1.5%, the sensitivity is 180–190 mV/A, the output reference voltage is 2.5 V, and the working temperature is 40–85°C. The detection circuit is shown in Fig 11.

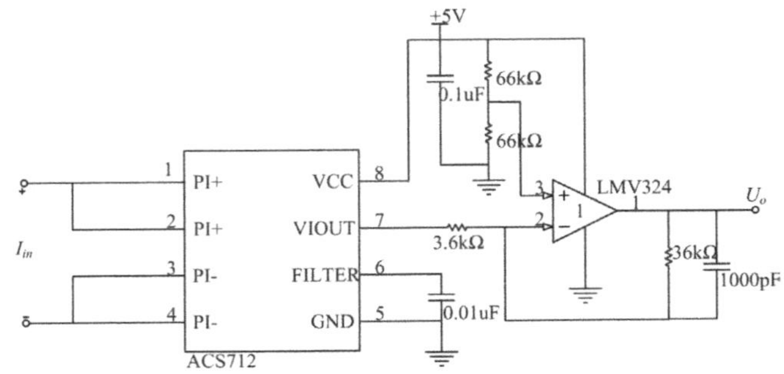

Fig 11: Current detection circuit.

http://dx.doi.org/10.1371/journal.pone.0156858.g011

Fig 11 shows that the amplifying circuit should be benchmarked against the 2.5-V standard. Because the design demand of the maximum electric current peak is 2 A, the output should be amplified by a factor of five for an output voltage range from 0.65 to 4.35 V, which can easily be collected by the A/D converter. In the diagram, PI+ and PI− are the input and output ports of the detected current, respectively. The analog signal output port is VIOUT.

Other Hardware Circuit Designs

The other hardware circuit designs include the communication alarm circuit, display circuit module, and voltage conversion module.

The communication and alarm circuit uses the general packet radio service (GPRS) communication methods, which apply the SIM900A module GPRS communication solutions. The 1602 LCD is chosen as the display module for real-time display of information such as the time and battery status. To meet the system power supply of each unit and the needs of the different modules on the power supply voltage, a system power supply circuit is designed and constructed. A 12-V battery is used as the main battery for the supply current. It supplies power to the motor drive module, display module, light-searching sensors, laser tube positioning, and realization of the MPPT circuit and FPGA minimum system using voltage-conversion chips 7808, 2940, and 1117–3.3–5.0.

SOFTWARE DESIGN

System Software General Structure Design

The following introduces the overall design of the software system:

Main working procedure of the system:

1. Initializing the program
2. Calculating the sun trajectory and determining the light intensity and battery level
3. Simultaneously controlling the first attitude and operating the MPPT
4. Connecting the GPRS network
5. Running the system

At the very start, the program initializes the processor registers and the system clock and sets the system performance period, standby period, discrete time interval, and the MPPT algorithm parameters. Next, the program initializes the A/D converter and GPRS, calculates the sun trajectory, and completes the determination of the current light intensity and battery level. The program roughly adjusts the sun tracking once more, precisely adjusts the position system attitude by photoelectric tracking, completes the first attitude adjustment, and runs the MPPT. While running, the system uses a discrete type of solar tracking with a tracking interval of 10 min, i.e., the system adjusts its attitude at set intervals. In other periods, each module of the tracking subsystem interrupts the power supply and remains in standby. During the running time, if the sun position identification sensor loses the position due to weather factors (such as rainy or cloudy conditions), the program would deal with the situation according to the change in the current light intensity. If the light intensity is higher than the sensory threshold value, the trajectory tracking methods will be chosen. If the light intensity is lower than the threshold power, the system will remain in standby. The charging mode of the system is three-phase charging, and MPPT is used in the large-current charging phase. The communication and alarm circuit varies the current-running state. It transmits an alarm when an exception occurs. The system software process is shown in Fig 12.

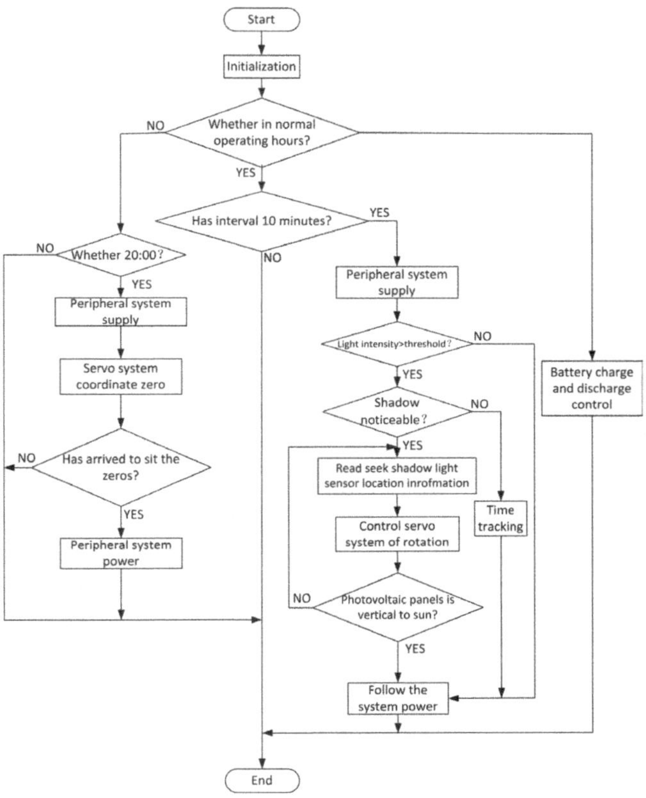

Figure 12: Program flow of the system software.

http://dx.doi.org/10.1371/journal.pone.0156858.g012

Software Implementation Process of the Improved Disturbance Observation Method

The software process of the improved disturbance observation method is shown in Fig 13. After initialization, the program first scans the battery level. Tracking of the maximum power point begins when the charging phase is identified as large-current charging. In the MPPT process, strong ambient light *S* is monitored in real time, and its effect on the algorithm is eliminated. Meanwhile, the scope of the duty cycle is delimited. When the duty-ratio adjustment reaches the limit of the duty-ratio boundaries, the system judgment is considered incorrect, and the duty ratio will be reset to return the system to its normal state. Proportional–integral–derivative (PID) control is used in the entire tracking progress. It adjusts the tracking step length, makes the output more stable, and avoids shock.

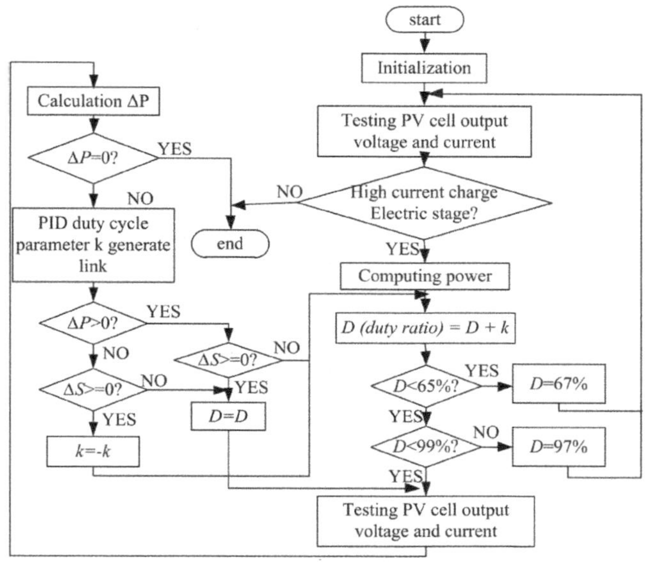

Figure 13: Software flowchart of the improved disturbance observation.

http://dx.doi.org/10.1371/journal.pone.0156858.g013

Duty-Ratio-Generated Subroutines

Duty-ratio-generated subroutines are used to transfer the duty-ratio values coming from the main program into the hardware-identifiable PWM signal. The duty-ratio-generated subroutine process is shown in Fig 14, which shows that the PWM signal cycle is determined by "time." The actual cycle is the product of the crystal vibration cycle and "time." The pulse width is determined by the reverse threshold "time PWM." The actual pulse width is the product of "time PWM" and crystal vibration cycle.]

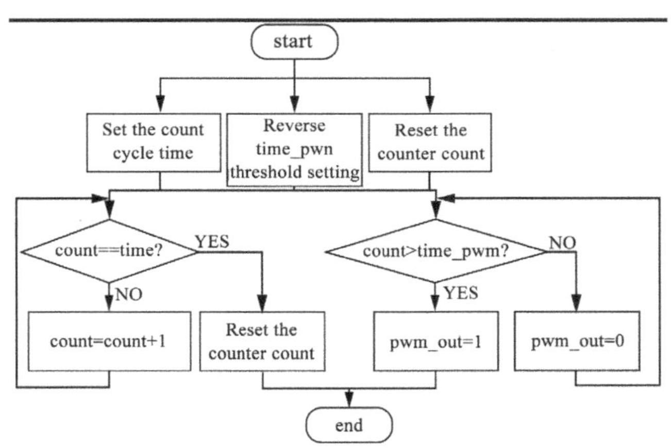

Figure 14: Duty cycle-generation software processes.

http://dx.doi.org/10.1371/journal.pone.0156858.g014

A/D Conversion Subroutine

The A/D converter is used to complete the module conversion function. The program first collects the output voltage of the PV cells, battery terminal voltage, and current through channel selection. Then, it proportionately restores the data coming from the A/D conversion into the corresponding voltage and current values. Finally, these data are output and used by the main program. The A/D conversion program is shown in Fig 15.

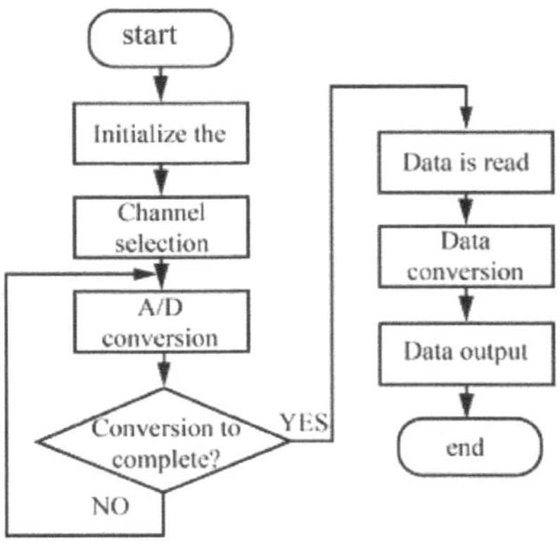

Figure 15: Program flow of the A/D conversion.

http://dx.doi.org/10.1371/journal.pone.0156858.g015

MODEL SIMULATION AND SYSTEM TEST

MATLAB/Simulink Simulation Model of Perturbation Observation

The MPPT simulation model is composed of the PV cell output current I_{pv} model, the MPPT model, which realizes the disturbance observation; the PWM model, which transfers the duty-ratio data into a PWM signal; and the basic circuit link modules. These components must be modeled in the Simulink environment. Each simulation model is modeled according to the equivalent mathematical model of each module. Fig 16 shows the basic mathematical model of the disturbance observation method to realize MPPT in which input parameters *T*, *S*, and*Vpv* are the environment temperature, light intensity, and output voltage of the PV cells, respectively. *Ipv* is the actual model current of the PV cells.

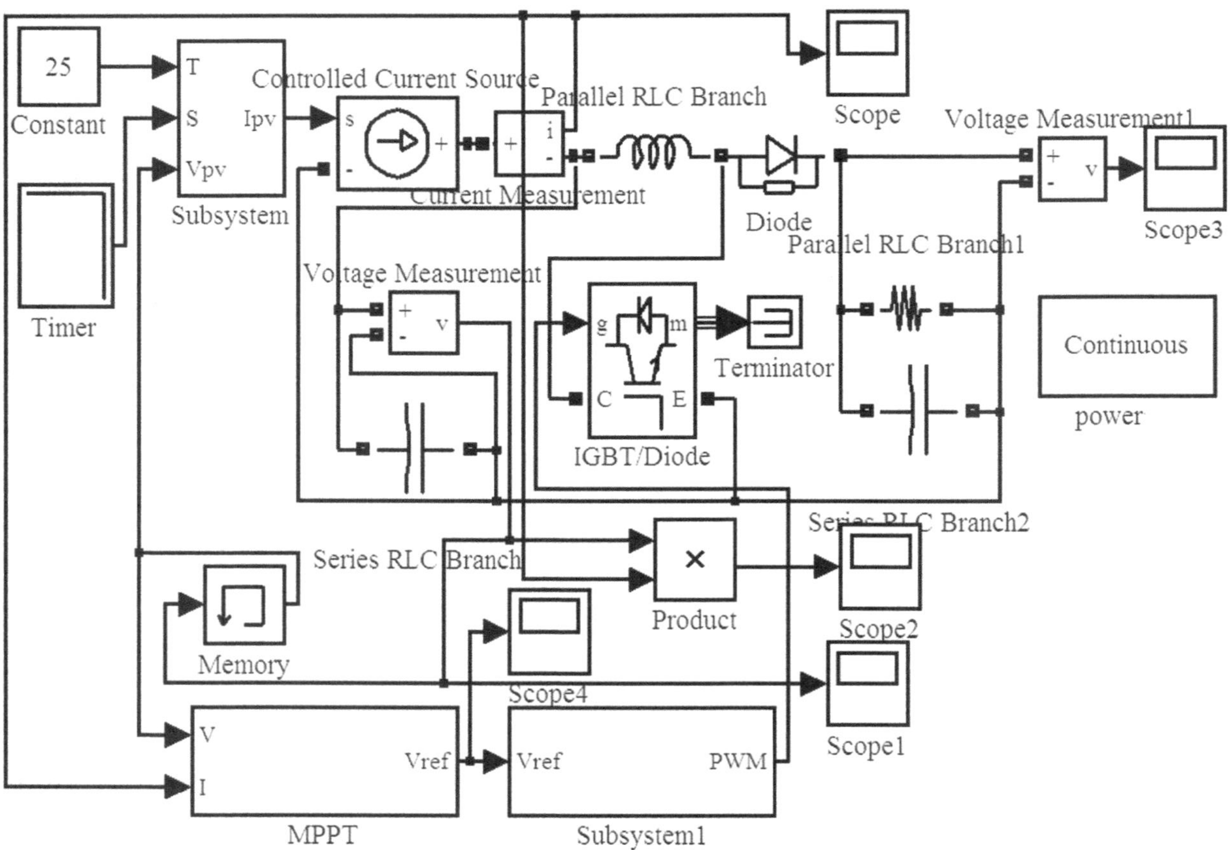

Figure 16: Disturbance observation of the MPPT Simulink model.

http://dx.doi.org/10.1371/journal.pone.0156858.g016

The PV cell output current (*Ipv*) model is determined by formula (1). Its Simulink simulation mathematical model is shown in Fig 17. *T*, *S*, and *Vpv* are quantities of the input control. *Ipv* is the output.

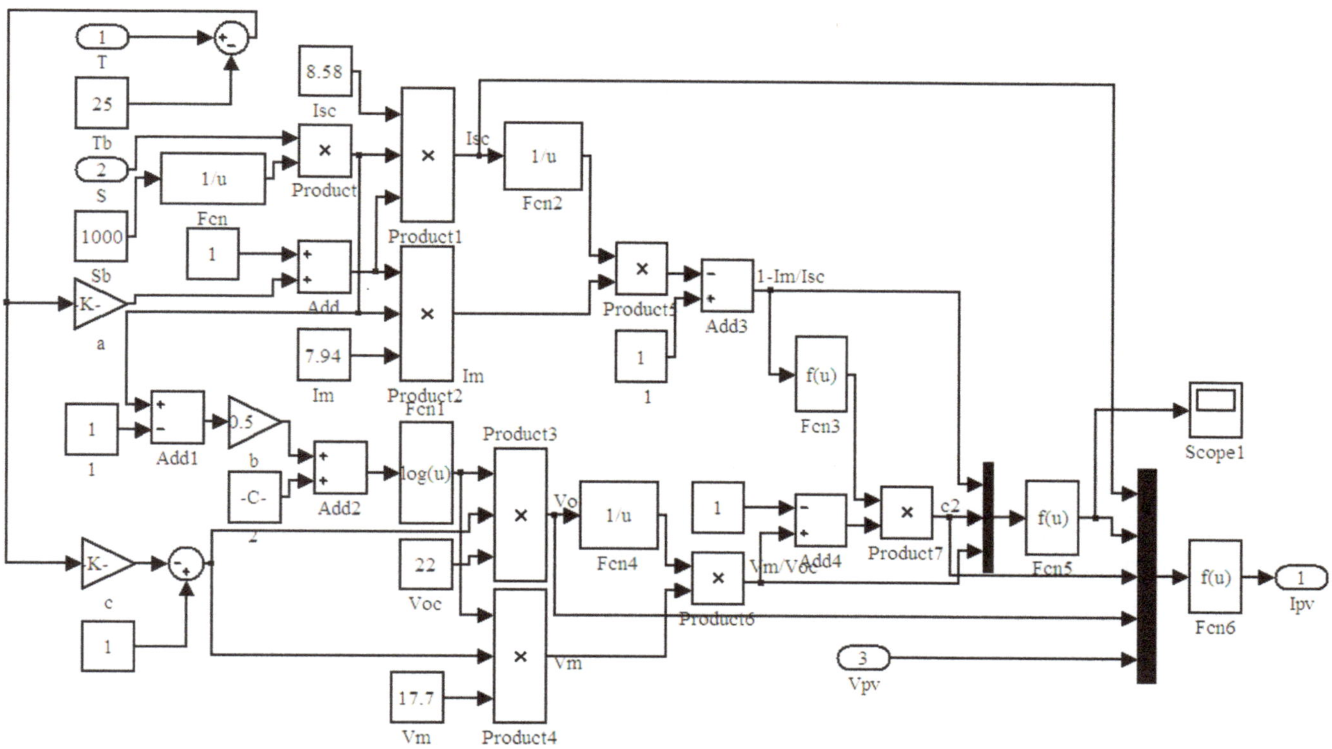

Figure 17: PV output current simulation model.

http://dx.doi.org/10.1371/journal.pone.0156858.g017

To realize the MPPT model of the disturbance observation, the current power calculation must be completed. Two values are needed: the difference with the previous state voltage and the difference with the previous state voltage. In addition, the duty-ratio adjustment direction should be determined. Finally, the adjusted duty-ratio value is determined. The Simulink model is shown in Fig 18. *V* and *I* are the input control quantity; *Vref* is the output.

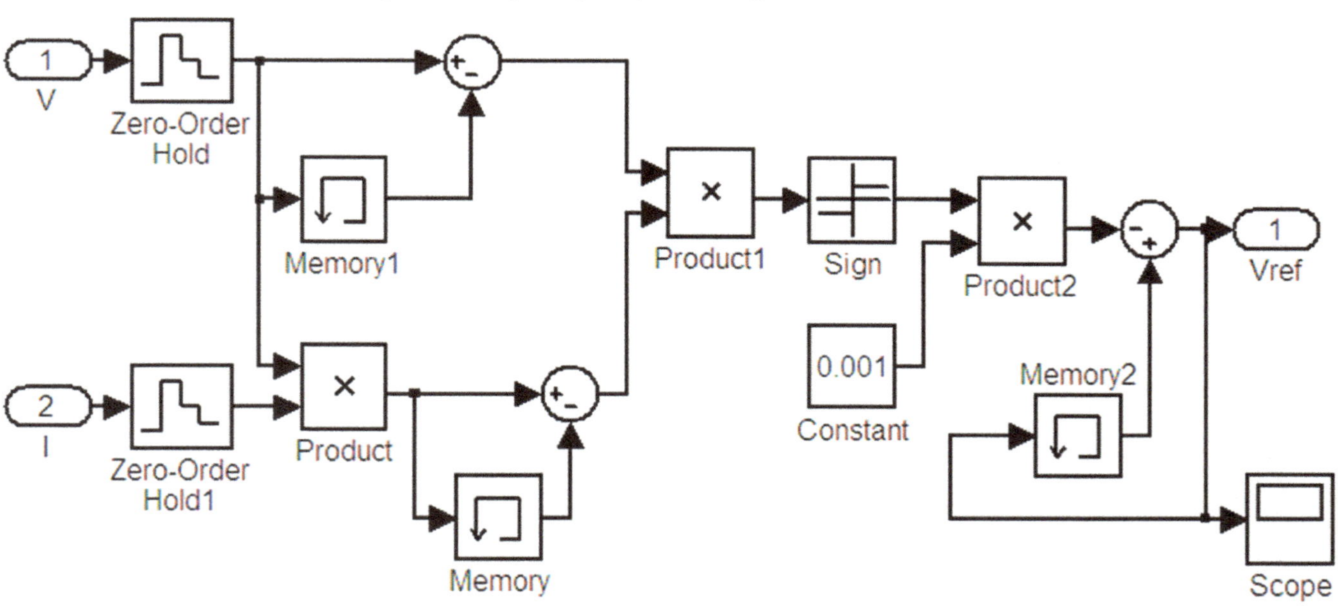

Fig 18: Model of the MPPT disturbance observation.

http://dx.doi.org/10.1371/journal.pone.0156858.g018

Fig 19 shows the PWM model that converts the duty-ratio data into a PWM signal. Its input variable *Vref* is the duty ratio. PWM represents the output PWM signal. The model is used to transfer the input duty ratio to the corresponding PWM signal output.

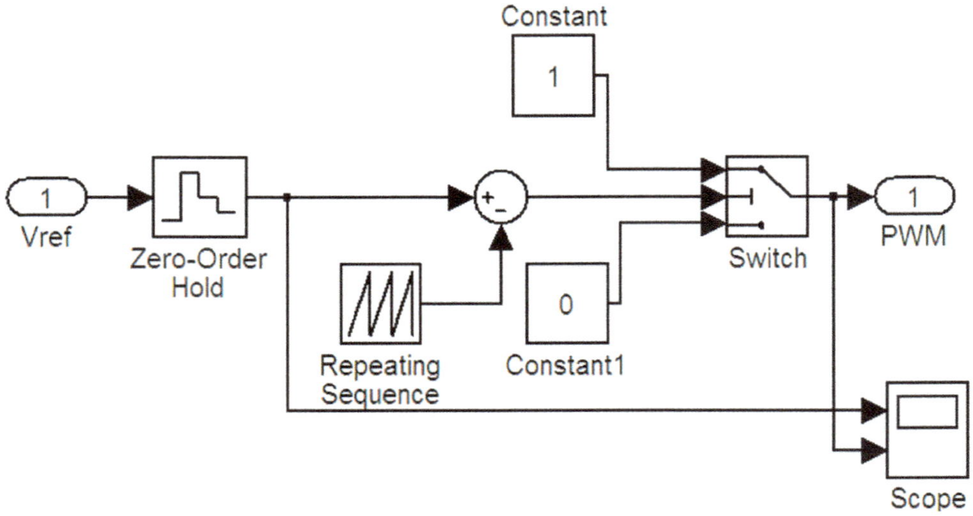

Figure 19: PWM generation model.

http://dx.doi.org/10.1371/journal.pone.0156858.g019

Analysis of the Disturbance Observation Method Simulation Results

In this design, the MPPT mathematical simulation is modeled using the Simulink software package, which comes from the MATLAB software, according to the mathematical model of the solar cells. MPPT is performed by a fixed-step-size perturbation observation method. The initial light conditions are set to be 1 kW/m², the sampling frequency is 1 kHz, and the step length is 5/1000. At the 0.1-s time point, the light intensity drops from 1 to 0.6 kW/m². The *Ipv*waveform changes, as shown in Fig 20. The duty-ratio curve is shown in Fig 21. The output voltage of the maximum power modulation is shown in Fig 22, and the data of Fig 22 .

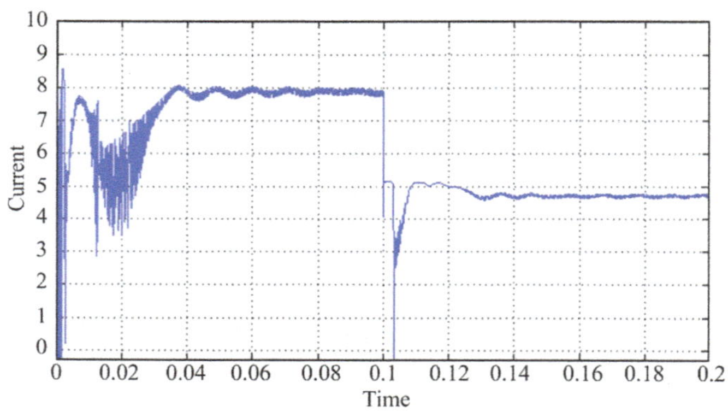

Figure 20: Solar cell output current *Ipv* **curve.**

http://dx.doi.org/10.1371/journal.pone.0156858.g020

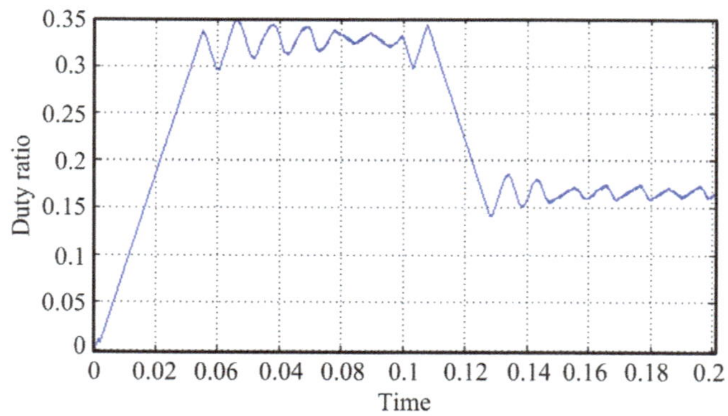

Figure 21: Duty-cycle curve.

http://dx.doi.org/10.1371/journal.pone.0156858.g021

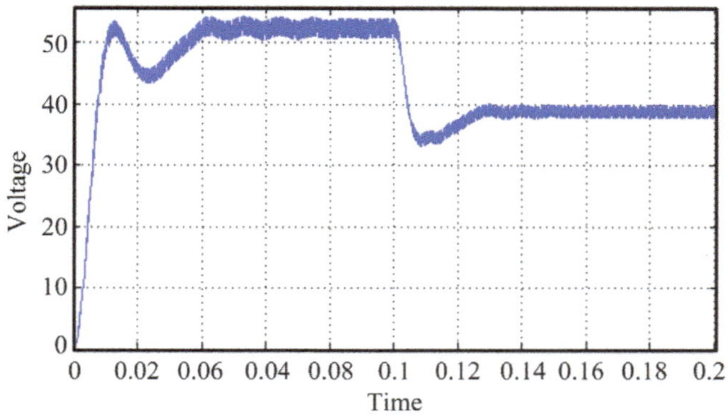

Figure 22: Tracking voltage curve of the maximum power point.

http://dx.doi.org/10.1371/journal.pone.0156858.g022

Improved MATLAB Disturbance Observation Model

The simulation model that uses a fixed-step-size perturbation observation to achieve MPPT suffers from some problems, i.e., the duty-ratio adjustment process overshoots, and after achieving dynamic balance, the duty ratio and the output voltage have a larger amplitude of oscillation. To overcome these shortcomings, the disturbance observation method is improved. By combining a PID control link, the duty-ratio-adjusted step length is gradually shortened as the maximum power point changes to reduce the overshoot and oscillation amplitude. Fig 23shows the improved disturbance observation method MPPT model. The PID control link is added to the Simulink simulation model, and its programming details are embedded in the file FUZZY_PID.fis. The duty-ratio curve of the improved model is shown in Fig 24 and the data ofFig 24 were in Table B. The improved MPPT voltage curve is shown in Fig 25.

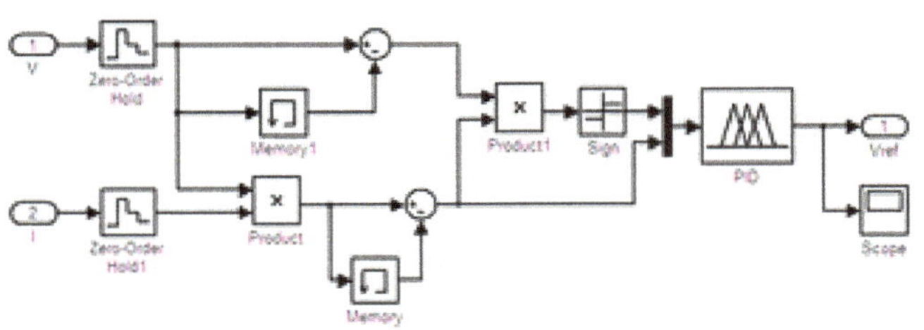

Figure 23: Improved disturbance observation model.

http://dx.doi.org/10.1371/journal.pone.0156858.g023

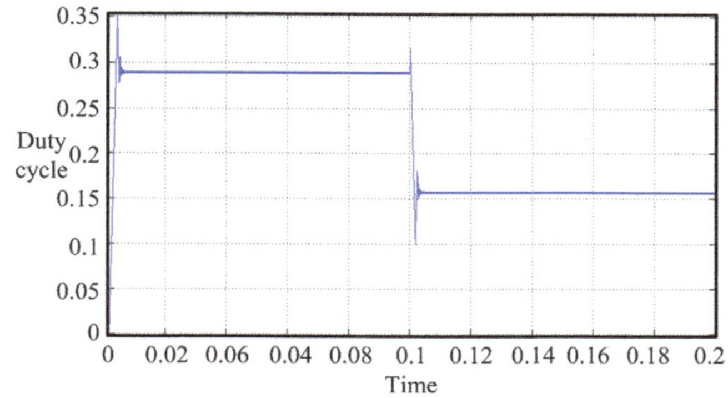

Figure 24: Duty-ratio curve of the improved model.

http://dx.doi.org/10.1371/journal.pone.0156858.g024

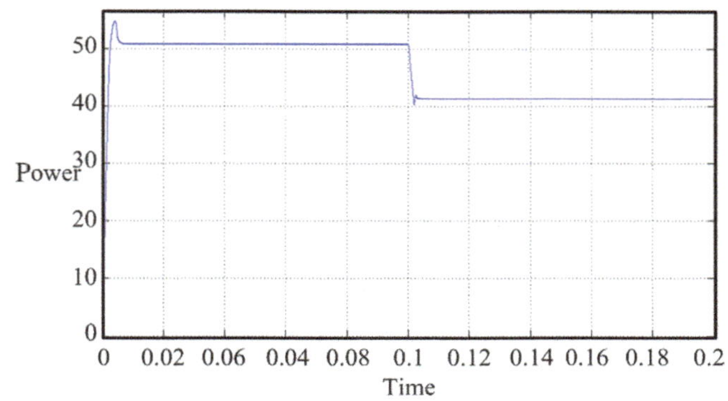

Figure 25: Improved MPPT voltage curve.

http://dx.doi.org/10.1371/journal.pone.0156858.g025

By comparing the improved simulation waveform with the unimproved one, we can see that after the improvement, the duty-ratio overshoot and steady-state oscillation amplitude obviously decrease. The duty-ratio value and output voltage of the improved model tend to be stable after the model reaches the steady state. The problems of vibration and overshoot are obviously reduced.

System Power-Consumption Test

To obtain the system power-consumption data, a measurement experiment was performed to measure the power consumption of each device module, as listed in Table 1.

Table 1: Average power consumption of the system device parts

http://dx.doi.org/10.1371/journal.pone.0156858.t001

Equipment	LCD	Sensors and laser tube	Level motor and drive circuit	Vertical motor and drive circuit
Power (mW)	824.2	134 8.9	106 1.7	158 8.7

doi:10.1371/journal.pone.0156858.t001

The system adopts a low-power design; hence, the energy consumption of each device is obtained by converting its actual operation time. The experimental result shows that the system mechanical structure level adjustment from zero to the maximum limit requires a time of 5 s, and the pitch adjustment from zero to the maximum limit requires 9 s, which means that every 10 min, the longest running time of the system is 9 s. The power consumption of each module is listed in Table 2.

Table 2: Average daily power consumption of the system equipment parts

http://dx.doi.org/10.1371/journal.pone.0156858.t002

System calibration frequency (s)	Horizontal maximum correction time (s)	Pitch correction time (s)	Motor and drive circuit power consumption (mW·h)	Sensors and laser tube power consumption (mW·h)	LCD power consumption (mW·h)	Total power consumption (mW·h)
72	5	9	538.5	242.8	148.5	929.65

doi:10.1371/journal.pone.0156858.t002

Table 2: Average daily power consumption of the system equipment parts

http://dx.doi.org/10.1371/journal.pone.0156858.t002

Measured System Charging Efficiency

The measurement experiment selected 30-W monocrystalline silicon solar panels with identical parameters. These solar panels were installed on a 45° south-facing fixed bracket with a pure tracking system and a tracking system that uses the MPPT technology. Each panel charged the same type of 12-V, 55-Ah battery, which was sufficiently discharged to the same state. When the experiment was in progress, the current entering the battery and the terminal voltage were detected in real time. The electric energy received by the battery at this moment was the net energy stored after considering the system power consumption. The current output power of the PV system and its energy storage can be obtained by calculation. The measurement experiment chose three discontinuous days, charging the system for 11 h from 06:00 to 17:00, and recorded a set of output data every 5 min for a total of 133 sets of data. A time power curve was then drawn, as shown in Fig 26, Fig27 and Fig 28, the data of Fig 26, Fig 27 and Fig 28 were in Tables C, D and E. The system daily output can be calculated by integrating the power over time, as listed in Table 3.

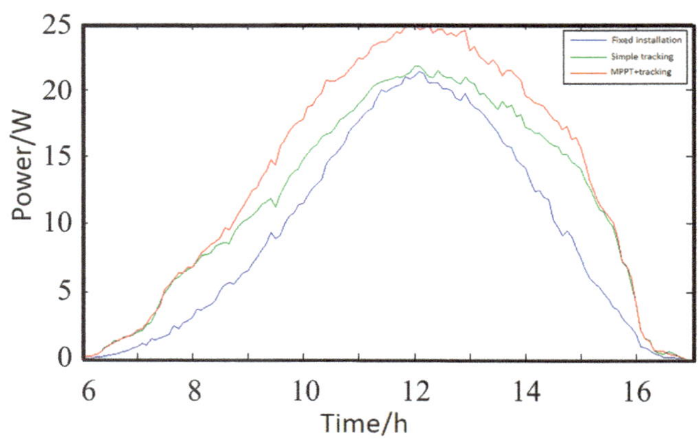

Figure 26: First comparison curves of the measured charging efficiency.

http://dx.doi.org/10.1371/journal.pone.0156858.g026

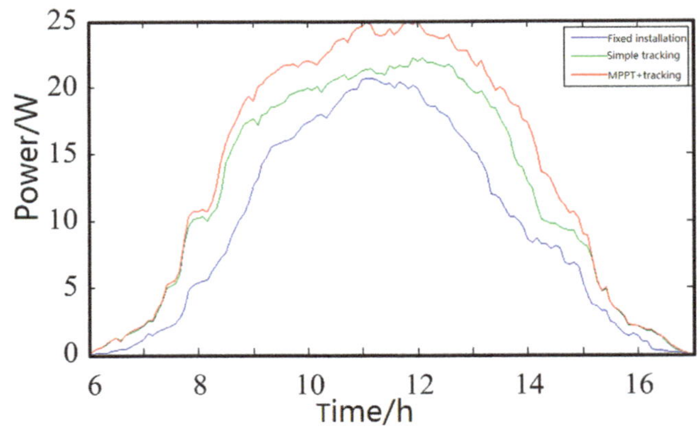

Figure 27: Second comparison curves of the measured charging efficiency.

http://dx.doi.org/10.1371/journal.pone.0156858.g027

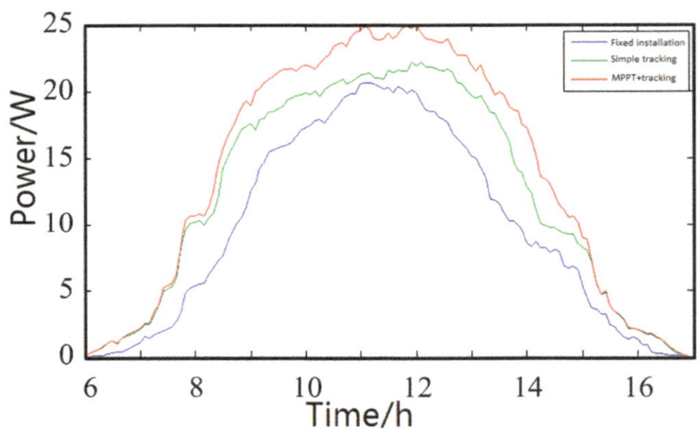

Figure 28: Third comparison curves of the measured charging efficiency.

http://dx.doi.org/10.1371/journal.pone.0156858.g028

Table 3: Comparison of the total power charging

http://dx.doi.org/10.1371/journal.pone.0156858.t003

State	Fixed installation	Simple tracking		MPPT	
	Total amount (W·h)	Total amount (W·h)	Relative fixed installation efficiency improvement	Total amount (W·h)	Relative fixed installation efficiency improvement
First test	103.17	130.32	26.32%	149.07	44.49%
Second test	105.79	134.76	28.33%	152.25	43.92%
Third test	90.54	107.37	18.59%	128.16	41.55%
Average	99.83	124.15	24.36%	143.16	43.40%

doi:10.1371/journal.pone.0156858.t003

Fig 26 shows the system charging efficiency comparison curves of the first measurement experiment from 06:00 to 17:00. The measurement conditions were as follows: temperature, from −9 to 2°C; relative humidity, 59%–79%; weather, fine. Fig 27 shows the system charging efficiency comparison curves of the second measurement experiment from 06:00 to 17:00. The measurement conditions were as follows: temperature, from −7 to 5°C; relative humidity, 60%–90%; weather, fine.

Fig 28 shows the system charging efficiency comparison curves of the third measurement experiment from 06:00 to 17:00. The measurement conditions were as follows: temperature, from −6 to 6°C; relative humidity, 60%–85%; weather, clear to overcast. Fig 28 shows that the weather markedly changed at approximately 08:30; the output curve when the sun was covered by clouds is circled in the figure.

The results of the three measurement experiments tended to be the same. This design of the PV system yielded the highest efficiency and could exhibit higher power generation efficiency improvement under different weather conditions. The system efficiency increase could reach 44.49% compared with that of the panels with fixed installation. The efficiency could still increase by 41.55% in a relatively bad rainy day, which proves that the influence of outside conditions is relatively small, and this system is stable and reliable.

CONCLUSION

This paper has presented the study and summary of the utilization and prospects of solar PV cells at home and abroad and has pointed out the significance of the PV power generation technology research. In addition, this paper has presented the profound study of the efficiency of the solar tracking and MPPT technology to enhance the mechanism. We designed the system capacity and provided a system technology target.

Related experiments and simulations to test the system performance as well as the parameters were designed. A Simulink simulation model of the disturbance observation method was designed. By comparing the performance of the model before and after the improvement, the superiority of the improved algorithm was proven. A performance-testing experiment on the sun-positioning recognition module was designed. This experiment tested the tracking accuracy and perspective of the module, the experimental verified that the photoelectric sensor resolution can reach 0.344° and the maximum tracking error was less than 2.5°. Finally, a charge–discharge and power-consumption testing experiment was designed, which proved the rationality of the system, the stability of operation, and the high generating efficiency,the experimental verified that the power consumption of system perihterals was under 1W·h a day, and the largest improvement in the charge efficiency could reach 44.5%.

AUTHOR CONTRIBUTIONS

Conceived and designed the experiments: YJW YBS. Performed the experiments: YJW XYY. Analyzed the data: XYY YJL. Contributed reagents/materials/analysis tools: YJW YBS. Wrote the paper: YJW YBS XYY.

REFERENCES

1. Zhao JZ, Shi YY, Ma J, Feng Y. Design and implementation of intelligent sun-tracking system[J]. Power System Technology, 2008, 32(24):93–97 (in Chinese).
2. Chen SY, Bao H, Wu CY, Han Y. Direct grid-tie power control method for distributed photovoltaic generation[J]. Proceedings of the CSEE, 2011, 31(10):6–11 (in Chinese).
3. Yang B, Li WH, Zhao Y, He XN. Design and analysis of a grid-connected photovoltaic power system[J]. IEEE Trans on Power Electronics, 2010, 25(4):992–1000 (in Chinese). doi: 10.1109/tpel.2009.2036432
4. Yu RJ. Optics and solar energy[J]. ActaOpticaSinica, 2009, 29(7):1751–1755 (in Chinese).
5. Yang XG, Wang YH, Ding N, Kou CR. Development of a system of solar LED street lamp lighting[J]. Transactions of China Electrotechnical Society, 2010, 25(6):130–136 (in Chinese).
6. Cao WH, Li MY, Chen X, Wu M. Novel high-efficient charge controller in independent photovoltaic system[J]. Journal of Zhejiang University:Engineering Science, 2010, 44(7):1260–1265 (in Chinese).
7. Bian XG, Yang FF, Xin QX. An automatic sun-tracking method in large range[J]. ActaEnerglae Solaris Sinica, 2010, 31(10):1298–1303 (in Chinese).
8. Du CX WP, Ma ZF, Wu YT. The application of day number in calculating the solar position[J]. ActaEnerglae Solaris Sinica, 2011, 32 (11):1640–1645 (in Chinese).
9. Hu XS, Li SB, Peng H. A comparative study of equivalent circuit models for Li-ion batteries[J]. Journal of Power Sources, 2012, 198, 359–367.
10. Hu XS, Sun FC, Zou Y. Online model identification of lithium-ion battery for electric vehicles[J]. Journal of Central South University of Technology, 2011, 18(5), 1525–1531.
11. Zhang L, Wang ZP, Hu XS, Sun FC, David DG. A comparative study of equivalent circuit models of ultracapacitors for electric vehicles[J]. Journal of Power Sources, 2015, 274, 899–906.
12. Zhang L, Hu XS, Wang ZP, Sun FC, David DG. Experimental Impedance Investigation of an Ultracapacitor at Different Conditions for Electric Vehicle Applications[J]. Journal of Power Sources, 2015, 287, 129–138.
13. Hu XS, Murgovski N, Johannesson LM, Bo E. Comparison of Three Electrochemical Energy Buffers Applied to a Hybrid Bus Powertrain With Simultaneous Optimal Sizing and Energy Management[J]. IEEE Transactions on Intelligent Transportation Systems, 2014, 15(3), 1193–1205.
14. Cai GW, Kong LG, Yang DY, Pan C, Sun ZL. Research on modeling and operation characteristics analysis of Large-Scale wind & light complementary electricity-generating system[J]. Power System Technology, 2012, 36(1):65–71 (in Chinese).
15. Liu LQ, Wang ZX. Variable-voltage MPPT algorithm and simulation of photovoltaic system[J]. Journal of South China University of Technology:Natural Science Edition, 2009, 37(2):140–146 (in Chinese).
16. Femia N, Petrone G, Spagnuolo G, Massimo V. A technique for improving P&O MPPT performances of double-stage grid-connected photovoltaic systems[J]. IEEE Transaction on Industrial Electronics, 2009, 56(11):4473–7782.
17. Jiao Y, Song Q, Liu H. Practical simulation model of

photovoltaic cells in photovoltaic generation system and simulation[J]. Power System Technology, 2010, 34(11):198–202 (in Chinese).

18. Neil SD, Luiz AC, Liu XJ. Comparative study of variable size perturbation and observation maximum power point trackers for PV systems[J]. Electric Power Systems Research, 2010(80):296–305.

19. Hohm DP, Ropp ME. Comparative study of maximum power point tracking algorithms[J]. Progress in Photovoltaic:Research and Applications, 2003, 11(1):47–62.

20. Rodriguez C, Amaratunga GAJ. Analytic solution to the photovoltaic maximum power point problem[J]. IEEE Trans on Circuit and Systems, 2007, 54(9):2054–2060.

21. Cao WH, Li MY Cen X. Independent photovoltaic system efficient charging controller design[J], Journal of Zhejiang University(Engineering Science), 2010, 44(7):1260–1265.

22. Rodriguez C, Amaratunga GAJ. Analytic solution to the photovoltaic maximum power point problem[J]. IEEE Trans on Circuit and Systems, 2007, 54(9): 2054–2060.

CHAPTER 14

AN AUTONOMOUS UNDERWATER RECORDER BASED ON A SINGLE BOARD COMPUTER

Manuel Caldas-Morgan, Alexander Alvarez-Rosario, and Linilson Rodrigues Padovese

Department of Mechanical Engineering, University of São Paulo, São Paulo, SP, Brazil

ABSTRACT

As industrial activities continue to grow on the Brazilian coast, underwater sound measurements are becoming of great scientific importance as they are essential to evaluate the impact of these activities on local ecosystems. In this context, the use of commercial underwater recorders is not always the most feasible alternative, due to their high cost and lack of flexibility. Design and construction of more affordable alternatives from scratch can become complex because it requires profound knowledge in areas such as electronics and low-level programming. With the aim of providing a solution; a well succeeded model of a highly flexible, low-cost alternative to commercial recorders was built based on a Raspberry Pi single board computer. A properly working prototype was assembled and it demonstrated adequate performance levels in all tested situations. The prototype was equipped with a power management module which was thoroughly evaluated. It is estimated that it will allow for great battery savings on long-term scheduled recordings. The underwater recording device was successfully deployed at selected locations along the Brazilian coast, where it adequately recorded animal and manmade acoustic events, among others. Although power consumption may not be as efficient as that of commercial and/or micro-processed solutions, the advantage offered by the proposed device is its high customizability, lower development time and inherently, its cost.

INTRODUCTION

Underwater acoustic research has acquired great importance in recent years, either in monitoring man-generated acoustic events or in assessing its impact on marine life. When long-term research activities are to be executed around remote locations where manned missions are difficult, this kind of monitoring can be logistically and economically daunting.

Since the water medium is a very efficient sound transmitter, underwater acoustic studies are suitable in conducting cetacean ecology research given that it allows the detection of acoustic events occurring as far as hundreds of kilometers away [1]. Typically; commercial Passive Acoustic Monitoring (PAM) devices, such as [2–5], often use hydrophones in conjunction with adequate hardware with the purpose of transmitting and storing acoustic events. In order to perform this task reliably; these devices are robustly engineered, they commonly use proprietary software (source-code not available to the public) and are not very flexible (impossible or very difficult to modify the hardware). Usually, their price range is in the order of thousands of dollars [4]. Despite the different manufacturer

Citation: Manuel Caldas-Morgan, Alexander Alvarez-Rosario, and Linilson Rodrigues Padovese, An Autonomous Underwater Recorder Based on a Single Board Computer, doi:10.1371/ journal.pone.0130297.

features (such as sample rates, maximum allowable depth, materials and storage capacity among others), autonomous underwater recorders commonly consist of at least one hydrophone; located on the external surface of a waterproof enclosure. The hydrophone output passes through a signal-conditioning board before entering into an analog to digital converter (ADC). Subsequently, data is saved on some sort of storage device (HDD, Flash drive, SD card). Usually, these systems are powered by a set of batteries, commonly located within the same enclosure as the electronics. These set of batteries have configurations that often vary from manufacturer to manufacturer. Even though it is difficult to establish a direct comparison between manufacturers using the information available in the market, it was estimated that the power consumption of the aforementioned recorders might lie somewhere between 450 mW to 720 mW.

In order to overcome the involved costs or to find alternative custom solutions for specific working conditions, several research groups around the world have developed their own versions of autonomous underwater recorders. Besides battery life, one of the most challenging aspects of building such a recorder is related to the electronics that handle the analog to digital data conversion and storage. Some authors [6–8] have proposed solutions based on microprocessors/microcontrollers that can be very energy-efficient. Others have favored ultra-high storage and high autonomy at the expense of energy-efficiency [9]. Nonetheless, the development of such systems requires a profound knowledge in both electronics and programming that might be well beyond the knowledge of non-engineering oriented users. An initiative to build an underwater autonomous recorder using a single board computer was previously attempted in [10]; however, besides the resolution being inferior to 16 bit (data was sampled by a 12 bit PCM-3718HG data acquisition module), the autonomy of about 4.5 hours appeared to be low given that the device was powered by a 35.2 Ah 16V lithium battery. There has been even a simpler, yet interesting initiative, in which the recorder was based on a commercial MP3 player running open source firmware [11]. A more comprehensive list composed of more than 30 solutions for underwater audio recording devices (available as of 2013), especially aimed at the studying and monitoring of marine mammals, can be found in [12].

The motivation of this work was to produce an autonomous underwater recording device for scientific research; with the intention of finding equilibrium between complexity, functionality, processing capabilities, power consumption, low-cost and quality. This device was based on the Raspberry Pi, a single board computer developed by the Raspberry Pi foundation in the United Kingdom; with educational purposes [13]. Due to its low-cost and great flexibility, this computer has become increasingly popular among scientists and hobbyists. Besides having built-in memory, a microprocessor, USB ports and being able to run a Linux based operating system; it is equipped with a set of general purpose input/output ports which further increase the interaction with several kinds of electronic sensors.

The proposed device was able to digitalize underwater acoustic signals coming from a hydrophone and routed through a custom built signal-conditioning board. The digitalization was handled by a commercial USB audio codec connected to a Raspberry Pi Model A. The device was powered by a set of batteries located inside the same containing enclosure of the electronics. After testing extensively, both the mechanical resistance of the containing enclosure and the power consumption of the electronics, the device was deployed at different locations along the eastern and southeastern coast of Brazil, in missions related to cetacean research and marine traffic monitoring.

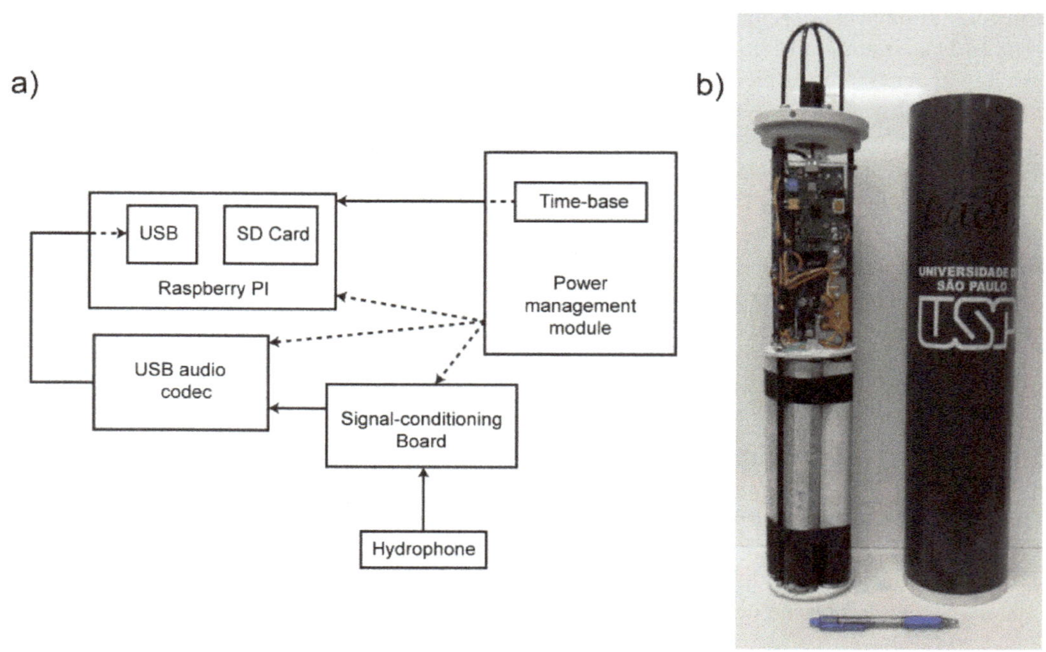

Figure 1: (a) The schematic representation shows the components of the autonomous underwater recording device and (b) the waterproof PVC enclosure, which contained the components.

MATERIALS AND METHODS

Before describing the procedures used for the development and testing of the device under laboratory conditions; it is worth mentioning that the field activities related to the testing of the device took place in areas considered of public access located in the eastern and southeastern coast of Brazil. According to the Brazilian law, no specific permissions are required as long as there is no transgression of areas that are considered private, located within conservation units or that are of any military importance. These field activities did not involve direct interaction with any endangered or protected species.

The main components of the proposed autonomous underwater recorder are illustrated in Fig 1a. Besides the basic hardware components depicted; namely the hydrophone, a signalconditioning board, the USB audio codec and the Raspberry Pi, there is a key set of components grouped inside the power management module (explained further ahead) that are related to a clock used to provide a time-base. With the exception of both the hydrophone and its stainless steel protection cage (which are mounted on the outer top surface), all of the hardware components were located inside a 50 cm long cylindrical PVC enclosure, having 114.30 mm of diameter and 9.52 mm of wall thickness (Fig 1b). This enclosure was tested inside a pressure chamber up to a gauge pressure of 10 bar (approximately 99.5 m of water depth), without showing signs of neither mechanical failure nor leakage.

Hydrophone and Signal-Conditioning

In principle, any kind of hydrophone can be used with the proposed device; from home-made models such as [14] to professional ones having superior specifications, such as the commercial model advertised in [15]. Notwithstanding that; the results presented in the this work were obtained with a custom-made hydrophone consisting of a polyurethane sealed PZT-5 piezoelectric cylinder (25 mm of diameter, 32 mm of length after embedding), with omnidirectional sensitivity of -198 dB re: 1 V/μPa and flat frequency response from DC up to 30 kHz within +/-2 dB. This hydrophone was connected to a custom-built signal-conditioning board. This board consisted of two cascaded non-inverting band pass filters with individual tunable gains and bandwidth. For the recordings mentioned further ahead in this work, the gain of the stages was set at 28 dB and 20 dB respectively. The combined gain of 48 dB was experimentally found to be well suited to the conditions at the selected deployment sites. This cascaded approach was chosen in order to preserve the gain-bandwidth product in circumstances where very high levels of gain are needed. Additionally, by using this configuration, it was possible to provide the signal-conditioning board with separate outputs coming from the first and the combined stages respectively. This feature could enable simultaneous two channel (stereo) recordings to be made at different gains using a single hydrophone; however, all the signals presented in this work were acquired using only the combined stage output. Fig 2 Illustrates the magnitude and phase response of a signal-conditioning board tuned to have a total gain of 48dB with a -3 dB bandwidth from 15 Hz to 20 kHz.

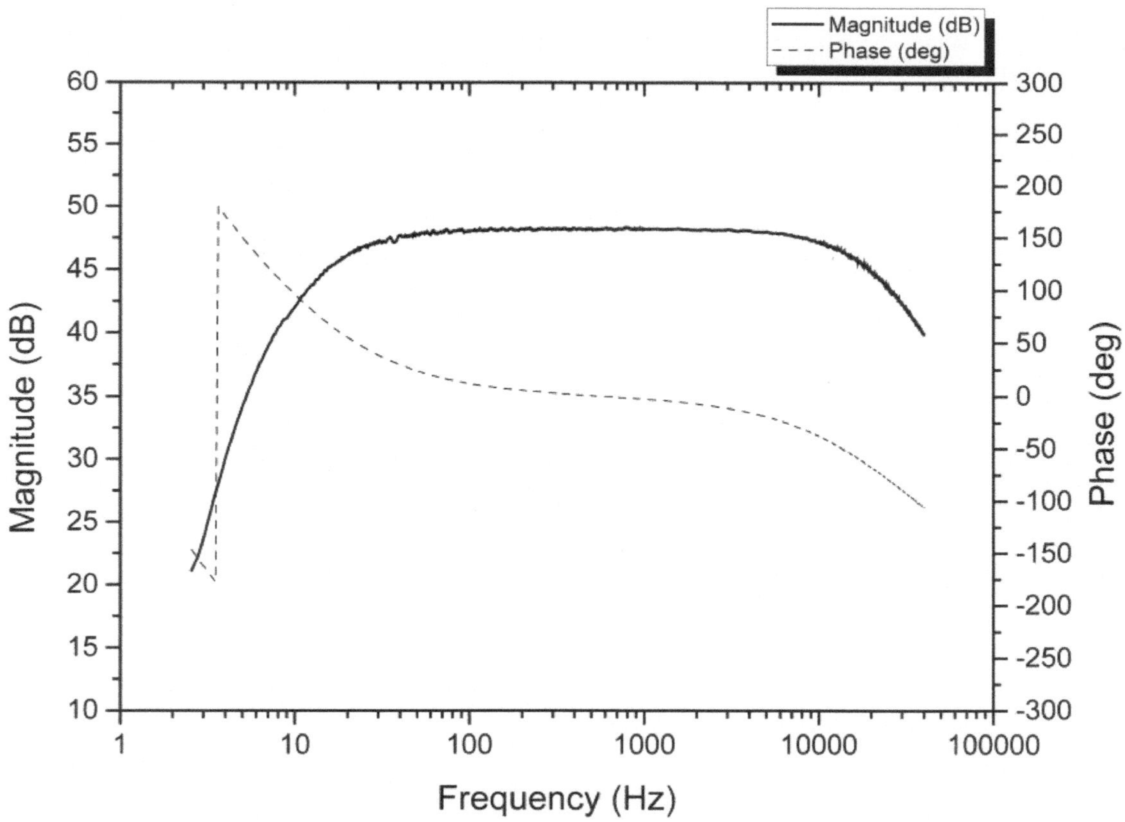

Figure 2: Experimental frequency response obtained with a particular tuning of the signal-conditioning board.

USB Audio Codec

After being filtered and amplified, the signal was routed into one of the line-in inputs of a Burr Brown PCM2902 audio codec (found in consumer electronics such as the Behringer UCA222 Audio Interface) connected to the USB port of the Raspberry Pi. This audio codec has an internal 16 bit Delta-Sigma ADC with hardware defined sampling rates of 8, 11.025, 16, 22.050, 32, 44.100 and 48 kHz [16] that allows the voltage coming from the signal-conditioning board to be digitalized and stored.

The vertical scale of the digitalized files (i.e. PCM WAV files) is known as Full Scale (Fs). Normally, when audio files are open in a computer, using any suitable software, the amplitude of the recorded waveform is represented by 216 Fs digital levels normalized between -1 and 1. Thus, a calibration procedure was needed in order to establish a relationship between these Fs levels and the voltage that produced them. This procedure involved the recording of a known pattern, which was made at a sample rate of 48 kHz in PCM WAV format. Preliminary tests showed that input waveforms with peak amplitudes as high as 2.25 V produced saturation in the recorded WAV file. Subsequently, a compatible calibration pattern was produced using an Agilent 33250A signal generator. This pattern was obtained by modulating a sine wave of 1 kHz and 2.25 V of peak amplitude with a crescent ramp function of 1 Hz of frequency. The rate of change of the amplitude of the Fs values was calculated by analyzing the obtained recording. As the rate of change of the peak input amplitude was known, the calibration constant could be found using the proportion between the two rates of change.

The use of a modulated signal of 1 kHz of frequency was judged adequate given that when sampling at 48 kHz, the digital filters of the ADC inside the USB audio codec provide a -3dB bandwidth between 3.5 Hz and 23.75 kHz [16].

The calibration results, in conjunction with the sensitivity of the hydrophone, allowed the acquired audio waveforms to be transformed into sound pressure level (SPL) variations relative to time; and accordingly, to gain a physical meaning. This was necessary in order to calculate the self-noise of the device, which was done by programming it to make a continuous recording (at 48 kHz) before positioning it inside an anechoic chamber. This chamber was supported by rubber feet, which helped to decouple it from low frequency vibrations. Eventually, the electric power supply of the testing facility was interrupted in order to avoid any possible electromagnetic interference. The self-noise of the device was estimated by calculating the RMS value of the resulting recording and expressing it in dB re: 1 V/μPa. Analogously, the background noise at the deployment sites was calculated by analyzing the quieter intervals registered during long-time recordings.

Single Board Computer

The USB audio codec was hosted by a Raspberry Pi Model A; a single board computer based on a BCM2835 system on a chip (SoC) with 256 MB of RAM, one USB port and no Ethernet port. This computer also features a 26 pin port, where 17 of those pins are GPIO (general purpose input/output), that allows for communication using protocols such as I2 C, SPI and serial [17]. The Raspberry Pi can be powered supplying 3.3 V and 5 V to pins 1 and 2 respectively; however, it is commonly powered just by 5 V as the board has an integrated 3.3 V linear dropout voltage regulator internally connected to pin 1. The Raspberry Pi is capable of running Linux, an open-source operating system that allows the user to perform some moderate to difficult tasks through a set of simple command lines based on routines (shell-scripts). The Linux distribution chosen to work with was Arch Linux, which is specifically designed for systems based on ARM technology. As this distribution lacks the drivers needed to interact with the USB audio codec, it was necessary to install the Advanced Linux Sound Architecture (ALSA) package in order to enable audio support. After having the adequate drivers installed, the next step was to install the SoX (Sound eXchange) package. This piece of software was used to configure the recording settings (i.e. sample rate, number of channels, bit depth), to choose the file format (between RAW binary or PCM WAV), to perform further operations such as splitting the files during long-term recordings (which, in the device, was configured to occur every 300 s by default) as well as filtering and resampling as needed. The complete list of SoX commands, with further information regarding its capabilities, is detailed in [18].

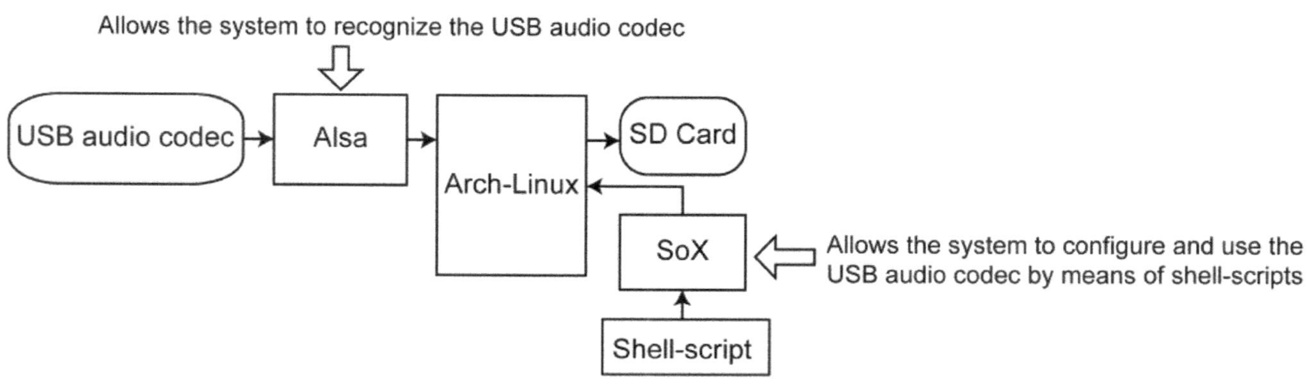

Figure 3: Schematic representation of the software/hardware interaction needed in order to capture the audio signals.

The interaction between the described software and the hardware, necessary in order to correctly store the acoustic data on the SD card, is shown schematically in Fig 3.

All the programming was written and tested using an additional Raspberry Pi Model B, which was accessed via Ethernet. After the programming was done, the SD card was transferred to the Raspberry Pi Model A installed in the device, which executed all the programmed operations equally well. Alternatively, the Raspberry PI Model A could be programmed directly using both a display and a keyboard; however, testing was more time consuming as this model only has one USB port, and had to be swapped between the keyboard and the USB audio codec. In order to retrieve the recorded files, only the SD card was needed. Data could be accessed by any computer capable of reading Linux partitions.

Given that the Raspberry Pi lacks an onboard clock, the time base was provided externally by a Maxim DS3231 real time clock (RTC), which communicates with the board via the GPIO pins 2 and 3 using the I2 C protocol. This RTC keeps time with an accuracy within ±2 minutes per year while being able to operate from -40°C to +85°C [19].

As storage capacity limits the amount of data that can be collected (and hence affects the autonomy), a 128 GB Kingston SDXC Class 10 card was used in the device. This is one of the largest commercially available SD cards compatible with the Raspberry Pi [20].

Power Management Module

Other factor that affects the autonomy is the power consumption. During testing, the device was powered by a battery pack formed by five D-size alkaline-manganese dioxide batteries (Duracell MN1300) wired in series. This pack was measured and had a total of 8 V (1.5 V nominal per battery, 1.6 V measured when brand new). Alternatively, as shown in Fig 4a, more packs could be connected in parallel in order to increase the capacity. The output of the battery pack was controlled by a pair of synchronous rectified step-down converters (MP2307) with an efficiency as high as 95% [21]. The first converter was set at 5 V, which is the operating voltage of the Raspberry Pi. The second converter, set at 3.3 V, was a more efficient alternative for the original onboard linear dropout voltage regulator (which was removed from the board).

Since the device was not equipped with neither an external switch nor a communication programming port to be turned-on; energy-wise, it was not convenient to have the device start recording before arriving at the deployment location, which could take hours, days or even weeks. Accordingly, the device was equipped with a Hall effect latch (Allegro A1250) that, under specific conditions, enabled a micro power relay (Omron G6L-1P 5V) to connect or disconnect the battery pack from the electronics. The Hall effect latch was controlled from the exterior of the PVC enclosure using a permanent magnet. When exposed to a south magnetic field, the Hall effect latch produced a logic HIGH output and vice versa.

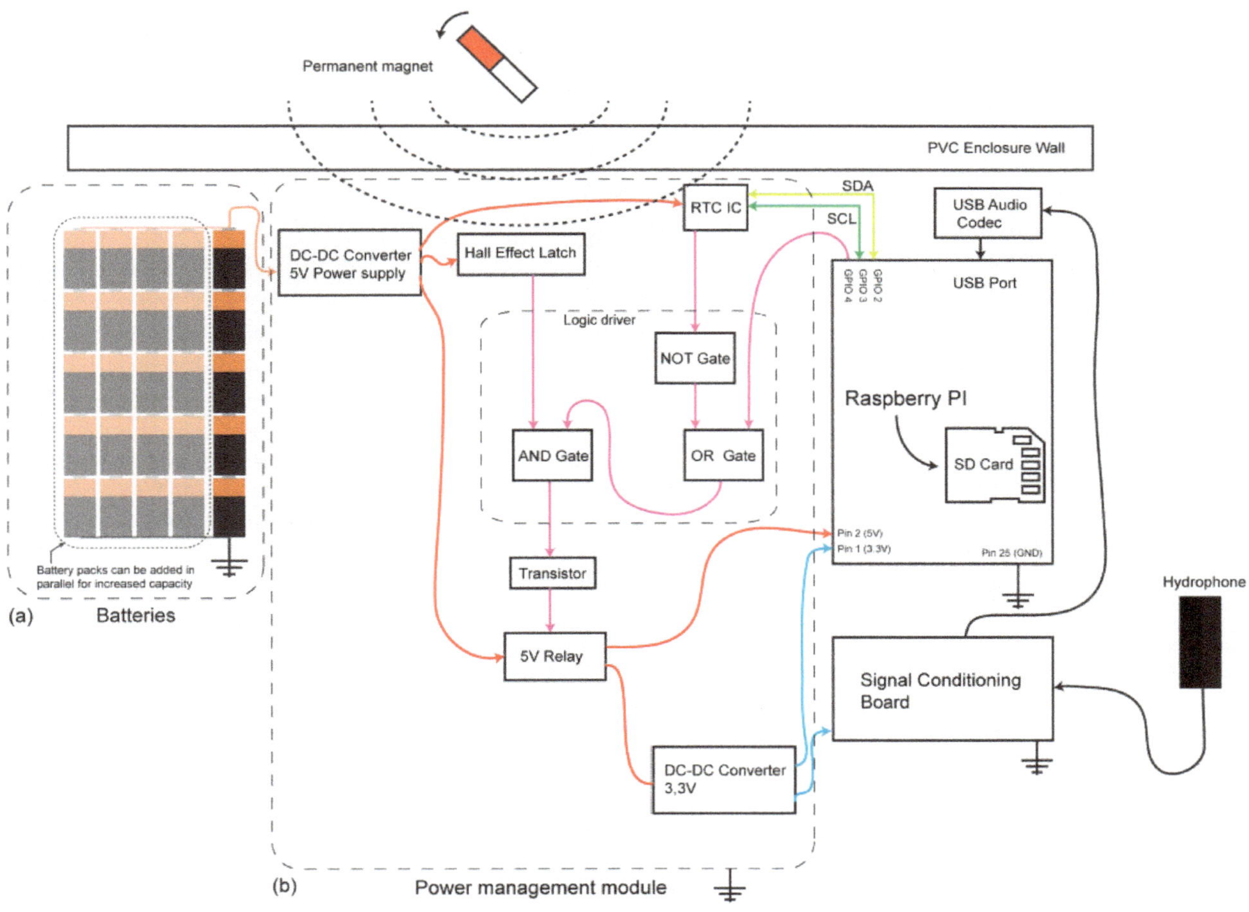

Figure 4: A schematic representation showing the connections made between the components of the underwater recording device.

Nonetheless, in an effort to further increase battery saving, a scheduling feature was implemented using the RTC; which allowed periodic recordings (e.g., during some specific set of hours per day). The RTC, which was directly powered by the 5 V DC-DC converter, is provided with a pin that outputs a digital alarm signal that can be software-programmed (RTC PIN 3). When not programmed or in standby, RTC PIN 3 stays on HIGH state, changing to LOW only when a programmed alarm goes off. This change of state could be used to polarize directly a small transistor serving as a switch to drive the relay. However, for the alarm to function more than once (e.g., an alarm set to be repeated daily), the RTC must be re-programmed each time an alarm is triggered. This was necessary in order to bring RTC PIN 3 to the standby state again. Usually, this can be done by means of a microcontroller, which might not be a straightforward solution for the average user. For the sake of simplicity, a workaround was implemented using a logic driver, which was built using a set of micro power logic gates that enabled the Raspberry Pi to do the re-programming on its own and turn-off the device at some scheduled time. This was achieved by only energizing the relay when certain combinations were input into the logic driver. The whole system is schematically shown in Fig 4.

The sequence of steps detailed in the flow chart of Fig 5 was used for re-programming the alarm. The combinations for the logic driver are shown in Table 1. The resulting logic state was used to drive the relay, by means of a general purpose NPN transistor (2N2222A).

For the procedure to be successful, it was necessary to use a RTC previously programmed to be in standby state (RTC PIN 3 HIGH). First, the magnet was used to activate the Hall Effect latch, which did not energize the relay directly but instead sent a signal to the NOT gate of the logic driver. As soon as certain programmed time condition was met, RTC PIN 3 changed its state to LOW. In sequence (as show in the truth table) the relay became active, powering the Raspberry Pi. After the boot process was completed, a command line instruction was issued in order to change the state of GPIO PIN 4 from LOW to HIGH (only to be turned-off after the designated recording time has elapsed). In sequence, another command line was issued with the purpose of reversing the state of RTC PIN 3 from LOW to HIGH, thus enabling the alarm for the next recording cycle. At this point, the system kept working until the recording time was reached, subsequently switching GPIO PIN 4 from HIGH to LOW in order to deactivate the relay (and accordingly turning the Raspberry Pi off). It remained off (device in sleep state) until the next programmed alarm reverted the state of RTC PIN 3 to LOW, which enabled the relay once more. This way, during a long-term mission (with the magnetic latch continuously in HIGH) the cycle would repeat either until the device is retrieved and manually turned-off or in the event a hardware/software failure (for example, when the batteries go below their cut-off voltage). The logic driver in conjunction with the RTC, the Hall effect latch, the step-down converters and the relay (including its driving transistor) compose the power management module, as illustrated in Fig 4b.

Power Consumption Testing

In order to discover any possible influence that the sampling rate might have over the autonomy of the device, a series of tests were run with the intention of monitoring both the instantaneous voltage and the total current drain. Initially, the device was energized with a benchtop power supply set to provide 7.5 V at constant current. Before performing the tests, all non-critical processes that occur during the Raspberry Pi boot cycle (such as video output) were disabled via software. First, with the intention of establishing the minimum power consumption levels, the average power consumption of the device was measured in standby state (i.e. all electronics were powered but not recording) and in sleep state (i.e. all electronics were disabled except for the 5 V DC-DC converter). Successively; 8 hour audio recordings were made, one at a time, using all the available sample rates offered by the USB codec in order to determine the average power consumption for each case. The average power at 8 kHz could not be measured since, at this rate, the recording always occurred at the next nearest value of 11.025 kHz; this might be a limitation either of SoX or the Raspberry Pi. Similar to the case of the USB audio codec, almost any off-the-shelve hand-held audio recording device allows to adjust the sample rate; however, usually offering fewer options to choose from.

However, besides being able to record at various sample rates, the Raspberry Pi has resources that allow for some post-processing to take place as well; for example, filtering or resampling (through the use of SoX commands). This kind of post-processing can be useful because storage space can be saved, as well as offline post-processing time, depending on the bandwidth of the target signals of interest. In order to detect a possible increase in power consumption due to post-processing, the initial 8 hour experiments were replicated using an additional low-pass filtering operation for each case (using a cut-off frequency equal to halfbandwidth of the corresponding sample rate).

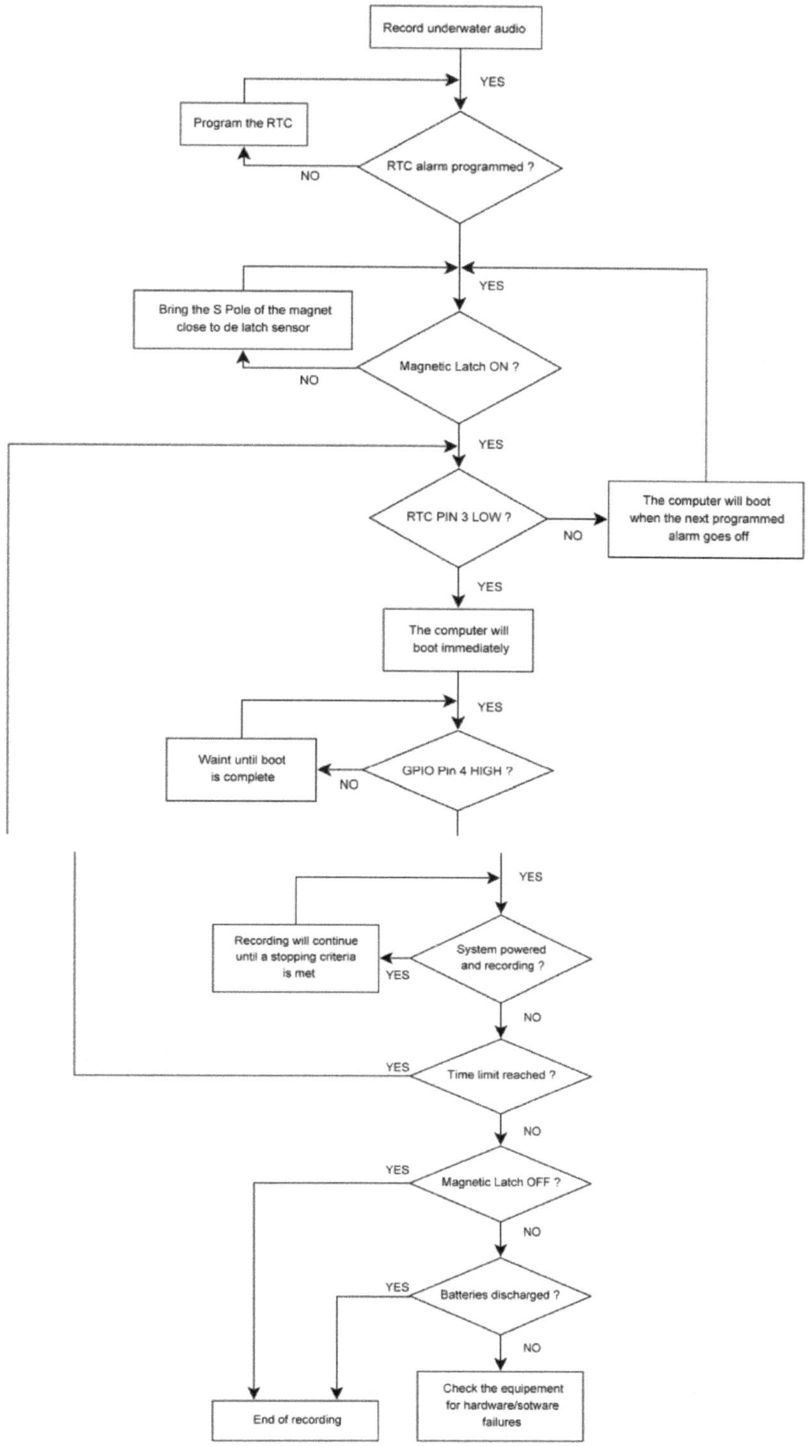

Figure 5: Flow chart of the procedure used to perform scheduled recordings using the resources of the RTC.

Table 1: Truth table showing the resulting state of the logic driver circuit

Magnetic Latch	RTC PIN 3	GPIO PIN 4	Relay State
HIGH	HIGH	LOW	DISABLED
HIGH	LOW	LOW	ENABLED
HIGH	LOW	HIGH	ENABLED
HIGH	HIGH	HIGH	ENABLED
LOW	HIGH	LOW	DISABLED
LOW	LOW	LOW	DISABLED

Finally, an endurance test was conducted both to determine the characteristic battery discharge behavior of the device and to estimate its autonomy. During this test, the device was powered by a single battery pack. Also, the recording was made at 48 kHz while re-sampling to 24 kHz instead of filtering, as done in the prior test. All the tests described in this section took place at a temperature of 25°C. All the files were recorded using a single channel (mono) at a 16 bit resolution, formatted as PCM WAV.

Data Management

As the size of collected data increases, it becomes more difficult to analyze. Long-term deployments can result in a huge amount of data. For instance, if the power supply was unlimited, a 128 GB SD card (discounting the 2 GB needed for the operating system) would be able to sore data produced by a continuous mono recording lasting approximately between 15.55 days (48 kHz sample rate) to 67.72 days (11.025 kHz sample rate).

Before deploying the device, the RTC is synchronized to the UTC time for the sake of organizing the successive 300 s recordings. They were named following the convention "YYYY.MM.DD_hh.mm.ss,sss.WAV". A database was consolidated using the MATLAB software in order to manage these files. This database allowed the selection of a portion of audio of interest (based on its initial and final date and time); making it possible to join its composing files as needed and process them as a single file. Using the calibration constant of the USB audio codec, it was possible to obtain the acoustic pressure waveform and subsequently to calculate parameters such as the SPL relative to 1 μPa. Using the SPL waveforms, it was also possible to calculate statistical parameters, based on window functions, such as the mean or the RMS value. In addition to time based processing, frequency domain calculations (such as the spectrogram) could be performed as well.

Field Deployments

Some autonomous underwater recorders underwent field tests at three different locations along the eastern and southeastern Brazilian coast with the purpose of evaluating their performance as well as the quality of the collected data.

The first deployment site was located in the southeastern coast of Brazil, near the city of Santos (24° 00' 31.1" S 46° 19' 34.8" W). A recorder was installed at a depth of 11 m with the aim of monitoring marine traffic in a public area situated near the entry of the port of Santos. Container-ship transit is steady because it is one of the busiest ports in Latin America. Nevertheless, some smaller ships might be found in the region; for example, outboard powered motorboats which serve as transportation for the local population. The equipment was programmed to record continuously at 22.050 kHz. Later, the sample rate was re-programmed to 16 kHz since this rate was better suited to the sounds being recorded, and it also contributed towards saving storage space.

The second deployment site was located in the eastern coast of Brazil, near the city of Ilhéus (14° 24' 30.35" S 38° 59' 51.87" W). The recorder was installed at a depth of approximately 20 m, with the purpose of registering local populations of marine mammals, particularly of the Megaptera novaeangliae species (Humpback whale). These animals are known to migrate from feeding grounds at high latitudes near Antarctica to breeding grounds at lower latitudes close to the Brazilian seaside during the winter [22]. Recordings were scheduled to take place daily between 7am and 5pm. The sample rate chosen for this mission was 11.025 kHz, which was judged adequate given that this species has a dominant frequency range that goes from 120 Hz to 4 kHz [23].

The third deployment site was located in the southeastern coast of Brazil, inside Sepetiba bay (22° 58' 26.4" S 44°00' 00.2" W). The device was suspended in the water column at a depth of 7 m. The recording sampling rate was set to 44.100 kHz as the goal was to detect the presence of members of the Delphinidae family.

RESULTS AND DISCUSSION

The results presented in this section provide a good benchmark of the power consumption under typical working conditions; which allow for estimating parameters such as the autonomy, the discharge characteristics of the battery pack and the storage space needed for a given recording scheme. These parameters can be used to optimize the equipment in order to provide the best performance according to the requirements of each specific mission. It was verified that the self-noise of the device was appropriate for recording at the selected field locations. This implies that the sample spectrograms presented at the end of this section are consistent.

Power Consumption and Data Management

The first parameters measured were related to the absolute minimum power consumption levels. It was observed that in average, the device consumed 1035.70 mW and 89.50 mW (at constant current) while in standby and in sleep respectively. Thus, by using the power management module during non-recording periods, it is estimated that an economy of up to 91.35% could be achieved. This would be crucial for long-term missions. Subsequently, the average power consumption (during recordings corresponding to each sample rate) was calculated. The results are shown on the lower data-set of Fig 6, where it can be noted that power consumption increases linearly with the recording sampling rate.

In contrast, during the second test (which included low-pass filtering) the power consumption increase was non-linear relative to the sample rate. Additionally, it can be observed that power consumption intensified due to the post-processing operation. The percent increase values, related to the corresponding standby state, are shown below the upper data-set of Fig 6.

Subsequently, while performing some experiments in preparation for the third test; it was perceived that whenever the supply voltage was adjusted, the current drain of the device varied as well (keeping the power constant). These observations were coherent with the behavior during the first portion of the endurance test, portrayed in Fig 7. It can be seen that for the first 40 hours (1.66 days) the battery pack discharged at a constant power of about 1150 mW. From this point, and for the next 21 hours (0.88 days), the battery pack began to discharge at an approximately constant current of 185 mA. The device continues to work fine until 61 hours (2.54 days) had elapsed. At that time, the batteries were no longer able to power the device and the Raspberry Pi stopped functioning. This point was highlighted by a dramatic power drop. The cut-off voltage was measured and had a value of 5.58 V (1.12 V per battery). Lastly, by integrating the power over the 61 hours of functioning time, it was found that the total energy delivered by the battery pack was 68.29 Wh; which gave an average power consumption of 1120mW. The integration of current over the same period of time yielded the capacity of the battery pack under the test conditions, which was of 10.79 Ah. However, it was expected that, at lighter working conditions (lower sampling rates, no post-processing, non-continuous recordings) the autonomy could be increased.

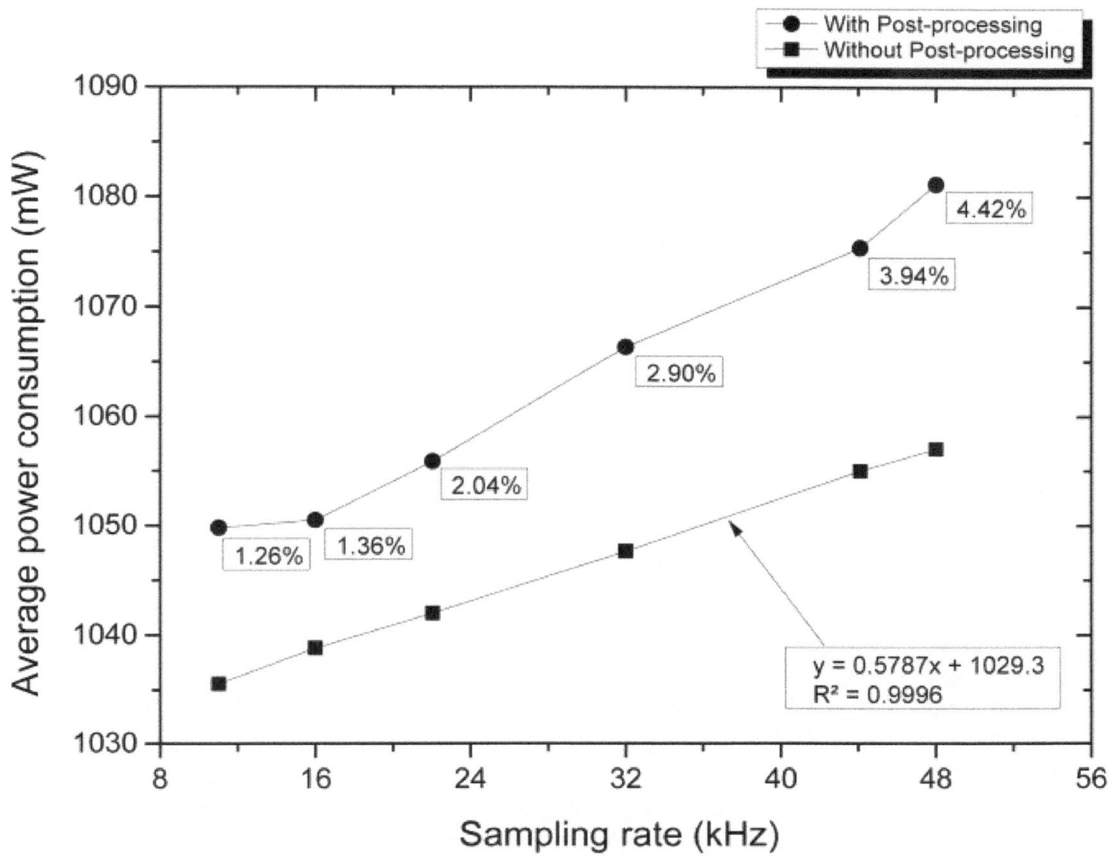

Figure 6: Average power consumption (at constant current) during continuous recordings made at different sample rates, with and without postprocessing.

Based on the energy delivered by the battery pack during this last test, It is predicted that by lowering the sample rate to 11.025 kHz (1035.71 mW, without post-processing) the autonomy would further increase to 65.93 hours (2.74 days) of continuous recording. By fitting the 5 battery packs allowed by the geometry of the PVC enclosure (as in Fig 4a), the energy capacity would be incremented to 53.95 Ah; bringing autonomy to 329.65 hours (13.70 days) while generating 24.98 GB worth of data.

Still under the last conditions, if scheduling was used instead (e.g. recording 1 hour per day), a good estimate of the expected autonomy could be made using the current drain measured (at constant voltage) both in sleep and standby (which was of 11.93 mA and 138.09 mA respectively). The battery capacity needed for the device to function 1 day is equal to 412.48 mAh (138.09 mA 1 hour + 11.93 mA23 hours); then, 5 battery packs would allow for 130.79 days (approximately 4.29 months) while generating 10.14 GB worth of data. Considering that the operating system needs around 2 GB of space, a 16 GB card would be adequate for this mission. Indeed, this kind of estimates could be useful to save costs. The predicted recording time could be used to choose smaller SD cards, coherent with the expected mission length.

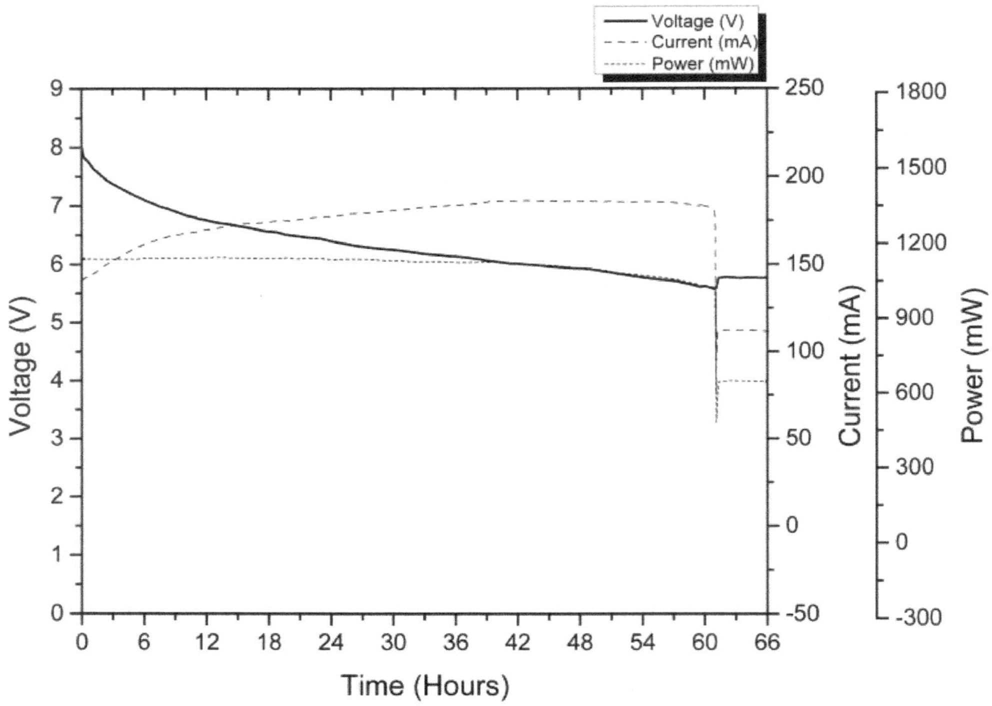

Figure 7: Battery discharge behavior registered while recording (at 48 kHz) and post-processing (resampling).

USB audio codec calibration and self-noise

As it can be seen in Fig 8, the recorded signal corresponded to the shape of the calibration pattern, until it saturated the internal ADC of the USB audio codec. The rate of change of the Fs with respect to time could be calculated outside of the saturated region and had a value of 1.26 Fs/s. As the peak amplitude of the calibration pattern changed at a rate of 2.25 V/s, the calibration constant was calculated and had a value of 1.79 V/Fs. The maximum input amplitude that could be registered before reaching the first clipping point was of 3.58 Vp-p (approximately 2.05 dBV).

Once the calibration constant was found, it was used in conjunction with the voltage sensitivity of the hydrophone in order to measure the lowest noise level of acoustic pressure that could be recorded. Following the procedure described in Materials and methods, the self-noise measured was of 84.22 dB_{RMS} re: 1 μPa.

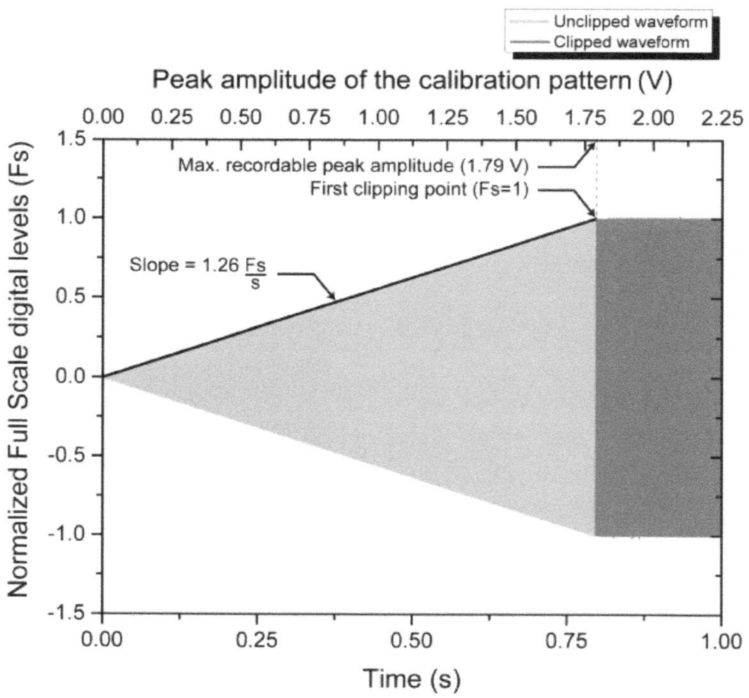

Figure 8: Recorded waveform of the calibration pattern (including clipping).

Audio Samples Collected During Field Deployments

The image depicted in Fig 9 corresponds to the spectrogram of a portion of audio recorded near the entry of the port of Santos, where the measured background noise was of 110.02 dBRMS re: 1 μPa. It mainly represents the sound produced by some small sized boats which seemingly were traveling close to the recorder. In this spectrogram, it can be seen that one of the boats (450 s into the recording) passed faster and that (probably) was closer to the device given that it produced a higher level of intensity (darker regions in the spectrogram). Complementing this kind of measurement with photographic records, would be very useful in order to create a database that could eventually lead to recognize ships by their sound signature.

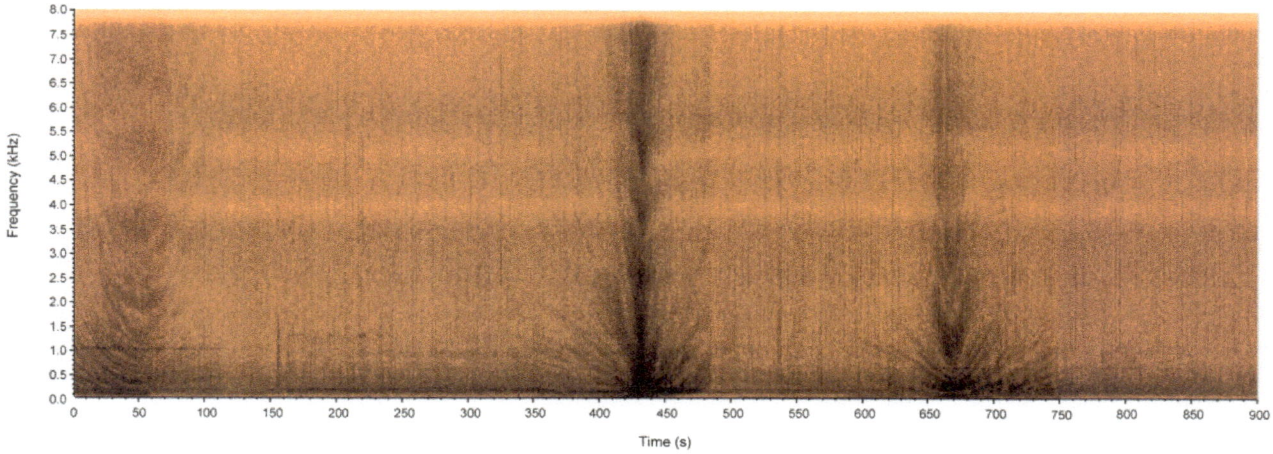

Figure 9: Spectrogram depicting marine traffic near the city of Santos, SP, Brazil.

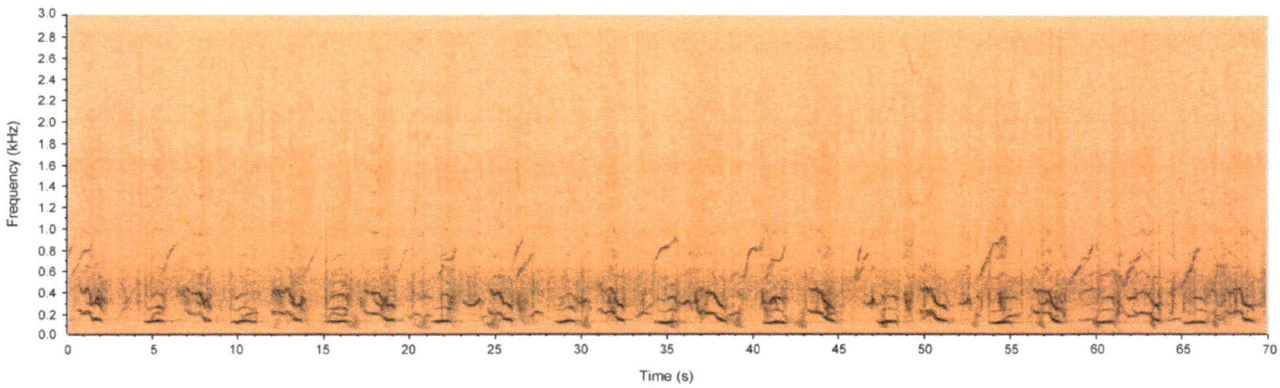

Figure 10: Spectrogram of Humpback whale vocalizations registered in the seaside of the city of Ilhéus, BA, Brazil.

An additional recording unit, installed near the coast of the city of Ilhéus, registered a portion of audio represented by the spectrogram of Fig 10. In this spectrogram, it is possible to identify the periodic components of a certain humpback whale vocalization. The spectrogram was zoomed to the lower 3 kHz; which evidenced that the greater components of those vocalizations were inferior to 500 Hz. At this location, the background noise measured was of 129.18 dB_{RMS} re: 1 μPa.

A third recording unit was installed at Sepetiba bay, where the background noise measured was of 118.52 dB_{RMS} re: 1 μPa. It registered some characteristic whistles that indicated the presence of Sotalia guianensis individuals (estuarine dolphins), as it can be detailed in Fig 11. Despite having a vocal range that goes above the 22.050 kHz band that was used, several vocalizations were captured (including the lower frequency portion of some echolocation sounds). This was enough to achieve the objective of detecting their presence.

CONCLUSIONS

This work presented an autonomous underwater recorder based on the Raspberry Pi (Model A) single board computer, which can be assembled and made functional without the need low level programming or specialized hardware, such as microcontrollers or microprocessors. However, some notion of programming is needed, as the user is allowed to perform configuration tasks by means of shell-scripts under a Linux operating system. Those tasks can allow, for example, setting the recording sample rate, scheduling the recordings and post-processing (e.g., resampling and filtering).

Acoustic audio signals were digitalized by means of a PCM2902 USB audio codec connected to the Raspberry Pi. This audio codec proved to be capable of digitalizing signals (having amplitudes as high as 2.05 dBV) with a maximum resolution of 16 bit, at sample rates of up to 48 kHz. The calibration procedure yielded a constant of 1.79 V/Fs.

It was determined that the maximum autonomy achieved using one single battery pack, while recording and post-processing a mono file at 48 kHz, was of 2.54 days. During that time, the battery pack delivered a total energy of 68.29 W.h. The measured capacity of the battery pack until reaching the cut-off voltage of 5.58 V (1.12 V per battery) was of 10.79 A.h. The average power consumed by the device was of 1120 mW. It was estimated that under lighter working conditions, by using the 5 battery packs allowed by geometry of the enclosure, the autonomy for continuous recording would be of 13.70 days; while a scheduled recording scheme of 1 hour per day would increase the functioning time to 4.29 months. Extensive tests of the prototypes showed that a compromise between storage space and long-term functioning can be achieved by means of fine adjustments according to the goal of the mission, as for example, selecting lower sampling rates or programming scheduled recordings.

Apart from the configuration used to test the performance of the device, other configurations might be possible without meaning that they would be less effective. For instance, by further taking advantage of the processing resources of the Raspberry Pi (or any extra resources offered by other suitable single board computer), it would be possible to implement features; such as, recording based on acoustic level detection. As only certain events within frequency and amplitude limits would trigger the recording, this measure would contribute with power saving and would also lower the amount of data to be stored and processed. Other configuration variations can include (but are not limited to), for example, another kind of hydrophone (with different sensitivity, dimensions, shape or directional pattern) or a different material for the enclosure (which could modify the depth rating).

The device proved to be useful for acquiring, processing and storing signals from biological and anthropogenic events at selected locations along the eastern and southeastern coast of Brazil. All field recordings were made with a hydrophone with a sensitivity of -198 dB re: 1V/μPa in conjunction with a 48 dB pre-amplifier. The self-noise with this configuration was of 84.22 dBRMS re: 1 μPa, which was lower than the background noise values measured at the deployments sites.

By altering the recording parameters (or by adding post-processing), the user can easily adapt the equipment accordingly to the nature of each mission. The cost of replication of the device was estimated to be around 500 USD. Given that basic hydrophone amplifiers, such as [24], are in the range of 500 USD [25] and that specialized underwater autonomous recorders have prices close to 6000 USD [4]; the device proposed in this work offers a good value for the money and provides an excellent low-cost alternative for research related activities.

ACKNOWLEDGMENTS

The authors wish to thank Isabel Gonçalvez, Marine Biologist, for the cooperation on the underwater acoustic monitoring campaign that took place in Ilhéus, BA, Brazil.

AUTHOR CONTRIBUTIONS

Conceived and designed the experiments: MCM AAR LRP. Performed the experiments: MCM AAR LRP. Analyzed the data: MCM AAR LRP. Contributed reagents/materials/analysis tools: MCM AAR LRP. Wrote the paper: MCM AAR LRP.

REFERENCES

1. Zimmer WMX. Passive Acoustic Monitoring of Cetaceans: Cambridge University Press; 2011.
2. Develogic GmbH. Sono. Vault Acoustic Recorder and Signal Analyzer [cited January 11, 2015]. Available from: http://www.develogic.de/wp-content/uploads/2014/05/SonoVault-Acoustic-Recorder-SignalAnalyzer-05 2014.pdf.
3. Multi-Électronique (MTE) Inc. AURAL-M2 (Autonomous Underwater Recorder for Acoustic Listening) Specification Sheet [cited January 11, 2015]. Available from: http://www.multi-electronique.com/files/ AURALspecsheet_EN.pdf.
4. Loggerhead Instruments. DSG-ST Ocean Acoustic Datalogger [cited January 11, 2015]. Available from: http://loggerhead.com/products/dsg-ocean-acoustic-datalogger.
5. Wildlife Acoustics Inc. Song Meter SM2M Marine Recorder User Manual [cited January 11, 2015]. Available from: http://media.nhbs.com/equipment/Song%20Meter%20SM2M%20Supplement.pdf.
6. Lammers MO, Brainard RE, Au WWL, Mooney TA, Wong KB. An ecological acoustic recorder (EAR) for long-term monitoring of biological and anthropogenic sounds on coral reefs and other marine habitats. The Journal of the Acoustical Society of America. 2008; 123(3):1720–8. doi: 10.1121/1.2836780 PMID: 18345859
7. Martinez JJ, Myers JR, Carlson TJ, Deng ZD, Rohrer JS, Caviggia KA, et al. Design and Implementation of an Underwater Sound Recording Device. Sensors. 2011; 11(9):8519–35. doi: 10.3390/ s110908519 PMID: 22164089
8. Wang C-C, Hsiao Y-H, Huang M-C. Development of MSP430-based ultra-low power expandable underwater acoustic recorder. Ocean Engineering. 2009; 36(6–7):446–55.
9. Wiggins SM, Hildebrand JA, editors. High-frequency Acoustic Recording Package (HARP) for broadband, long-term marine mammal monitoring. Underwater Technology and Workshop on Scientific Use of Submarine Cables and Related Technologies, 2007 Symposium on; 2007 17–20 April 2007.
10. Ming-Hao C, Chau-Chang W, Jin-Yuan L, Chia-Wei C, editors. Design and Application of Autonomous Underwater Acoustic Recorder. Underwater Technology and Workshop on Scientific Use of Submarine Cables and Related Technologies, 2007 Symposium on; 2007 17–20 April.
11. Hamilton LJ, Lees C. A low cost (near expendable)

underwater sound logger using an mp3 player: testing in Sydney harbour. Acoustics Australia. 2009; 37(2):57–60.

12. Sousa-Lima RS, Norris TF, Oswald JN, Fernandes DP. A Review and Inventory of Fixed Autonomous Recorders for Passive Acoustic Monitoring of Marine Mammals. Aquat Mamm. 2013; 39(2):205–10.
13. Raspberry Pi Foundation. What is a Raspberry Pi? [cited January 11, 2015]. Available from: http:// www.raspberrypi.org/help/what-is-a-raspberry-pi/.
14. Romero Vivas E, León López B. Construction, calibration, and field test of a home-made, low-cost hydrophone system for cetacean acoustic research. Proceedings of Meetings on Acoustics. 2012; 11 (1):010001.
15. Brüel & Kjær. Standard measuring hydrophone with integral 10m cable terminated BNC plug [cited January 11, 2015]. Available from: http://www.bksv.com/Products/transducers/acoustic/hydrophones/ 8104.
16. Texas Instruments. Stereo audio codec with USB interface, single-ended analog input/output and S/ PDIF [cited January 11, 2015]. Available from: http://www.ti.com/lit/ds/symlink/pcm2900.pdf.
17. Raspberry Pi Foundation. GPIO—An introduction to GPIO and physical computing on the Raspberry Pi [cited January 11, 2015]. Available from: http://www.raspberrypi.org/documentation/usage/gpio/.
18. SoX—Sound eXchange. Documentation [cited January 11, 2015]. Available from: http://sox. sourceforge.net/Docs/Documentation.
19. Maxim Integrated. DS3231 Extremely Accurate I2C-Integrated RTC/TCXO/Crystal [cited January 11,
20. 2015]. Available from: http://datasheets.maximintegrated.com/en/ds/DS3231.pdf.
21. Embedded Linux Wiki. RPi SD cards [cited January 11, 2015]. Available from: http://elinux.org/RPi_ SD_cards.
22. Monolithic Power Systems. MP2307 3A, 23V, 340KHz Synchronous Rectified Step-Down Converter [cited January 11, 2015]. Available from: https://www.monolithicpower.com/DesktopModules/ DocumentManage/API/Document/GetDocument?id=583.
23. Zerbini AN, Andriolo A, Heide-JøRgensen MP, Moreira SC, Pizzorno JL, Maia YG, et al. Migration and summer destinations of humpback whales (Megaptera novaeangliae) in the western South Atlantic
24. Ocean. Journal of Cetacean Research and Management (Special Issue). 2011;(3:):113–8.
25. Richardson WJ, Thomson DH. Marine Mammals and Noise: Academic Press; 1995.
26. DolphinEar Hydrophones. DE200 Series: Complete, Ready-to-go [cited January 11, 2015]. Available from: http://www.dolphinear.com/de200.html.
27. DolphinEar Hydrophones. Products [cited January 11, 2015]. Available from: http://www.dolphinear.com/products.html.

INDEX

P

Q

R

S

T

V

W